DESIGN OF
MOS VLSI CIRCUITS
FOR
TELECOMMUNICATIONS

DESIGN OF
MOS VLSI CIRCUITS
FOR
TELECOMMUNICATIONS

Edited by

Yannis Tsividis
Columbia University
New York, New York, USA

Paolo Antognetti
University of Genova
Genova, Italy

Prentice-Hall, Inc.
Englewood Cliffs, New Jersey 07632

Cover photo courtesy of Texas Instruments, Inc.

ISBN: 0-13-200643-X

Library of Congress Catalog Card Number: 85-60438

Printed in the United States of America

10 9 8 7 6 5 4 3 2 1

ISBN 0-13-200643-X 01

Prentice-Hall International (UK) Limited, London
Prentice-Hall of Australia Pty. Limited, Sydney
Prentice-Hall Canada Inc., Toronto
Prentice-Hall Hispanoamericana, S.A., Mexico
Prentice-Hall of India Private Limited, New Delhi
Prentice-Hall of Japan, Inc., Tokyo
Prentice-Hall of Southeast Asia Pte. Ltd., Singapore
Editora Prentice-Hall do Brasil, Ltda., Rio de Janeiro
Whitehall Books Limited, Wellington, New Zealand

CONTENTS

CONTRIBUTORS

APFEL, R. J.
    Advanced Micro Devices, Sunnyvale, California, U.S.A.

BAKER, P.
    Filter Design Consultants Ltd., Lincolnshire, U.K.

CASTELLO, R.
    University of California, Berkeley, California, U.S.A.

DEMOULIN, E.
    Centre National d'Etudes des Telecommunications (CNET),
    Meylan Cedex, France

FISCHER, T.
    ITT-Intermetall, Freiburg, West Germany

GRAY, P. R.
    University of California, Berkeley, California, U.S.A.

HANSON, K.
    Texas Instruments, Inc., Houston, Texas, U.S.A.

HODGES, D. A.
    University of California, Berkeley, California, U.S.A.

HOSTICKA, B. J.
    University of Dortmund, Dortmund, West Germany

KLAASSEN, F. M.
    Philips Research Laboratories, Eindhoven, The Netherlands

LERACH, L.
    Siemens AG, München, West Germany

PFEIFER, H.
    ITT-Intermetall, Freiburg, West Germany

SCHWEER, R.
    ITT-Intermetall, Freiburg, West Germany

SEDRA, A. S.
    University of Toronto, Ontario, Canada

SENDEROWICZ, D.
    SGS-Ates, Agrate Brianza, Italy

SWANSON, E. J.
    AT&T Bell Laboratories, Reading, Pennsylvania, U.S.A.

TSIVIDIS, Y.
    Columbia University, New York, New York, U.S.A.

ULBRICH, W.
    Siemens AG, München, West Germany

VITTOZ, E. A.
    Centre Suisse d'Electronique et de Microtechnique,
    S.A. (CSEM), Neuchâtel, Switzerland, and Swiss Federal
    Institute of Technology, Lausanne, Switzerland

PREFACE

Telecommunications systems are growing explosively, both in terms of the number of users and in terms of the level of circuit sophistication. Systems with complex features are increasingly found in data acquisition, conversion, processing, networking, and transmission. Thus low cost, small size, low power consumption, and high performance are all goals of prime importance, and in many cases they must be attained simultaneously. Very Large Scale Integration is the means to achieve these goals; the numerous techniques that are involved in the process are the subject of this book.

The material covered is devoted exclusively to MOS technology, which has proven to be best suited for VLSI. The emphasis is on relevant, working techniques that can make the difference between academic knowledge and the design of successful chips. Since computational circuits and basic logic circuit design are extensively covered in numerous sources, these subjects are largely left out in the present book (except for considerations applying to the specific systems discussed). Instead, extensive material is included on new or relatively new subjects that have not been adequately described before in a book of this level, or even in technical journals. The trend of merging digital and analog functions on the same chip is evident throughout most of the book. The material is divided in three parts. Part I covers fabrication technology and device modelling. Part II deals with the detailed design of building blocks, ranging from amplifiers to filters. Part III covers representative systems which use the circuits discussed in Part II.

The book is addressed to technically sophisticated readers. This reflects the fact that individual chapters have grown out of lectures given at the Advanced Course on the Design of MOS VLSI Circuits for Telecommunications, held June 18-29, 1984 at Scuola Superiore Guglielmo Reiss Romoli (SSGRR) in L'Aquila, Italy. The course was supported by the European Economic Community and was organized by these editors. Drafts of the chapters in this book were provided to all course participants, who came from several companies and universities in various countries.

Many thanks are due to our expert authors whose high quality work on the chapters matches that of the lectures they gave in the course. Thanks are due to the course attendees, too, many of them experienced integrated circuit designers, for their active participation; they made the course a stimulating experience, and their comments and suggestions are reflected in this book. The editors would also like to thank Prof. C. Colavito, Dr. S. Breccia,

and all the staff of SSGRR for their support. They provided an
excellent setting and organization; without their cooperation the
course, and thus the present book, may not have been possible.
Finally we owe an explanation to our readers, who may find parts of
the book not completely polished in terms of style, or not totally
free from typos. Such imperfections are normally corrected during
the detailed editing process by the publisher. However, we insisted
that the publisher forgo this time-consuming process for the present
book. This was the only way we could make sure that the book would
be as timely as demanded by the rapid changes in the field of
Telecommunications VLSI.

                                    Yannis Tsividis

                                    Paolo Antognetti

PART I

FABRICATION TECHNOLOGY AND DEVICE MODELLING

INTRODUCTION

The first part of this book covers two subjects that form an essential foundation for state-of-the-art circuit design. The fabrication of MOS integrated circuits is discussed by Eric Demoulin in the first chapter; he describes several alternative processes, and provides quantitative information on process parameters element values and element value matching. The second chapter covers the modelling of MOS devices, and is written by Francois Klaassen. He considers models both for CAD and for approximate "hand" analysis.

1

FABRICATION TECHNOLOGY OF MOS IC's FOR TELECOMMUNICATIONS

Eric DEMOULIN
CENTRE NATIONAL D'ETUDES DES TELECOMMUNICATIONS (CNET)
BP 98
F-38243 MEYLAN CEDEX
FRANCE

1. INTRODUCTION

Over the last decade, the world of Telecommunications has been intensively innovated by the technology of Integrated Circuits. Every domain of public and private Telecommunications systems now relies to some degree on integrated circuits : this is true for transmission and switching systems as well as for the peripherals of telephone systems. As a matter of fact, the market share of Telecommunications ICs came to around 20 % of the total IC consumption in 1983. The Telecommunications business still shows evidence of good health, explained by the flourishing associations of computers and communications systems and the promising advent of videocommunications. It is therefore expected that the share of Telecommunications in the IC market may rise to 25 or 30 % in the coming years. This bright future has led all the IC companies to form dedicated divisions for Telecom ICs and to launch R & D activities first, on the design of specialized ICs for Telecom systems, and also, where necessary, on specially adapted processes for these circuits.

In a general public communications system (fig. 1) integrated circuits are to be found in all the subsystems : switches, interfaces to the subscriber loop, telephone sets, etc. These circuits fall into two broad categories : standard and specific circuits. Among the first, the Telecommunications industry makes wide use of general purpose microprocessors and associated peripherals, digital signal processors, various kinds of memories (RAMs, ROMs), gate arrays, PLAs, etc. The latter are dedicated circuits developed specially for Telecommunications sytems, even though some of them are fast becoming standard products. These circuits have undergone very rapid growth, particularly since the introduction of specialized circuits in switching, the integration of an increasing number of functions in telephone sets, and the development of new functions and systems to be installed in subscribers' offices and homes.

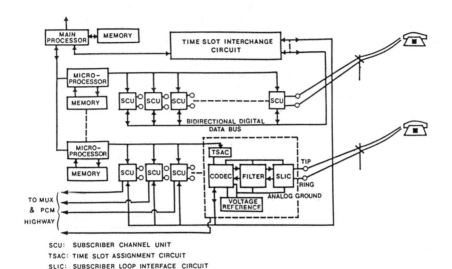

SCU: SUBSCRIBER CHANNEL UNIT
TSAC: TIME SLOT ASSIGNMENT CIRCUIT
SLIC: SUBSCRIBER LOOP INTERFACE CIRCUIT

Fig. 1.- Typical public communication system (1)

The advent of LSI MOS technologies for digital applications, introducing the possibility of integrating a large number of devices per chip at very low cost, has allowed circuit designers to adopt solutions that would have been unfeasible using discrete devices or SSI circuits. Likewise, the emergence of MOS designs for basic analog functions (operational amplifiers, filters, etc.) offering performances similar to their bipolar counterparts (2), has attracted a large number of designers in the Telecommunications industry. The above-mentioned dedicated circuits include both digital and analog functions. They are : SLICs (Subscriber Loop Interface Circuits), CODECs (Coders/Decoders), Filters, Echo Cancelers, Circuits for MODEMS, and so on. The analog parts they contain are implemented using the basic analog components provided by the "analog" technology (2) : precision resistors, precision capacitors, analog switches and operational amplifiers. The techniques for making high quality samples of the first three components will be described later in this chapter ; the design and optimization of operational amplifiers will be reviewed in subsequent chapters.

Even if the trend is to push the barrier towards "less analog, more digital" in order to take advantage of the latest advances in VLSI, the necessity of handling some analog signals will remain. Therefore there exists a definite trend towards mixing digital and analog parts on a single chip, i.e. towards maximum benefit from the progress in both worlds. Such

3

mixing has been made possible by the development of hybrid MOS processes, suitable for the realization of both high speed digital subcircuits and high performance analog parts on a single chip. These MOS processes are usually named "TELECOMMUNICATIONS MOS" processes, and result from a compromise between the requirements for high speed digital signal processing and high speed analog signal processing.

High voltage capability is sometimes quoted as also being an attribute of a Telecom process. This is not completely general. Even if a few circuits must either withstand very high surge voltages or drive high voltage lines for particular functions (the ringing in a telephone set, for instance), far more Telecom circuits do not require voltages higher than 5 or 10 volts. Low-voltage digital/analog circuits are therefore in a class by themselves, a class for which developing a dedicated process really makes sense. Where necessary and economical, an adaptation of this process to extend the range of working voltages will allow the realization of high-voltage parts. Note, however, that the number of such parts will probably drop off as copper lines are replaced by optical fibers.

In any case, the MOS technologies have shown their intrinsic qualities and advantages for the integration of large Telecommunications circuits. Both single (NMOS) and double channel (CMOS) technologies have been used extensively in developing advanced Telecom products ; they also have become an industrial reality with many parts in mass production today. These NMOS and CMOS processes will be reviewed in this chapter, together with the basic techniques on which they rely.

2. A BASIC NMOS PROCESS

A modern Silicon-gate NMOS process aims at realizing integrated circuits designed with both enhancement and depletion N-channel MOS transistors. It usually provides capability for making direct contacts between N^+ doped regions and the polycrystalline silicon (called "polysilicon") of the gate, the so-called "buried contacts". One or two layers of metallic interconnection are generally available, depending on the complexity of the circuit and the accepted process cost.

Figure 2 presents a schematic cross-section of an enhancement and a depletion transistor showing a buried contact in the latter, and a single layer of metal. Expected characteristics for these devices are, among others :
- low sheet resistance for the sources and drains (N^+ doped regions) as well as the gates (polysilicon) ;
- low resistance, high reliability contacts between the silicon and the metal ;
- perfect insulation and good stability and reliability for the gate dielectrics (silicon dioxide) ;
- low leakage currents and small capacitances in all the diodes ;
- perfect isolation between adjacent devices (no current flows below the field region) ;
- small parasitic capacitances.

4

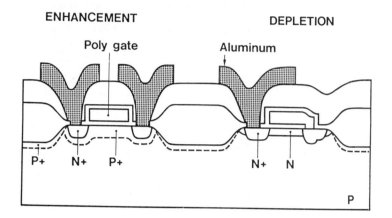

ENHANCEMENT   DEPLETION

Poly gate   Aluminum

Fig. 2.- Cross-section of an E/D NMOS inverter.

These characteristics have to be taken into account when the process is
being designed.

      Whereas only a few VLSI-like large memories and microprocessors are
currently being produced using advanced scaled-down processes, it should be
noted that a 3 to 4 micron process is typical of today's analog circuits.
For reasons of complexity, noise, and cost, analog and analog-intensive
Telecommunications circuits are most often produced using a 3 to 6 micron
technology. A basic 3.5 micron process will now be reviewed in some detail.

2.1. An NMOS process flow

      The key steps in an NMOS process flow are illustrated in figure 3, and
can be described as follows.

      One starts with <100> oriented, P-type silicon substrates slightly
boron doped to give small junction capacitances. They are first oxidized,
then covered with a thin silicon nitride layer, which is patterned in a
first photolithography step (fig. 3.a). The silicon nitride prevents the
active regions from being oxidized in a subsequent step. The nitride-free
field regions are first implanted with boron atoms to increase the dopant
concentration near the surface and, consequently, increase the threshold
voltage of the parasitic field transistor. The field oxide is then grown

locally in a so-called LOCOS (for LOCal Oxidation of Silicon (3)) process (fig. 3.b) to a thickness typically of 1 to 1.2 microns.

Next, the silicon nitride and thin silicon dioxide layers are removed from the active regions. After careful cleaning of the surface, the gate oxide is grown under rigorously controlled conditions to a thickness of about 60 nm.

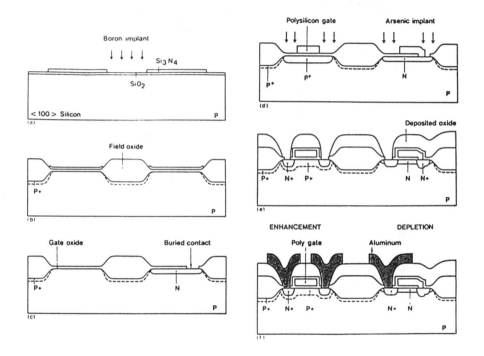

Fig. 3.- Sequence of steps in an NMOS silicon gate process.

Two implantations are then performed to adjust the threshold voltages of the transistors :
- an unmasked boron implantation to set the enhancement threshold voltage at about 0.7 Volts ;
- a masked phosphorous or arsenic implantation to set the depletion threshold voltage at the desired value in the range of -1.5 to -3 Volts.

Windows are then opened in the thin oxide (fig. 3.c) to allow for the buried contacts. Then comes the deposition of a thin polycrystalline silicon layer (typically 0.4 to 0.5 microns) which is subsequently subjected to heavy phosphorous doping to give it a low resistivity (typically 20 to 50 ohms/square). The lithography step which follows defines, at one and the same time the gates of the transistors and the first level of interconnection lines (fig. 3.d). A high dose arsenic implantation forms the N^+ sources and drains, which are self-aligned with respect to the gates, thereby minimizing the interelectrode capacitances. These N^+ regions typically exhibit a sheet resistance in the range of 15 to 40 ohms/square.

After a short re-oxidation step, an isolating oxide layer is deposited to ensure good electrical insulation between the underlying silicon and the metal to be deposited. Contact holes are then opened in a lithography step (fig. 3.e) to give access to both N^+ diffused and polysilicon regions. An aluminum layer is deposited and delineated lithographically to form the required interconnect patterns (fig. 3.f). Finally, passivation oxide layer containing windows opened over the aluminum bonding pads is deposited over the whole circuit.

Such a process requires 7 to 9 lithography steps, most of which include etching of the underlying material. These techniques and as others (oxidation, deposition of polysilicon or metal, etc) will be briefly reviewed in Section 3.

2.2. Typical results

Figure 4 shows a micrograph of the cross-section of an NMOS transistor observed by means of a Scanning Electron Microscope (SEM) (4). Some parameters and objectives for the dimensions and electrical characteristics of a typical 3.5 micron silicon gate NMOS process are summarized in Table I. It must be mentioned that an actual process yields results that are scattered around the targetted central values. This scattering arises from non-uniformities in the basic processes (across a wafer, from wafer-to-wafer in a lot) and from the sometimes poor reproducibility of a few techniques. As a first consequence, circuit designers must take these parameter fluctuations into account in their designs, and, secondly, a specialized MOS process must be built including basic techniques which minimize the scattering of the specific result to be obtained.

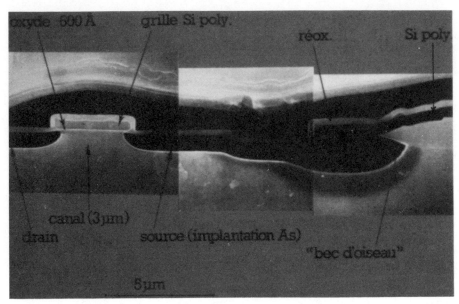

Fig. 4.- SEM cross-section of an NMOS transistor (4).

TABLE I		
Resistivities	Substrate	10-20 ohms cm
	N^+ regions	15-40 ohms/square
	Polysilicon	20-50 ohms/square
Thicknesses	Gate oxide	60 nm
	Field oxide	1 micron
	Deposited oxide	0.8 micron
	Polysilicon	0.5 micron
	Aluminum	1.1 micron
	N^+ junctions	0.5 micron
Minimum size	Polysilicon	3.5 microns
	Contact holes	4.5 microns
	Aluminum	6 microns
Voltages	Threshold :	
	– enhancement	+ 0.7 Volts
	– depletion	– 1.5 to – 3 Volts
	– field	> 12 Volts
	Breakdown	> 12 Volts

2.3. The scaling rules

The scaling down of MOS devices has been shown to have a large potential for greatly improving the performances of digital integrated circuits. The original work demonstrating this (5) introduced the concept of "constant field" (CF) scaling according to which proper adaptation of the dimensions, doping, and voltages keeps the electric field constant in the devices. More recently, two scaling laws have been proposed (6) : the "constant voltage" (CV) and the "quasi constant voltage" (QCV), respectively. The CV law takes into account the difficulty of having a large voltage supply in small devices as well as problems associated with hot carrier-related reliability and operation close to breakdown conditions in the junctions and thin oxides.

The scaling laws can be defined as a set of multiplying factors based on a dimensionless factor k (> 1) which are applied to the device parameters. Table II gives the main factors for the CF, CV, and QCV scaling laws. It should be mentioned that these scaling rules simply provide guidelines for designing a scaled process the final parameters will also depend on product-related and technological considerations. For example, following the CV law, the supply voltage has pinned at 5 Volts in spite of a dramatic reduction in dimensions from 4 microns down to 1.5 micron. However, the occurrence of excess currents related to impact ionization in one micron and submicron devices will probably lead to a QCV scaling for reasons of reliability.

Up to now, the definite trend towards reduced dimensions has strongly affected digital MOS processes. The question now arises among circuit designers as to whether or not it will also affect the analog and Telecom circuits. This point will be discussed in Section 6.

	TABLE II		
Device parameters	CF	CV	QCV
Horizontal dimensions	k^{-1}	k^{-1}	$k^{-1/2}$
Vertical dimensions	k^{-1}	$k^{-1/2}$	k^{-1}
Doping	k	k	k
Operating voltages	k^{-1}	1	$k^{-1/2}$

3. THE BASIC TECHNOLOGIES

Any MOS process is nothing but a putting together in the right way of the basic steps outlined in the brief description of the NMOS process. They are : oxidation, dopant introduction and diffusion, microlithography, etching, deposition of thin layers of dielectric material (oxide, nitride), deposition of conductive material (polycrystalline silicon, aluminum). These steps will now be briefly reviewed with emphasis on their recent evolution in the production of scaled down devices.

3.1. Oxidation

Typical oxide thicknesses in MOS devices are in the range of 100 to 20 nm for the gate insulator and of 1 micron to 0.5 micron for the field regions. These oxides result from the oxidation at high temperature of the silicon material, in an oxygen or steam ambient. They are called <u>thermally grown</u> oxides. As the oxide thickness has decreased in compliance with the scaling rules, the oxidation temperature has dropped from the 1100°C down to the 900-950°C range. The oxidation time is selected to yield the desired thickness, as determined by typical kinetics curves (fig. 5).

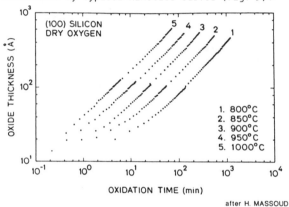

after H. MASSOUD

Fig. 5.- Oxidation of <100> Silicon in dry oxygen : oxide thickness vs. time for various oxidation temperatures (7).

The key parameters to be controlled are the oxide thickness and the charges at the interface. The scattering of these values is responsible for part of the distribution in the threshold voltages of MOS devices. The thickness distribution over a long period of time typically shows an overall standard deviation of 3 % while the distribution in a lot shows an average standard deviation of less than 1 %. No degradation is observed when the oxide thickness is reduced.

Figure 6 illustrates typical breakdown characteristics for a 25 nm thermally grown oxide (8). Low-field breakdowns refer to pathological devices, and are significative of gross defects in the capacitors. Their number gives the defect density for an oxidation process. High-field breakdown sets the useful limit of MOS capacitors. Typical results lie in the 10-12 MV/cm range. A 60 nm oxide withstands 50 to 60 V without disrupting ; no degradation occurs when the thickness is decreased, since a 25 nm oxide exhibits a breakdown voltage of over 25 volts.

The previous description holds for thin oxides thermally grown on single-crystal Silicon. Oxide layers are also thermally grown on doped polycrystalline silicon (the so-called <u>polyoxides</u>). The rugosity of the polysilicon surface, resulting from the granular nature of this material

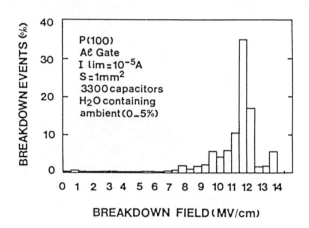

Fig. 6.- Histogram of the percentage of breakdown events of Al gate MOS capacitors on P(100) silicon (Oxide thickness = 25 nm).

(see 3.6), is transferred during oxidation, producing a sometimes highly inhomogeneous oxide layer. Figure 7 shows an SEM micrograph of the cross-section of a polyoxide grown under conditions that enhance the inhomogeneity (9). Even optimized growth conditions yield oxides which display worse thickness distributions and lower high-field breakdowns than oxides grown on single-crystal silicon.

3.2. Dopant introduction and diffusion

In modern technologies, dopants are introduced in the silicon substrate mainly by ion implantation, by far the most accurate and flexible technique. In MOS processes, ion implantation is used to adjust the threshold voltages of both active and field devices. It is also used to create the N^+ or P^+ heavily doped regions required for N^+P, P^+N junctions and interconnections. Finally, by implantation of the proper ion dose, the desired sheet resistivity can be obtained in silicon for use in resistance formation.

For N^+ regions, phosphorous and arsenic are the implanted species used for deep and shallow junctions respectively. For P^+ regions, only boron is

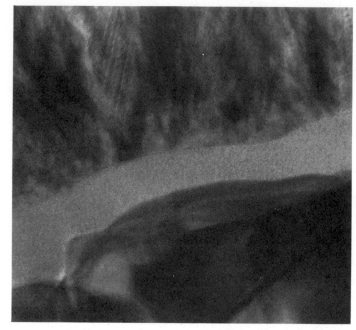

Top polysilicon

Polyoxide

Bottom
polysilicon

Fig. 7.- SEM cross-section of an inhomogeneous polyoxide (9).

available and normally makes deep junctions. It must be pointed out that the ion implants yielding shallow doped layers have to be followed by a high temperature step (at least 30 min at 900-1000°C) which electrically activates the implanted dopants, for the implant damages the crystal lattice and the dopants themselves are electrically inactive). During this anneal step, the impurities move away from the surface causing the increase in junction depth. This is mainly true of boron and phosphorous which diffuse much faster than arsenic.

This now classical technique allows junctions to be made in devices scaled down to 0.25 or 0.3 micron N^+P junction depths and 0.4 micron P^+N junction depths. For shallower junctions, new techniques such as transient or flash annealing (using higher temperatures, 1000 to 1150°C, for very short times, 10 to 20 sec) are currently under investigation.

A crucial question for N^+P and P^+N junctions is the leakage current. Typical values for accepted currents at room temperature are 0.01 to 1 fA/square micron over the area of the diode, and 0.1 to 10 fA/micron along its perimeter. These currents are the not well understood result of many factors among which can be mentioned : the quality of the starting substrate, the thermal history of the wafer, the density of defects such as metallic impurities and dislocations, etc. In essence, the leakage currents are highly sensitive to the process sequence itself and to many of the individual steps.

For the sake of completeness, it should be said that the "old" furnace doping technique is still used for heavy doping of the polysilicon layers, which are deposited undoped (see 3.6).

3.3 Lithography

Lithography means transfer of a pattern onto a thin (about 1 micron thick) layer of photosensitive material deposited on the wafer (fig. 8). This material, called photoresist, must be able to withstand the subsequent etching without uncontrolled damage, so that the pattern generated in the resist is conformally transferred onto the underlying layer.

Three main characteristics help in assessing a lithography process, especially in its exposure step. The first is the resolution or minimum feature size obtained in the resist ; the second is registration accuracy

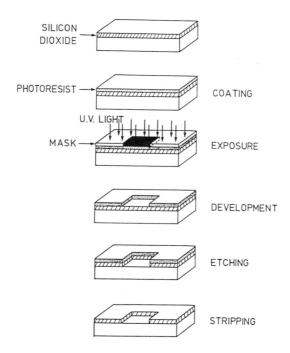

Fig. 8.- Basic steps in a lithography process.

which indicates how well a lithography level has been aligned with respect to the previous one. These parameters are essentially exposure equipment dependent. The third characteristic is the defect density and results from the overall cleanliness of the process.

Figure 9 presents the different strategies used in lithography today. Contact printing and proximity printing were and are still extensively used for geometries in the 8 to 4 micron range, which remains typical of some Telecom products. Their main drawback is the high level of defect density which restricts them to the production of low complexity circuits. 1:1 projection printing is today the most widely used lithography technique in the range of 5 to 2.5 microns : since there is no contact between the mask and the resist, the defect density can be kept at a low level, making this technique a good choice for volume production of high complexity circuits. The registration accuracy shows values of 1.5 to 0.5 micron at three sigma.

The pressure for reduction of feature size in VLSI products such as memories and microprocessors has led to their being produced according to the most advanced, but also the most expensive, optical printing technique, namely direct stepping on the wafer. In this case, each dye on the wafer can be aligned individually. The resolution is in the range of 1.5 to

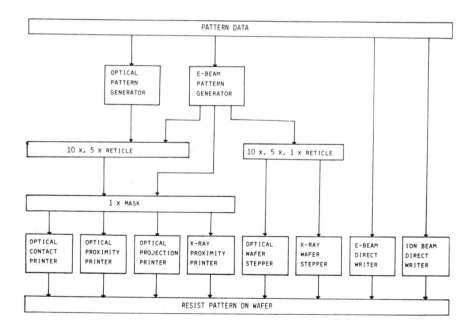

Fig. 9.- Printing techniques for IC lithography.

14

0.8 micron, while the registration accuracy should reach values at three sigma between 0.6 and 0.2 micron. It is likely that advanced Telecom products will soon benefit from the improvements arising from wafer stepper lithography, but the cost of this technique will limit it at first to specific applications.

3.4. Etching

The transfer of the resist pattern onto the underlying layer is done during the etching step. The original processes involved using chemicals in a so-called <u>wet etching</u> technique : the material is dissolved in a chemical solution (ex. : SiO_2 in HF, Al in H_3PO_4, etc.) which does not attack the protective resist layer. Since wet etching is isotropic, some lateral etching occurs, making dimension control rather difficult. <u>Dry etching</u> techniques are based on the etching action of species created in a plasma. Classical plasma conditions yield isotropic etching, while reactive ion etching (R.I.E.), which adds a "bombardment" component, provides essentially anisotropic etching.

Actually, the best dimensional control is obtained under R.I.E. conditions (fig. 10). In a poly layer, the combination of standard plasma etching and 1:1 projection printing brings about an average dimension loss of 0.82 micron and a standard deviation per wafer of 0.12 micron. The same printing technique associated to reactive ion etching gives an average loss of 0.56 micron and a standard deviation of 0.07 micron. Finally, the use of a wafer stepper and R.I.E. brings these values down to 0.4 and 0.03 micron respectively.

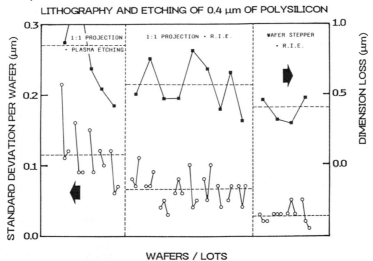

Fig. 10. - Loss of dimensional control in a lithogaphy/etching process (full squares, right-hand scale) and standard deviation per wafer (open circles, left-hand scale) (10).

3.5. Dielectric deposition

In cases where silicon dioxide cannot be thermally grown from silicon, it can be deposited by supplying both silicon and oxygen from a gas mixture. Chemical vapor deposition (CVD) of silicon dioxide results from the pyrolitic decomposition of silane (SiH_4) in the presence of oxygen either at atmospheric pressure (AP-CVD) or low pressure (LP-CVD). The deposition temperature is chosen between 400°C (Low Temperature Oxide - LTO) and 900°C (High Temperature Oxide - HTO) ; the deposition is sometimes enhanced by plasma (PE-CVD). These CVD oxides exhibit large variations in thickness (long term standard deviation over 5 or 6 %) and a density of defects greater than that of thermal oxides. They also suffer from charge-voltage hysteresis effects when deposited on silicon, thereby making them poor candidates for MOS capacitors.

Silicon nitride can also be deposited in an LP-CVD reactor via the decomposition of dichlorosilane (SiH_2Cl_2) and ammonia (NH_3) at a temperature of 800°C. The nitride thickness can be controlled easily and exhibits a typical long term standard deviation of from 3 to 3.5 %.

3.6. Conductive layers

Polycrystalline silicon deposition is normally performed in an LP-CVD reactor and occurs as a result of pyrolysis of silane (SiH_4). Typical layers are 0.3 to 0.6 micron thick ; long term standard deviations are in the 2 % range. The polysilicon can be doped in various ways : either in-situ by adding, for example, a phosphorous compound to the gas mixture, by ion implantation, or by furnace predeposition of dopants. The cost effectiveness of this last technique justifies its use for creating low resistivity material (sheet resistances of 15 to 30 ohms/square are obtained by predeposition of phosphorus at about 1000°C). Ion implantation is used when a low and/or accurate dopant concentration is required, for example when realizing high value or precision resistors. The final grain size of the polysilicon layer depends on a number of factors : the initial grain size (the film is amorphous if deposited below 600°C, but has a grain size of a few tens of nanometers when deposited at the standard temperature of 620°C), the doping mode, the doping level, the thermal history of the film (11). Typical results give an average grain size of a few hundreds of nanometers.

In high speed digital circuits the resistance of polysilicon interconnections introduces delays which may limit the operation frequency of the circuit. In order to lower the resistivity of polysilicon, a sandwich layer, the so-called polycide (12), is now being used extensively : a thin layer (0.2 microns) of standard polysilicon is covered with a film of a refractory metal silicide (WSi_2, $MoSi_2$, etc). The final sheet resistance reaches the 1 to 5 ohms/square, providing a tenfold improvement in propagation delay on the line.

Low resistivity interconnects rely almost exclusively on aluminum thin films, which are deposited in vacuum systems, either by e-gun evaporation, or, as is most often the case, by sputtering. The resistivity of aluminum is about 3×10^{-6} ohms/cm. One of the difficulties with aluminum involves the realization of reliable, low-resistance contacts with silicon. Interdiffusion of aluminum and silicon tends to create short circuits in shallow P-N junctions, thereby limiting the annealing temperature or imposing a diffusion barrier at the interface. Typical specific contact resistances are in the range of 200 to 400 ohms x sq. microns for present day technologies.

4. TELECOMMUNICATIONS CMOS

A "Telecommunications process" has already been defined previously as one capable of realizing both digital and analog functions. In MOS, the analog functions are more easily implemented with complementary devices than with single channel devices ; in particular, the basic operational amplifier offers larger gains in CMOS than in NMOS, while taking up less space on a chip. CMOS has indeed become the first choice for Telecom ICs where its superior behavior in power dissipation turns out to be a strong advantage, especially for the applications with remote supply. Even if clever designs succeed in keeping attractive the NMOS parts, the future looks brighter for CMOS than NMOS, particularly since the difference in complexity between both processes has been greatly narrowed down.

4.1. The basic digital CMOS

Figure 11 shows a schematic cross-section of a basic CMOS structure in which the P-channel transistor has been placed in a diffused N-well. The issue of well type has been discussed extensively over recent years and is still a matter of controversy. The points to be considered in making the choice for a digital CMOS are (13) : latch-up resistance, performances of the N-channel vs the P-channel transistor, number of N-channel transistors, resistance to input overvoltage, availability of an NMOS process, etc. No definite argument exists in favor of a single solution that would apply in all circumstances. The choice of well type has to be made case by case, on the basis of product considerations. However, economical or historical reasons often bias the selection process.

An N-well CMOS process offers several advantages. It can be built on the same substrate as an NMOS process, from which many basic steps can be adopted. The N-channel transistors made in a lowly doped substrate exhibit high mobility, low substrate capacitances, and low body-effect and are therefore superior to their P-well counterparts made in a higher doping region. An N-well process can thus yield faster circuits especially if their designs make use of more N-channel than P-channel transistors.

Typical parameters for a 3.5 micron N-well CMOS process are presented in Table III.

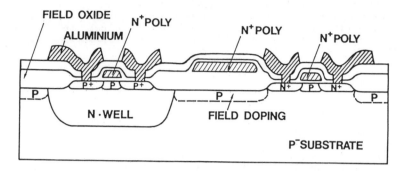

P·CHANNEL MOST   N·CHANNEL MOST

FIELD OXIDE   N⁺POLY   N⁺POLY   N⁺POLY
ALUMINIUM

Fig. 11.- Cross-section of a N-well CMOS structure

	TABLE III		
Resistivities	Substrate	5-10 ohms cm	
	N^+ regions	15-40 ohms/square	
	P^+ regions	50-100 ohms/square	
	Polysilicon	20-50 ohms/square	
	Well	4-6 k ohms/square	
Thicknesses	Gate oxide	60 nm	
	Field oxide	1 micron	
	Deposited oxide	0.8 microns	
	Polysilicon	0.5 microns	
	Aluminum	1.1 microns	
	N-well	5 microns	
	N^+ junctions	0.5 microns	
	P^+ junctions	0.6 microns	
Minimum sizes	Polysilicon	3.5 microns	
	Contact holes	4 microns	
	Aluminum	5 microns	
	N^+P^+ spacing	10 microns	
Voltages	Threshold		
	− N-channel	+ 0.7 Volts	
	− P-channel	− 0.7 Volts	
	− field	> 12 Volts	
	Breakdown	> 12 Volts	

4.2. Specific components for analog CMOS circuits

Precision resistors and capacitors, as well as switches are the key basic components required for the design of MOS analog circuits (2, 14). In most cases, the absolute value of the resistor or capacitor is not as important as the ratio of such elements which provide the precision components in the critical parts of these circuits.

4.2.1. Resistors : In a silicon-gate CMOS process, resistors can be formed from the N^+ or P^+ source-drains diffusions, the diffused well or the polysilicon lines. Figure 12 shows different resistor types in such a process. High resistances values (up to several hundred kilohms) are often required in filter design and realized via "serpentine" layouts. The absolute resistance value is given by :

$$R = \rho_\square \frac{L}{W}$$

where $\rho_\square$ represents the sheet resistance of the material, and L and W are the length and width of the resistor, respectively. If these parameters are assumed to be statistically independent, the standard deviation of the resistance value is given by :

$$(SD)_R = \left[\left(\frac{\delta\rho_\square}{\rho_\square} \right)^2 + \left(\frac{\delta L}{L} \right)^2 + \left(\frac{\delta W}{W} \right)^2 \right]^{1/2} .$$

In most cases, L is much larger than W ; therefore, the previous expression is simplified as :

$$(SD)_R = \left[\left(\frac{\delta\rho_\square}{\rho_\square} \right)^2 + \left(\frac{\delta W}{W} \right)^2 \right]^{1/2} .$$

Generally speaking, the first term is smaller
- for ion implant than for furnace predeposition ;
- for bulk silicon than for polysilicon (the increased variation being due to the granular structure of this material).

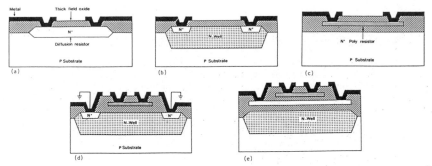

Fig. 12. - Resistor types in an N-well silicon-gate CMOS process : (a) N$^+$ (or P$^+$) source-drain regions ; (b) diffused N-well ; (c) polysilicon ; (d) polysilicon with substrate shield ; (e) a double-polysilicon distributed RC structure. After (14).

The second term has been greatly reduced with the advent of plasma etching techniques, especially RIE, which have replaced wet etching. Typical widths were 5 microns but smaller values can now be accepted with advanced lithography techniques.

The resistance ratio is the precision component is some designs. Symmetrical, interdigitated layouts are then used to compensate for long-range gradients in sheet resistance and width. Relatively large widths (20 to 50 microns) give precision resistor pairs that typically match to within about 0,25 % (15).

Temperature coefficients and voltage coefficients are in the range of 200 to 10 000 ppm/ºC and 20 to 20 000 ppm/Volt, respectively (see (14), Table I, p. 970).

Parasitic capacitance with respect to the substrate may add high-frequency noise to the signal flowing through the resistor. Electrically shielded resistors (cases (d) and (e) from fig. 12) present an advantage over other designs.

4.2.2. Capacitors : Capacitors in analog MOS circuits are mostly based on the use of silicon dioxide for the dielectric layer ; some technologies use a sandwich of silicon dioxide/silicon nitride, and thus take advantage of the higher dielectric constant of silicon nitride. Deposited oxides are generally not suitable for precision capacitors because of their large variations in thickness (see section 3.5). Thermally grown oxides can be formed from either single-crystal silicon or polycrystalline silicon (polyoxides). A variety of capacitor configurations is illustrated in figure 13.

The important characteristics of MOS capacitors are : absolute accuracy, ratio accuracy, voltage and temperature coefficients, parasitic capacitances. Ideally, the value of a capacitor is given by :

$$C = A \frac{\varepsilon_{ox}}{t_{ox}} = WL \frac{\varepsilon_{ox}}{t_{ox}}$$

where A is the capacitor area (W and L are the side dimensions of a rectangular capacitor), ε_{ox} is the permittivity of the oxide, and t_{ox} is the oxide thickness. If these variables are statistically independent, the standard deviation of the capacitance value will be :

$$(SD)_C = \left[\left(\frac{\delta W}{W} \right)^2 + \left(\frac{\delta L}{L} \right)^2 + \left(\frac{\delta \varepsilon_{ox}}{\varepsilon_{ox}} \right)^2 + \left(\frac{\delta t_{ox}}{t_{ox}} \right)^2 \right]^{1/2} .$$

W and L are generally chosen equal ; the corresponding errors depend on the quality of the lithography and etching steps (see sections 3.3 an 3.4) and are called <u>edge effects</u>. The last two <u>oxide effects</u>. Edge effects dominate in small capacitors, while oxide effects determine the accuracy of large capacitors ; the crossover point occurs for dimensions ranging from about 20 to 50 microns (14). A discussion on the accuracy of capacitors accuracy can be found in references (16) and (17).

Note that, in the case of ratio accuracy, critical for switched-capacitor filters and A/D converters, the effect of edge variation can be minimized by building ratioed capacitors from varying numbers of unit

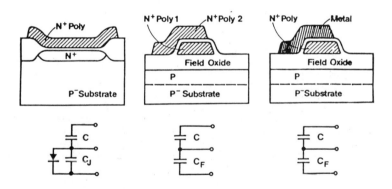

Fig. 13.- Capacitor types in an N-well CMOS process involving either one or two polysilicon layers.

capacitors. Systematic variation is effectively cancelled, but some random variation still remains. Similarly, systematic variations in oxide thickness are rather well cancelled by the choice of common centroid layouts, while random variations will depend on the particular technology.

- The thin gate oxide of MOS transistors can be used for making MOS capacitors. Gate oxides are of higher quality than polyoxides as far as thickness control and breakdown characteristics are concerned (see section 3.1). However, the formation of the lower electrode requires additional process steps : a selective high-dose implant has to be performed in the substrate to form the P^+ (or N^+) layer before the deposition of polysilicon. The polysilicon layer then serves simultaneously for the transistor gates and the upper electrode of the capacitors. The main drawback of this configuration is the relatively large parasitic capacitance occurring between the P^+ (or N^+) layer and the substrate.

- The polyoxide must be sufficiently thick to ensure breakdown voltages larger than the maximum voltage in the circuit ; typical values are 100 to 150 nm. Owing to the poorer oxide quality, the ratio accuracy of capacitors made on polysilicon is usually poorer than that obtained from oxides grown on single-crystal silicon. Nevertheless, the better isolation with respect to the substrate is beneficial ; the parasitic capacitance between the polysilicon and the substrate is about ten times smaller than the useful capacitance. In addition, capacitors can be electrically shielded from the substrate by grounding the silicon below the polysilicon lower electrode.

When the first polysilicon layer is used as the lower electrode, the upper electrode can either be an aluminum or a second polysilicon layer. The former solution yields asymmetrical capacitors (polysilicon on one side, metal on the other). The latter requires extra processing steps (polysilicon deposition, doping and patterning) but provides and additional interconnection layer. An interesting feature arises from the highly symmetrical structure : both positive and negative voltages give rise to the same capacitance behavior.

The ratio accuracy of capacitors on a given chip depends mainly on their area ratio (especially if we try to minimize the capacitors' area in order to increase the integration density). Improvements in lithography and etching (see sections 3.3 and 3.4) will improve the overall capacitor ratio accuracy and/or allow reduced dimensions of the minimum size unit capacitor. Typical results in ratio accuracy range from about 1 % for small geometries in a double polysilicon structure (2) to about 0.03 % for large geometries where the oxide has been grown on single-crystal silicon (16). Note that the matching characteristics of capacitors are strongly dependent on the processing conditions and, therefore, must be determined for each MOS process as well as for each processing facility.

The voltage coefficient is small, especially when MOS capacitors are made using silicon plates that are heavily doped to approximately the same doping level. Such capacitors display voltage coefficients in the range of 10 to 200 ppm/Volts (16, fig. 6). The temperature coefficient is usually very low, in the range of 1 to 10 ppm/°C (16, fig. 12). The associated capacitance variations are small enough to be insignificant in most applications.

4.2.3. MOS transistors : In analog circuits, enhancement MOS transistors are typically used as charge switches and amplification devices. MOS switches are characterized by their ON resistance, OFF leakage currents, and the parasitic capacitances. A minimum-sized MOS transistor (W = L   5 microns, for example) exhibits an ON resistance of a few kilohms ($R_{ON} = 1/g_m$ where g_m denotes the transconductance of the device). From the standpoint of technology, the gate oxide should be as thin as possible, while the mobility should be maximized ; this is exactly the trend being followed in the scaling down of devices for digital applications. The leakage current to the substrate must be kept to a low level (see section 3.2) so as to be negligible in most applications ; for a unit diode currents of about 10 fA are easily obtainable in MOS technologies. Finally, one must minimize parasitic capacitances by reducing the dopant concentration in the substrate as far as possible . Values in the range of 5 to 20 fF can be obtained, and must be compared to the values for the capacitors which range from 1 to 30 pF .

In operational amplifiers, the above likewise holds true. In addition, the gain in amplification, which in MOS transistors depends on the output impedance, forces the designer to use long-channel devices, thereby increasing the die area. Reduced geometries are suitable, provided the MOS transistors present flat output characteristics in the saturation region, or if some circuit tricks are used to obtain an equivalent result. The noise figure of the devices is also of critical importance in analog circuits. The low-frequency noise (or 1/f noise, or flicker noise) depends on the device processing conditions. Therefore, the noise parameters must be experimentally determined for each device at each specific operating point. N-channel transistors exhibit larger 1/f noise levels than P-channel devices, especially if the implanted dose in the channel is larger (14). As far as thermal noise is concerned, N-channel devices seem to be slightly better than the P-channel ones.

4.3. Telecommunications CMOS process

As defined previously, a dedicated Telecommunications CMOS process mixes the digital aspects exposed in section 4.1 with the analog components described in section 4.2. The digital section uses a standard 5V power supply, while the analog circuit is operated at the usual - 5V and + 5V levels. Reports have been published on CMOS processes, digital and analog, exhibiting P-well, N-well, and Twin-well configurations.

As for purely digital circuits, the choice of the well type for Telecom CMOS is an endless matter of discussion. Nevertheless, the N-well structure has been adopted in many designs. N-channel transistors exhibit superior characteristics, as pointed out in section 4.1 for the digital blocks. High-gain N-channel devices are also necessary for linear circuits and fast I/O buffers, and for driving heavily loaded lines. In some designs for operational amplifiers (18), the input transistors are put in a well which is tied to the sources of the input transistors. In an N-well structure the input transistors are P-channel devices, which is beneficial in terms of noise ; this probably arises from the fact that the P-channel transistor in an N-well process is "slightly depleted", since the V_T adjust boron implant forms a conductive channel away from the $Si-SiO_2$ interface. The main drawback of the N-well configuration is that, the P substrate being tied to the most highly negative power supply (- 5V, for example), the digital blocks operate between - 5V and 0V. A voltage translation is thus required to interface these circuits to the 0-5V external world.

Precision capacitors have been realized either between a heavily doped substrate and N^+ polysilicon or between two layers of polysilicon. The respective advantages and drawbacks of the two approaches have already been summarized in section 4.2.2. The lower parasitic capacitance inherent in the latter configuration, and its ability to provide capacitor shielding, has generally led to the adoption of the double polysilicon structure.

As a consequence, the N-well double polysilicon CMOS process is now widely accepted as a typical Telecom CMOS process. A cross-section of the basic devices is shown in figure 14. The process flow for designs of about 4 microns is as follows : the P-type substrate is similar to the substrate used in the afore-described NMOS process, but is usually less resistive. The N-well regions are lithographically defined and implanted with the proper dose of phosphorous ions (in the 10^{12} atoms/cm^2 range) such that, at the end of the process, well depth will be about 5 microns and surface concentration of a few 10^{16} atoms/cm^3. The active regions are then defined by a LOCOS technique, just as in NMOS. The P field region is boron-implanted to increase the field threshold voltage, whereas the N field region over the N-well does normally not require this extra doping. The next steps include formation of the gate oxide (about 60 to 80 nm), boron implant to adjust threshold voltage, deposition and doping of polysilicon, and definition of the gates. The N^+ and P^+ source-drains are formed by selective ion implantation of arsenic or phosphorous, and boron respectively.

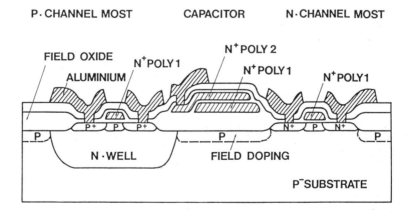

Fig. 14.- Cross-section of an N-well double polysilicon CMOS structure.

Typical doses are in the range of 1 to 5 x 10^{15} atoms/cm^2. A careful oxidation step then provides the dielectric layer for precision capacitors on top of the polysilicon ; typical thicknesses are around 100 nm. The second polysilicon layer is deposited, doped and patterned. The contact and interconnect operations are then performed ; note that contacts to N$^+$ and P$^+$ regions must be made simultaneously. A total of 10 to 13 lithography steps are required to complete such a process.

The dopant concentrations in both active and passive devices result from trade-offs between contradictory requirements. For example, dopant concentration in the substrate must be low, to maximize mobility and minimize the junction capacitances, and thereby comply with the high speed requirements. At the same time, it must be as high as possible to decrease the resistive paths in the bulk and, consequently, improve latch-up immunity. Dopant concentration in the well is selected with the following guidelines in mind : it must, 1. yield the right P-channel threshold voltage after the V_T adjusting boron implant (Note that a single implantation step suffices for adjustment of the threshold voltages in both N- and P-channel active devices.) ; 2. provide a high enough threshold voltage over the N field region ; 3. minimize the junction capacitances ; 4. minimize the resistive paths to increase the resistance to latch-up. The well depth must be large enough to prevent complete well depletion (when the source/drain to well depletion region reaches the well to substrate depletion region) and decrease the current gain of the vertical NPN transistor (to kill the latch-up susceptibility). But it must be small enough to limit lateral diffusion and allow for tight N$^+$ to P$^+$ spacing for obvious reasons of packing density. In active devices, he channel dopant concentration must be adjusted to give the right threshold voltages (+ 0.7 volts and - 0.7 volts for the N- and P-channel transistors respectively), adequate subthreshold characteristics, and clean behavior even when drain-to-source voltages are in excess of 10 volts.

4.4. Hybrid CMOS/bipolar process

Bipolar devices present an advantage over MOS transistors in terms of drive capability, especially in the output buffer stages. They can also operate at higher voltages with standard off-the-shelf technologies. Finally, as they have been the basic devices for analog circuits in the past, a wide variety of such designs is available ; hence the interest in combining both MOS and bipolar devices on a single chip. Specific processes have been designed and optimized to get the best of both worlds (20). Figure 15 shows the cross-section of a CMOS/bipolar structure. The gain in performance is unfortunately counterbalanced by the cost of the technology. Nevertheless, a few IC companies already propose this solution for specific high-voltage applications (21).

5. HIGH VOLTAGE TECHNOLOGIES

The line interface circuits in a digital switching system provide several primary functions, well known as the BORSHT functions (battery feed, overvoltage protection, ringing, supervision, hybrid or two-wire to

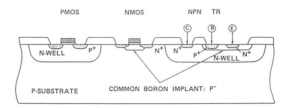

Fig. 15.- Cross-section of a CMOS/bipolar structure (20).

four-wire conversion, test). The SLIC has to be able to handle high voltages (ringing, power-voltage connection) and withstand large foreign potentials, such as 50 or 60 Hz potentials induced from power lines or lightning. Most of the actual implementations involve a low-voltage integrated circuit and high-voltage components to ensure the high-voltage functions. High-voltage realizations have been reported in both the dielectric-isolated technology (22) and the junction-isolated technology (23). In both approaches, the active devices are bipolar transistors.

   The MOS technology has been proposed for high-voltage applications through the use of DMOS transistors which combine high blocking capabilities with good current driving characteristics (24). A N-channel DMOS structure is realized by simultaneous diffusion of P type and N type dopants from the source window (Fig. 16). The P-type surface concentration near the source determines the threshold voltage. The N-drift region between this P-type region and the N^+ drain can absorb sharp drops in voltage, thereby allowing the device to be operated at high voltages. This device can be isolated from the substrate either by reverse biasing a PN junction (25) or by using a dielectric-isolated tub for 400 V blocking capability (26). Both technologies feature integration of the low-voltage control and interface circuitry, and of the high-voltage functions on the same substrate.

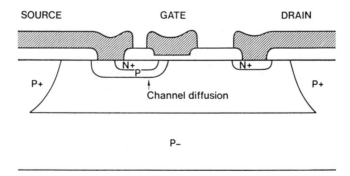

Fig. 16.- Cross-section of a high-voltage DMOS structure.

26

6. SCALED DOWN TELECOM PROCESSES

Device scaling based on the constant field, constant voltage, and quasi-constant scaling laws (see section 2.3) has been successfully applied to digital circuits with notable gains in speed and power consumption. Clearly, analog designers would also appreciate taking advantage of the higher speed and device integration resulting from device scaling. Typical operational amplifiers indeed use MOS transistors having channel lengths of more than 6 microns, whereas the high speed digital circuits currently being designed have channel lengths of less than 2 microns.

The performances of scaled analog circuits have been evaluated by weighing such key parameters as : gain, bandwidth, mismatch, and signal-to-noise ratio (27-29). The scaling relationships of the scaling parameters are shown in Table IV where k is the scaling factor defined in section 2.3. The analog performance parameters are invariant to the application of CF scaling, with the exception of speed which increases while the signal-to-noise ratio (S/N) on the other hand is severely degraded. Nonconstant field scaling laws give substantial improvements in speed and frequency response. Nevertheless CV scaling is affected by a large voltage

TABLE IV			
	CF	CV	QCV
Gain stage in CMOS, A_V	1	$k^{-1/2}$	$k^{-1/4}$
Operational amplifier in CMOS:			
- A_V, tot	1	k^{-1}	$k^{-1/2}$
- Power/Area	1	$k^{5/2}$	$k^{3/2}$
- GBW product	k	k^2	$k^{3/2}$
- Slew rate/V_{DD}	k	k^2	$k^{3/2}$
- S/N	k^{-1}	$k^{-1/2}$	$k^{-1/2}$

gain degradation and increased power dissipation per unit area. Lastly, QCV scaling constitutes an acceptable compromise offering limited degradation in voltage gain, power dissipation and signal-to-noise ratio.

Experimental results on three NMOS op amp designs are given in (29), for typical 8.5 micron, 4.5 micron and 2.5 micron design rules, respectively. They are summarized in Table V. The transient response as well as the integration density are greatly improved by decreasing the device dimensions, whereas the signal-to-noise ratio is only somewhat reduced. Part of the expected degradation, however, is compensated for a lowering by 1/f noise levels. This has been demonstrated in recent technologies and is seen as a result of better process control and cleaner techniques.

The minimum dimensions in analog circuits have been reduced progressively over the past years with respect to the 5 to 8 microns initially adopted for most designs. Scaled-down CMOS processes have been used for :
- an analog adaptative equalizer IC for wide-band digital communication networks (30) : 4,5 and 2,5 microns were the minimum gate lengths in the analog and digital parts, respectively.

TABLE V			
CMOS OPERATIONAL AMPLIFIERS	8.5 microns	4.5 microns	2.5 microns
Power supplies	$^+$ 7.5 Volts	$^+$ 5 Volts	$^+$ 5 Volts
Area	0.156 mm^2	0.062 mm^2	0.035 mm^2
Gain Bandwidth product	1.5 MHz	9.6 MHz	13.0 MHz
Positive slew rate (+ 1 Volt step input)	2.15 V/usec	13.8 V/usec	16.4 V/usec
Low frequency open loop gain	65 dB	66 dB	60 dB
Input offset voltage			
- mean	- 5.8 mV	0.90 mV	- 2.4 mV
- standard deviation	3.4 mV	5.05 mB	3.87 mV
Signal-to-noise (1/f) ratio for voltage follower	104.1 dB	103.0 dB	96.6 dB

- a fully integrated active RC filter (31) based on 3.5 micron design rules in a twin-tub CMOS process (32).

- a switched-capacitor adaptative line equalizer for a high-speed digital subscriber loop (33) operated at 200 kb/sec ; operational amplifiers designed in a 3 micron CMOS process work on a single 5 Volt power supply.

- a switched-capacitor variable line equalizer (34) in a 2.5 micron CMOS technology (5 volts single power supply is used for the operational amplifiers).

- 2 micron CMOS switched-capacitor circuits for analog video signal processing (35), likewise operated at a single 5 Volt power supply.

It must be pointed out that the trend towards shorter channel lengths is necessarily accompanied by a reduction in supply voltage. Drain-induced barrier lowering or avalanche multiplication give undesirable current increase in saturation for short channel transistors. Moreover, overcoming this will call for a careful optimization of the doping concentration in the channel and around the drain regions.

7. CONCLUSIONS

Today, the MOS technologies are used extensively for the fabrication of analog circuits and Telecommunications circuits that contain both analog and high speed digital functions. This family of technologies and circuits has now reached a good level of maturity. Recent improvements in the basic techniques for MOS processes will provide even better performances and allow for the monolithic integration of more complex functions. It is expected that the specific Telecommunications circuits will move progressively into the VLSI world by taking advantage of the breakthroughs accompanying the evolution of digital IC technologies.

8. REFERENCES

(1) C.A.T. SALAMA, IEEE J. Solid-State Circuits, SC-16, 253-260 (1981)

(2) D.A. HODGES, P.R. GRAY and R.W. BRODERSEN, IEEE J. Solid-State Circuits, SC-13, 285-294 (1978)

(3) J.A. APPELS et al., Philips Res. Rep., 25, 118 (1970)

(4) Courtesy of J.P. GONCHOND

(5) R.H. DENNARD, F.H. GAENSSLEN, H.N. YU, V.L. RIDEOUT, E. BASSOUS, and A.R. LEBLANC, IEEE J. Solid-State Circuits, SC-9, 256-268 (1974)

(6) P.K. CHATTERJEA, W.R. HUNTER, T.C. HOLLOWAY, and Y.T. LIN, IEEE Electron Dev. Lett., EDL-1, 220-223 (1980)

(7) H.Z. MASSOUD, Ph. D. Dissertation, Stanford University, Technical Report N° G502-1 (1983)

(0) P. DELPECH, private communication

(9) Courtesy of C. D'ANTERROCHES

(10) J. PERCHET, private communication

(11) L. MEI, M. RIVIER, Y. KWART, and R.W. DUTTON, Semiconductor Silicon 1981, ed. by H.R. Huff et al., The Electrochemical Society, Proceedings Vol. 81-5, 1007-1019 (1981)

(12) B.L. CROWDER and S. ZIRINSKI, IEEE Trans. Electron Dev., ED-26, 369-371 (1979)

(13) D.L. WOLLESEN, J. HASKELL, and J. YU, IEEE 83 IEDM, Dec. 5-7, Washington, Technical Digest, 155-158, (1983)

(14) D.J. ALLSTOT and W.C. BLACK Tr., Proceedings IEEE, 71, 967-986 (1903)

(15) G. NELSON, H.H. STELLRECHT and D.S. PERLOFF, IEEE J. Solid-State Circuits, SC-6, 396-403 (1973)

(16) J.L. Mc CREARY, IEEE J. Solid-State Circuits, SC-16, 608-616 (1981)

(17) J.L. Mc CREARY, ECCTD 83, Stuttgart, Technical Digest, 26-32 (1983)

(18) P.R. GRAY and R.G. MEYER, IEEE J. Solid-State Circuits, SC-17, 969-982 (1982)

(19) M. AOKI, Y. SAKAI, and T. MASUHARA, IEEE Trans. Electron Devices, ED-29, 296-299 (1982)

(20) J. MIYAMOTO, S. SAITOH, H. MOMOSE, H. SHIBATA, K. KANZAKI, and S. KOHYAMA, IEEE 83 IEDM, Dec. 5-7, Washington, Technical Digest, 63-66 (1983)

(21) For example : MIETEC N.V., data sheet on 70 V BIMOS technology (1983)

(22) T. OHNO, T. SAKURAI, Y. INABE, and T. KOINUMA, IEEE 1984 ISSCC, San Francisco, Digest of Technical Papers, 230-231 (1984)

(23) Electronics, March 22, 72-73 (1984)

(24) M.J. DECLERCQ and J.D. PLUMMER, IEEE Trans. Electron Devices, ED-23, 1-4 (1976)

(25) P. SHAH, N. BATRA, and M. GILL, Electronics, April 5, 145-149 (1984)

(26) For example : MIETEC N.V., data sheet on 400 V HVMOS technology (1983)

(27) S. WONG, and C.A.T. SALAMA, Dig. Tech. Papers, 1982 VLSI Syposium, Tokyo, 48-49 (1982)

(28) S. WONG, and C.A.T. SALAMA, IEEE J. Solid-State Circuits, SC-18, 106-114 (1983)

(29) T. ENOMOTO, T. ISHIHARA, M. YASUMOTO, and T. AIZAWA, IEEE J. Solid-State Circuits, SC-18, 395-402 (1983)

(30) T. ENOMOTO, M. YASUMOTO, T. ISHIHARA, and K. WATANABE, IEEE J. Solid-State Circuits, SC-17, 1045-1053 (1982)

(31) M. BANU, and Y. TSIVIDIS, IEEE J. Solid-State Circuits, SC-18, 644-651 (1983)

(32) L.C. PARRILLO, R.S. PAYNE, R.E. DAVIS, G.W. REUTLINGER, and R.L. FIELD, Tech. Dig., '80 IEDM, Washington, 752-755 (1980)

(33) Y. KURAISHI, Y. TAKAHASHI, K. NAKAYAMA, and T. SENBA, Tech. Dig., CICC, Rochester (1984)

(34) T. SUZUKI, H. TAKATORI, H. SHIRASU, M. OGAWA, and N. KUNIMI, IEEE J. Solid-State Circuits, SC-18, 700-706 (1983)

(35) K. MATSUI, T. MATSUURA, and K. IWASAKI, 1982 Int. Symp. on Circuits and Systems, 241-244 (1982)

2

MOS DEVICE MODELLING

F.M. KLAASSEN
PHILIPS RESEARCH LABORATORIES
EINDHOVEN, THE NETHERLANDS

INTRODUCTION

In order to achieve IC's of greater complexity structural
dimensions of MOSFET devices have been reduced considerably
during the last decade. On a laboratory scale we have now
entered the submicron range. An unfortunate consequence of this
continuing effort is that multidimensional physical and parasitic effects are making it increasingly difficult to describe the
characteristics of small devices. Although most of these effects
can be investigated more suitably by numerical methods, for efficient circuit analysis programs we need (preferably explicit)
analytic expressions of the characteristics with parameters that
can easily be determined.

With the aid of exact numerical results and physical insight as a guideline, accurate CAD models can be made. Since a
very high accuracy is not necessary for most applications and
additionally may conflict with the efficiency of the simulation,
a hierarchy of models is preferred. Successively we discuss
general purpose models for small enhancement transistors, similar models including subthreshold conduction, charge models, a-c
models and models for depletion devices. Throughout the text it
is assumed that we are dealing with n-channel devices unless otherwise stated. Finally this chapter is closed with a discussion
of parameter determination.

1. ENHANCEMENT DEVICE CURRENT MODELS

In present MOS IC processing two types of enhancement devices are made. The first, well-known type (fig. 1a) is characterized by a threshold voltage adjustment implant of the same type
as the substrate. In the second type (fig. 1b) an implant of the
opposite type is used. Since the latter implant is very shallow,
the surface layer is depleted at zero gate bias (normally-off depletion transistor). When the threshold voltage is adjusted to
practical values, current conduction occurs at the interface, and
this type of transistor can be modelled in the same way as the
usual enhancement type [1].

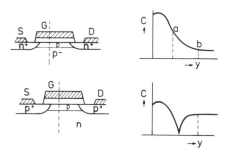

Fig. 1. Cross-section and doping profile of a MOSFET

1.1. Threshold voltage

Essentially the threshold voltage is determined from a balance of the charges at the gate, in the gate insulator and in the channel area. Since the first-mentioned charge is proportional to the voltage drop over the insulator, this balance is expressed by

$$C_{ox}WL(V_T - V_{FB} - u_F) + Q_D = 0 , \tag{1}$$

where V_{FB} is the flat-band voltage, u_F is the band bending, Q_D is the associated depletion charge, C_{ox} is the insulator capacitance per unit area, W is the channel width and L the channel length. Usually in the above balance the mobile charge is much smaller than Q_D and can be neglected in eq. (1).
Since the depletion charge is also shared by the source and drain junction and the channel stop area (compare fig. 2a and 2b) the threshold voltage becomes geometry-dependent in small devices. Generally V_T decreases with decreasing channel length and increases with decreasing channel width. Although the approach is questionable, a useful geometrical correction formula has been proposed [2,3] .
In addition the depletion charge is determined by the

threshold implantation, which is widely used in modern MOS processes. Taking into account that the latter implant can be adjusted to full depletion at the onset of channel formation, and that the above geometry effects are of second order in properly scaled devices [4] , the threshold voltage is given by

$$\begin{aligned}
V_T &= V_{Tx} + k(V_o + V_{SB})^{\frac{1}{2}} &, \\
V_{Tx} &= V_{FB} + u_F \pm D_{imp} C_{ox}^{-1} &, \\
k &= k_\infty - k_L L^{-1} + k_W W^{-1} &.
\end{aligned}\Biggr\} \qquad (2)$$

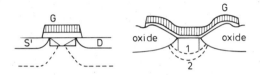

Fig. 2. Charge sharing (grey area) in a small MOSFET.

In this relation D_{impl} is the total implantation dose (with a plus sign for normal enhancement devices [5] and a minus sign for normally - off depletion devices [1]). k_∞ is the classical long channel body effect, which is mainly determined by the substrate doping, an anti-punch-through implant, or a well doping in a CMOS process. k_L and k_W characterize the geometry effect associated with electrical channel length and channel width. V_0 is a voltage constant smaller than u_F for an enhancement device and larger than u_F for a normally - off depletion device [1] . Obviously, when the above mentioned conditions are not satisfied, more complex relations for V_T have to be used (compare [6]).

1.2. Mobility reduction effects

It is well known that the mobility in the inversion layer decreases owing to the combined action of the normal field at the interface [7] , and the lateral field, which causes velocity saturation effects [8] . Although more exact semi-empirical relations have been suggested [8,9] , we prefer to model these effects by the simple relation

33

$$\mu = \mu_0[1 + \theta_A(V_{GS} - V_T) + \theta_B V_{SB}]^{-1} [1 + E/E_c]^{-1} , \tag{3}$$

where θ_A, θ_B and μ_0 are process parameters. The first term between brackets not only represents the influence of the normal field increase with gate bias, but has the additional advantage of representing effects due to the source series resistance. The second term represents an additional decrease of mobility with back-bias. Usually this effect is neglible in n-channel devices, but it is important in p-channel devices [1] . The third term has been shown to be sufficiently accurate to represent velocity saturation and leads to a closed-form relation for the saturation voltage of the drain current [10, 11] . Generally θ_A increases with decreasing channel length L.

1.3. Saturation mechanism

When a drain voltage is applied to the conducting channel, the increasing channel potential V(x) causes an increase of the depletion charge towards the drain. Generally this increase can be expressed in terms of $k(V_0+V(x))^{1/2}$. Consequently the drain current can be written at moderate drain voltages in the form

$$I_{DS} = \frac{\beta[f(V_G,V_S) - f(V_G,V_D)]}{2[1 + \theta_A(V_{GS} - V_T) + \theta_B V_{SB}](1 + \theta_c V_{DS})} , \tag{4}$$

where

$$f(V_G,V) = (V_G - V_{Tx} - V)^2 + \frac{4}{3} k(V + V_o)^{3/2} , \tag{5}$$

$\beta = \mu_0 C_{ox} W/L$ is the gain factor and $\theta_c = 1/LE_c$.
With a further increase of the drain voltage the channel is pinched off, but a correct calculation of the saturation voltage generally leads to a cubic equation. From the many approximations for V_{DSS} [1,11,12] we consider here the most useful approach. Since with modern processes the factor k has a modest value and the last term in eq. (5) is less important than the first one, we approximate the numerator of eq. (4) by

$$f(V_G, V_S) - f(V_G, V_D) \approx 2(V_{GS} - V_T)V_{DS} - (1 + \delta)V_{DS}^2 \, , \tag{6}$$

in which

$$\delta \approx \frac{0.5k}{(1 + V_o + V_{SB})^{\frac{1}{2}}} \, . \tag{7}$$

As shown by numerical calculations the above approximation only causes unacceptable errors for drain voltages in excess of ten Volts.

When a high drain voltage is applied, the normal field in the insulator close to the drain is inverted, and beyond a point x_p in the channel, mobile carriers are pushed into the bulk and move towards the drain with a saturated velocity. This situation is illustrated in fig. 3. Here x_p corresponds to the classical pinch-off situation. Numerical calculations show that the carriers have reached velocity saturation already at $x_c < x_p$. In addition between x_c and x_p the so-called gradual channel approximation, upon eq. (4) is based, becomes invalid. Assuming that the above approximation is still valid at x_c, andassuming that the potential and lateral field at x_c are V_{DSS} and E_{xc}, respectively, the saturated current at the right hand side of x_c can be expressed by

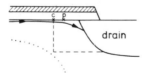

Fig. 3. Current saturation region

$$I_{DSS} = \frac{\beta[V_{GS} - V_T - (1 + \delta)V_{DSS}]L\,E_{xc}}{[1 + \theta_A(V_{GS} - V_T) + \theta_B V_{SB}](1 + E_{xc}/E_c)} \cdot \qquad (8)$$

Since I_{DSS} at the left-hand side of x_C is still given by eqs. (4) and (6), we obtain by equating an expression for the saturation voltage:

$$V_{DSS} = \theta_c^{-1}\left[\left\{\frac{2\theta_c(V_{GS} - V_T)}{1 + \delta} + 1\right\}^{\frac{1}{2}} - 1\right] \cdot \qquad (9)$$

Owing to the fact that $E_{xc} \gg E_c$ it will be seen that E_{xc} has disappeared from eq. (9).
For drain voltages in excess of V_{DSS} the drain current is in the saturation mode and can increase only by a static feedback effect and a slight shift of x_C towards the source. This will be a subject of the next section.

1.4. Saturation mode

When the average distance between the conducting drain and the channel becomes small, an increase of the drain bias induces some excess mobile charge in the channel. The presence of this effect has been demonstrated numerically for a MOSFET in saturation [10] and has been estimated analytically [13]. Generally the average effect on the carrier density can be represented by a term

$$\langle \Delta n \rangle \sim V_{DS}/L^n , \qquad (10)$$

where $1 < n < 1.5$
To take into account the above effect on the drain current we simply modify the gate driving voltage in the numerator of eq. (6):

$$V_{GT} = V_{GS} - V_T + \gamma V_{DS} \ . \tag{11}$$

In this way the extra voltage source γV_S remains active in the saturation region. Since γ depends on the insulator thickness, the drain doping profile and the threshold implant profile, we consider the above quantity as a process parameter.

In addition to the above static feedback drain effect I_{DSS} may increase with V_{DS} by a slight shift of the saturation point x_c towards the source. This effect is known as channel length shortening. Actually an exact calculation of the above effect is not easy, since it requires a two-dimensional solution of Poisson's equation for the channel saturation area indicated by the interrupted line of fig. 3. A pseudo two-dimensional analysis has been given in implicit form [14], but the result is limited to long channels. For short-channel devices more approximate solutions have been proposed [1,15,16]. According to the last approach cited, the shortening in channel length is expressed in terms of the lateral surface field E_{xc} (V_{DS}), where

$$E_{xc}(V) = [\alpha^2(V - V_{DSS})^2 + E_c^2]^{1/2} \tag{12}$$

and

$$\Delta L = \alpha^{-1} \ln \left[\frac{\alpha(V_{DS} - V_{DSS}) + E_{xc}(V_{DS})}{E_c} \right] \ . \tag{13}$$

Although α can be expressed in terms of the insulator thickness and junction depth [16], we prefer to consider this factor as a parameter. One reason is that in practical situations an additional voltage drop occurs over the drain region. In order to guarantee continuity of the drain conductance a slight modification of the saturation voltage V_{DSS} is required when the electrical channel length L in eq. (8) is corrected for the above shortening.

1.5. Simple long-channel model

For first-order estimations of circuit performance often the long-channel current expression is used. This expression can be easily derived from eqs (2) through (9) by neglecting all terms inversely proportional to L and W. In this case the threshold voltage can be written in the form

$$V_T = V_{To} + k\left[(V_{SB} + u_F)^{1/2} + u_F^{1/2}\right] , \tag{14}$$

where V_{To} is the zero-bias threshold voltage and k is the body coefficient given by

$$k = \frac{(2\epsilon q N_B)^{1/2}}{C_{ox}} , \tag{15}$$

where N_B is the substrate or well doping. For the triode region the drain current reads

$$I_{DS} = \beta_o \frac{W}{L} \left[(V_{GS} - V_T)V_{DS} - \frac{1}{2}(1 + \delta)V_{DS}^2\right] . \tag{16}$$

where δ is given by eq. (7).
However for the saturation mode ($V_{DS} > (V_{GS} - V_T)/1 + \delta$) this current is given by

$$I_{DSS} = \frac{1}{2}\beta_o \frac{W}{L} \frac{(V_{GS} - V_T)^2}{1 + \delta} . \tag{17}$$

When the gate driving voltage has a moderate value ($\theta_c(V_G - V_T) < 0.10$) the above relations even apply to short-channel devices.

1.6 Shift of threshold voltage with drain bias

Since an increase of the drain bias affects the depletion layer associated with the formation of an inversion layer, the threshold voltage of a short-channel device will be subjected to a further decrease (also called drain-induced barrier lowering [17]). Since the charge division rules of the so-called charge-sharing models [2] become questionable [18] , a rigorous solution of the two-dimensional Poisson equation is necessary for predicting the ΔV_T (V_{DS}) effect. Although such an analysis has been given in terms of a Fourier series expression [19] , the result overestimates the voltage dependence of ΔV_T when compared to experimental data. Since the subthreshold current depends critically on an accurate threshold voltage shift, for

38

the time being use has to be made of a semi-empirical expression of ΔV_T ($V_{SB}V_{DS}$).

1.7. Narrow width effects

Owing to two-dimensional effects in narrow width devices, the depletion layer spreads into the channel stop diffusion area (compare fig. 2b). Since the latter areas are often highly doped to increase the parasitic field thresholds, the excess charges cannot be neglected. Unfortunately the problem is aggravated by unsatisfactory knowledge of two-dimensional doping profiles. Therefore modelling of effects on the characteristics has only been approximate [20]. When the channel is not too narrow, the charge spreading is relatively small and the effect leads to an increase of the threshold voltage and of the body effect factor inversely proportional to the channel width W (compare eq. (2)).

An additional effect, which has not been modelled in the literature, occurs with increasing back-bias. In this case the depletion layer can hardly expand in the channel stop region and consequently shows a very curved shape (see fig. 2b; broken line no 2). Because of this the field lines no longer cross perpendicular to the interface in a large part of the channel. Consequently the inversion charge decreases from the centre of the channel towards the edges and causes the electrical width effectively to decrease with the back-bias [21].
The above effect is observed in the characteristics of narrow-width devices as a strong decrease in the conductance at higher back-bias.

1.8 Subthreshold mode

When a gate bias below threshold is applied, a potential barrier exists between the channel area and the source. In this case current flow is possible by means of carriers which are able to pass the above barrier and transport takes place by diffusion rather than by drift. In order to guarantee a continuous transition between the subthreshold mode and the inversion mode a single drain current expression has been derived in terms of the surface potential $\emptyset_S$ [15,22] . Unfortunately I_D can be calculated only numerically. Therefore much is to be said for a simpler approach if it is sufficiently accurate.

Since the drain current is of the diffusion type, it can be written in a form [22]

$$I_{DS} = WL^{-1}qD_nN(o)(1 - \exp - qV_{DS}/kT) ,$$

where $N(o)$ is the number of carriers that can pass the potential barrier $\varnothing_{SO}$ close to the source. By solving the charge balance equation it can be proved that $\varnothing_{SO}$ is proportional to the gate bias. Consequently the subthreshold current can be written in the form [23]

$$I_{DS} = WL^{-1}I_o \exp q(V_{GS} - V_T)/mkT \; , \qquad (18)$$

in which I_O is a current constant and the slope factor m is given by

$$m = m_o + \frac{1}{2} k(V_{SB} + V_o)^{\frac{1}{2}} \; , \qquad (19)$$

where mo is a parameter (dependent on the surface state density) and k is the body effect coefficient.
In the transition region near threshold it is necessary to limit the value of (18) by multiplying by an empirical limiting function. By adding the combined relation to the strong inversion equation, a sufficiently accurate, continuous current relation is obtained for all bias regimes.

1.9 Temperature relations

Although the drain current in the strong inversion regime according to eq. (18) may vary with temperature owing to a variety of temperature-dependent factors, in practice only three major factors are present: the band bending u_F, the mobility μ_0 and the saturation parameter Θ_C. For the temperature dependence of u_F it follows from the physics that

$$\frac{\partial u_F}{\partial T} = T^{-1}(u_F - E_G/q) \; , \qquad (20)$$

where E_G is the zero-temperature band gap.
The mobility decreases with increasing temperature according to

$$\mu(T) = \mu(T_0) \, [T/T_0]^{-1.6} \qquad (21)$$

where T_O is room temperature.
As the mobility varies with temperature, but the saturation velo-

city hardly at all, the saturation field E_C becomes temperature-dependent. Therefore θ_C decreases with increasing temperature. From measurements it has been found that

$$\theta_c(T) = \theta_c(T_0) - \Delta\theta_c(T - T_0) \ . \tag{22}$$

Obviously the subthreshold current varies strongly with temperature owing to the exponential factor. This causes the current to vary at lower temperatures with a steeper slope.

2. CHARGE MODELS

Compared to the notice given to the MOST current in a variety of papers, the modelling of charge and capacitance has attracted little attention. This is understandable because these quantities do not vary much with the bias condition and are often eclipsed by parasitic capacitances. Many models use a few small signal capacitance values derived for a long-channel device, disregarding the body effect [14] . This approach not only leads to errors when some capacitances of short channel devices, but generally charge is not conserved. [25, 26] . This may cause large errors in simulating dynamic circuits or switched capacitor circuits. Taking into account the important charge conservation rule

$$Q_G + Q_S + Q_D + Q_B = 0 \ , \tag{23}$$

we have to calculate the charges Q_S, Q_D and Q_B as functions of the node voltages. Such a calculation has been given for a long-channel MOST [25] and later corrected for a short-channel device [26] . However in the first case the partitioning of channel charge into Q_S and Q_D is incorrect and in the latter case the calculation of the channel charge is very approximate.

None of the above problems exist in a model based on the calculation of the charge in the channel and the depletion layer from a double integration of the basic equation leading to the current expression (4). The only approximation is in the necessary split of the channel charge into parts Q_S and Q_D attributed to the source and drain terminal, respectively [27] . The above charges are given by

$$Q_S = -\frac{1}{2} C_{ox} WL[V_{GS} - V_T + (1 + \delta)V_{DS} F(V_{DS})]$$

$$Q_D = -\frac{1}{2} C_{ox} WL[V_{GS} - V_T - 3(1 + \delta)V_{DS} F(V_{DS})]$$

$$(24)$$

where F is a function depending on the bias condition

$$F = \frac{V_{GS} - V_T - \frac{2}{3}(1 + \delta)V_{DS} - \frac{1}{6}(1 + \delta)\theta_c V_{DS}^2}{2(V_{GS} - V_T) - V_{DS}(1 + \delta)} . \tag{25}$$

Note that all voltages refer to the source voltage and that only the unit area insulator capacitance C_{ox} turns up as a new parameter. With the above charge relations all relevant capacitances can be calculated. However this is beyond the scope of this chapter. Generally the above expressions lead to an asymmetrical capacitances matrix, i.e. $C_{GD} \neq C_{DG}$. The split between Q_S and Q_D has been made such that for a long-channel device ($\theta_c \approx 0$) Q_D according to (24) is zero in the saturation region. This corresponds to the physical situation of channel pinch-off. For a short-channel device no real pinch-off occurs and therefore $Q_D \neq 0$. From the evaluation of the charge derivatives it turns out, however, that $\partial Q_G / \partial V_{DS}$ becomes zero in saturation.

To the active charges in the above-threshold region an expression has to be added for the bulk charge in the subthreshold depletion mode and in the accumulation mode. Approximating the threshold implant by a box profile it can be shown [26] that in the subthreshold region

$$Q_B = C_{ox} WL \frac{K^2}{2} \left[\left\{\frac{1 + 4(V_{GB} - V_{FB})}{K^2}\right\}^{\frac{1}{2}} - 1\right] \tag{26}$$

and that in the accumulation region

$$Q_B = C_{ox} WL(V_{GB} - V_{FB}) \tag{27}$$

Since K is determined by the peak value of the threshold implant it has to be considered as a parameter. In addition, since the Q_B expression (26) has to be kept constant in the active region ($V_{GS} > V_T$), some numerical manipulation is needed to avoid a discontinuity in the gate capacitance at $V_{GS} \approx V_T$.

Since parasitics have a marked effect on circuit delay times, it is absolutely necessary to model important parasitics with care. Among these quantities we mention the gate overlay capacitances, the bias-dependent junction capacitances to the channel stop areas, and the interwiring capacitances. As the value of most of these capacitances is determined by spreading effects too, special test configurations are required. A four-terminal charge presentation of the MOSFET is given in fig. 4.

circuit model

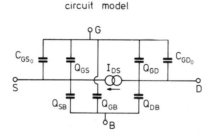

Fig. 4. Circuit model of a MOSFET

3. SMALL-SIGNAL MODELS

For analog design often a small-signal model is preferred. The large-signal charge model is considered as too complex and inaccurate for the above application. Compared to the scheme of fig. 4 the current generator is replaced by a transconductance g_m with a drain conductance g_d in parallel and the charges have been replaced by capacitances. However usually only three small--signal capacitances are taken into account: C_{GS}, C_{GD} and G_{GB}.

In practice such a scheme is only suitable for first order estimations of circuit performance. In most cases, it has been shown that the drain conductance in saturation [32] and the transconductance of narrow width devices are poorly modelled [21]. The first defect, for instance, causes large errors in calculating the voltage gain of CMOS amplifiers. In addition the

mentioned three capacitances are inadequate to represent all
signal currents, which pass the device terminals. In fact in
several bias regions other transcapacitances like C_{DG}, C_{SB}
and C_{DB} are equally important [33] .

Regarding the above shortcomings, an accurate a.c. model
requires a scheme, which is hardly simpler than that of fig. 4.
Therefore for analog design the model of fig. 4 can be used just
as well. It should be noted, however, that an appropriate
description of g_d requires the parameters α and γ to be
determined from mixed measured values of drain current and drain
conduct-tance [32] .

For first-order estimations of the frequency dependence and
the voltage gain of amplifier stages the underlying simplified
equivalent circuit model can be applied.

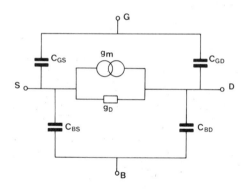

Fig. 5. Small-signal circuit representation of a MOSFET

All capacitance terms in this scheme can be derived from the
charge expressions (24) and (25) in case of a long channel
device. For reason of simplicity only major components for the
active saturation mode have been taken into account and capacitance terms for the subthreshold region have been neglected. In
addition several important parasitics have been added to the
intrinsic terms, like the drain-gate overlay capacitance per unit
width (C_{DGO}) and the source (C_{JS}) and drain (C_{JD})

junction capacitance. Successively we have

$$C_{GS} = \frac{2}{3}\,C_{ox}WL$$

$$C_{GD} = C_{GDO}\,W$$

$$C_{BS} = \frac{2}{3}\,C_{ox}WL\delta + C_{JS} \tag{28}$$

$$C_{BD} = C_{JD}$$

The above equations also apply to smaller devices, provided that the gate driving voltage is not too high.

The transconductance term can be easily derived from the current expressions of section 1.5 and the drain conductance from eqs (8), (9) and (11). Thus we obtain

$$g_{ms} = \beta_o\,\frac{W}{L}\,\frac{V_{GS} - V_T}{(1 + \delta)\,(1 + \theta_c(V_{GS} - V_T))} \tag{29}$$

$$g_{DO} = \gamma g_{ms}$$

In these equations δ is given by (7). Since a simple analysis of the saturation mode is impossible, the above expression of g_{DO} has a limited value. When $L < 10$ μm, the accuracy of g_{DO} is satisfactory, but for longer channels the error may become 30%.

4. DEPLETION DEVICE MODELS

In modern depletion devices a channel of a type opposite to the substrate material is implanted and the implant dose is selected such that a conducting layer remains at zero gate bias.

This layer is indicated as a grey area in the concentration

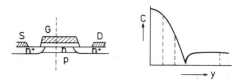

Fig. 6. Cross section and doping profile of a depletion MOSFET

profile of fig. 6. When the device is used only as a load the
characteristics can be satisfactorily described by an enhancement
model with an appropriate threshold shift. We shall return to
this later. However, to account for the depletion mode of operation, a wider utilization in IC's requires a separate model.
Usually the model equations are based on a constant channel
doping [28, 29,30]. At moderate drain bias the device may operate in two possible modes (compare fig. 7a and 7b): the depletion
mode occurring at gate bias values close to threshold

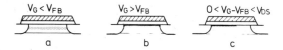

Fig. 7. Modes of operation of a depletion MOSFET

and the accumulation mode. Since the conductance in the depletion mode is smaller than in the accumulation mode the characteristics is of a type shown in fig. 8. In addition to the

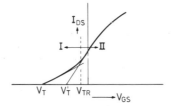

Fig. 8. Characteristics of a depletion MOSFET

above modes a third mixed mode of operation occurs at higher values of the drain voltage. This is shown in fig. 7c. The associated high value of the electron potential in the channel causes
the accumulation charge in part of the channel to disappear in
favour of a depletion charge.

Owing to the above three modes of operation a further evaluation of the basic transport equation has led to a rather complicated set of analytic expressions to describe the complete
characteristics [29]. When the results are compared with measured characteristics of shallow implanted structures, several
deficiencies are observed. The accuracy is rather poor, in particular for small devices. Determination of characteristic parameters is rather complicated. The current is not expressed
explicitly in terms of a major factor like V_T. Most often velocity saturation is neglected and cannot be incorporated successfully owing to mathematical complications. The measured tail
region of the depletion mode is often less pronounced.
The latter is caused by the channel profile, which in practice
decreases gradually towards the substrate.
Hence from a practical viewpoint it is attractive to formulate an
alternative model, which takes into account all the above deficiencies, but requires some approximations to obtain characteristics expressed explicitly in basic parameters.

In fact, by assuming a linear channel doping profile, it can
be shown that the depletion mode conductivity g_D can be described by [34]

$$g_D \approx \mu_n C_i W (V_{GS} - V_T) \ , \tag{30}$$

where V_T is the real threshold voltage, μ_n is the bulk mobility
and C_i is an effective gate capacitance given by

$$C_i \approx [\frac{2d_i}{3\epsilon_s} + \frac{t_{ox}}{\epsilon_{ox}}] . \tag{31}$$

On the other hand, when the device operates in the accumulation mode ($V_{GS} > V_{TR}$), the conductivity g_A is given by

$$g_A \approx \mu_n r C_i W(V_{GS} - V_T') , \tag{32}$$

where V_T' is an effective threshold voltage (compare fig. 8) given by

$$V_T' = \frac{V_T + (r_0 - 1) V_{TR}}{r_0} \tag{33}$$

and the gain ratio constant r can be modelled by

$$r = r_0 [1 + \theta_A (V_{GS} - V_{TR})]^{-1} . \tag{34}$$

Usually $1 < r_0 < 2$, owing to a larger effective gate driving capacitance in the accumulation mode.
Note that the form of eq. (32) makes it possible to model the characteristics in the accumulation mode as a pure enhancement device with an apparent threshold voltage V_T'. This is often done in practice. In the more general case eqs. (30) and (32) are still similar to the basic equations of an enhancement model and therefore effects of velocity saturation and of p-n junction space charge can be easily implemented. In fact, using a smoothening function for the gate driving voltage to correct for the difference in slope at the boundary V_{TR}, the drain current can be presented in a form similar to eq. (4). Compared to the pure enhancement transistor we need two extra parameters: V_{TR} and r_0.

48

5. PARAMETER DETERMINATION AND COMPARISON TO MEASUREMENTS

5.1. Enhancement devices

Although most parameters can be obtained simultaneously from curve-fitting routines based on least squares optimization, it turns out that it is more practical to split the parameters into four groups. In table I the parameters have been classified. The first group is determined from the strong inversion regime at lowdrain bias, the second group from the general I_D-(V_{GS}, V_{DS}, V_{SB}) characteristics in strong inversion, the

I	II	III	IV
V_{Tx}	θ_C	m_0	C_{ox}
k	γ	I_0	V_{FB}
β	α	SH	K
θ_A		P	
θ_B		n	
V_o			

table I

third group from the subthreshold characteristics, and the last group from the gate bulk capacitance in the subthreshold and accumulation modes.

Fig. 9 gives the measured conductance characteristics (fully drawn lines) at low drain bias for a p-channel MOST. This device is of the normally-off depletion type. For comparison the simulated characteristics (dots) and associated parameters are also given. The latter values have been calculated from a least squares optimization routine. Generally the accuracy is satisfactory (usually the average deviation is less than 2%). Note that owing to the relatively high value of θ_B the conductance slope decreases at higher back-bias. This effect is rarely observed in n-channel devices. In addition the zero bias threshold voltage $V_{TO} = V_{TX} + kV_o^{1/2}$ is given as a parameter. From a practical viewpoint the latter procedure is often preferred.

In order to establish correct geometry relations for several parameters in the case of small devices, the effective electrical channel length (L) and channel width (W) have to be determined additionally. For this purpose the conductance of several geometry combinations has to be measured. From plots of β against the drawn or mask dimensions L_M^{-1} and W_M, L and W are obtained via offset values. Next the above values of L and W may be used to obtain the correct geometry relations for other parameters. As an example we present in fig. 10 the body effect

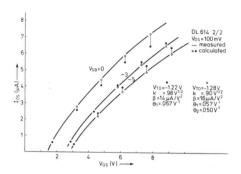

Fig. 9. Conductance of a p-channel MOSFET

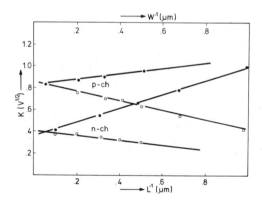

Fig. 10. Body coefficient vs. device dimensions

coefficients as a function of geometry. Note the linear
dependence down to submicron dimensions.
    For an n-channel MOST with L=1.8 μm, fig. 11 gives the
measured high current characteristics at a back-bias of 2.5V
(dots) together with the simulated values (fully drawn lines).
In addition the associated parameters have been indicated. Most

accurate results (relative errors within 5%) are obtained from optimizing θ_C, γ and α with a set of measured values at several values of back-bias. Note that for simplicity α has been taken as zero. In reality the value is non-zero, but compared to γ, α has only a minor influence on the characteristics of short-channel MOSFETs.

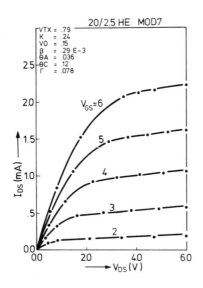

Fig. 11. Characteristics of enhancement MOSFET at V_{SB}=2.5Volt

Finally in fig. 11 the subthreshold characteristics have been plotted. Usually the average deviation between measured and simulated results is larger than in former cases (between 10 and 15%). The effect of an additional threshold voltage shift with drain bias is clear. Note the small anomaly in the calculated curves, which arises from the summation of the two expressions (4) and (18). However this anomaly does not cause a continuity problem.

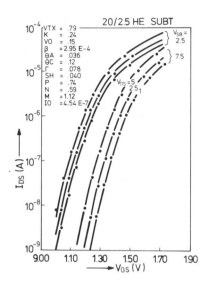

Fig. 12. Subthreshold characteristics of enhancement device.

5.2. Depletion devices

For the description of depletion devices two additional parameters are needed: the transition voltage V_{TR} and the gain
ratio r_O.
Similar to the procedure for enhancement devices the parameters
have to be classified in two groups: V_T, k, β, V_{TR}, r_O and
θ_A, which are determined from conductance measurements and θ_C
and γ , which are determined from the general characteristics.
Obviously, for the first group an optimization routine gives the
most accurate results.
In fig. 13 measured characteristics are compared with calculated ones. The parameter values obtained from this comparison
are indicated. In addition, these values have a physical meaning.

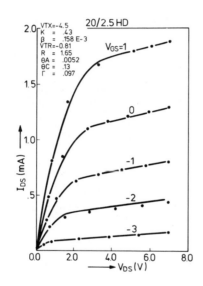

Fig. 13. Characteristics of depletion MOSFET at V_{SB}=2.5Volt

References

1. F.M. Klaassen and W. Hes, "The compensated MOSFET", accepted for publication in Solid-State Electronics.
2. L.D. Yau, "A simple theory to predict the threshold voltage of IGFETS", Solid-State Electronics, vol. 17, pp. 1059-1063, 1974.
3. G. Merckel, "Simple model of threshold voltage in small MOSFETS", Solid-State Electronics, vol 23, pp. 1207-1213, 1980.
4. R. Dennard, F. Gaensslen, E. Walker and P. Cook, "A 1 μm MOSFET VLSI technology, part II", IEEE Transactions on Electron Devices, vol. ED-26, pp. 325-333, 1979.
5. V.L. Rideout, F. Gaensslen and A. LeBlanc, "Device design considerations for ion-implanted MOSFETS", I.B.M. Journal of Research and Development, vol. 19, pp. 50-59, 1975.
6. N. Herr, B. Garbs and J.J. Barnes, "A statistical modelling approach for simulation of MOS VLSI circuit designs" Digest IEDM 82, pp. 290-293, 1982.
7. A.G. Sabnis and J.T. Clemens, "Characterization of the electron mobility in the inverted 100 Si surface" Digest IEDM 79, pp. 18-21, 1979.
8. G. Merckel, J. Borel and N.Z. Cupcea, "An accurate large-signal MOST model", IEEE Transactions on Electron Devices, vol. ED-19, pp. 681-698, 1972.
9. D.L. Scharfetter and H.K. Gummel, "Large-signal analysis of a Si Readdiode oscillator", IEEE Transactions on Electron Devices, vol. ED-16, pp. 64-77, 1969.
10. F.M. Klaassen and W.J. de Groot, "Modelling of scaled-down MOS transistors", Solid State Electronics, Vol. 23, pp. 237-242, 1980.
11. E. Sun, "Short channel MOS modelling for CAD", Digest IEDM 79 pp. 493-499, 1979.
12. N. Murphy, F. Berz and I. Flinn, "Carrier mobility in Si MOSTs", Solid State Electronics, vol. 12, pp. 775-786, 1969.
13. T. Poorter and J.H. Satter, "A d.c. model for a MOS transistor in the saturation region", Solid State Electronics, vol. 23, pp. 765-772, 1980.
14. Y.A. El Mansy and A.R. Boothroyd, "A new approach to the theory of IGFETs", IEEE Transactions on Electron Devices, vol. ED-24, pp. 254-270, 1977.
15. P.P. Guebels and F. van de Wiele, "A small geometry MOSFET model" Solid State Electronics vol. 26, pp. 267-273, 1983.
16. P.K. Ko, "Hot electron effects in MOSFETs", PhD thesis, University of Berkely, 1982.
17. R.R. Troutman and A.G. Fortino, "Simple model for threshold voltage in a short-channel IGFET, Transactions on Electron Devices, vol. ED-24, pp. 1266-1268, 1977.

18. T.N. Nguyen and J.D. Plumer, "Physical mechanisms responsible for short-channel effects in MOS devices", Digest IEDM, vol. 81, pp. 596-599, 1981.
19. K.N. Ratnakumar and J.D. Meindl, "Short-channel MOST threshold voltage model", IEEE Journal of Solid-State Circuits, Vol. SC-17, pp. 937-948, 1982.
20. C.R. Ji and C.T. Sah, "Analysis of the narrow-gate effect in submicrometer MOSFETs", IEEE Transactions on Electron Devices, vol. ED-30, pp. 1672-1677, 1983.
21. F.M. Klaassen, "Narrow-width effects in MOSFETs", to be published.
22. J.R. Brews, "A charge-sheet model of the MOSFET", Solid-State Electronics, vol. 21, pp. 345-355, 1978.
23. M.B. Barron, "Low level currents in IGFETs", Solid-State Electronics, vol. 15, pp. 293-302, 1972.
24. J.E. Meyer, "MOS models and circuit simulation", RCA Review, vol. 32, pp. 42-63, 1971.
25. D.E. Ward and R.W. Dutton, "A charge-oriented model for MOST capacitances", IEEE Journal in Solid State Circuits, vol. SC-13, pp. 703-707, 1982.
26. P. Yang, B. Epler and P.K. Chatterjee, "An investigation of the charge conservation problem for MOSFET simulation", IEEE Journal of Solid State Circuits, Vol. SC-18, pp. 128-138, 1983.
27. F.M. Klaassen, "Charge modelling of the MOSFET", to be published.
28. G. Merckel, Process and Device Modelling, ed. Noordhoff, Leyden, pp. 677-710, 1977.
29. Y.A. El Mansy, "A nonlinear CAD model for the depletion-mode IGFET" Digest IEDM 78, pp. 20-24, 1978.
30. P.E. Schmidt and M.B. Das, "DC and high-frequency characteristics of built-in channel MOSFETs", Solid State Electronics, vol. 21, pp. 495-505, 1978.
31. F.M. Klaassen and W.J. de Groot, "Modelling of the depletion-mode IGFET", ESSDERC 80, Europhysics Conference Abstracts ECA 4H, pp. 107-108, 1980.
32. Y.P. Tsividis and G. Masetti, "Problems with precision in modelling of the MOS transistor for analog applications", IEEE Transactions on Computer-Aided Design, vol. CAD-3, pp. 72-79, 1984.
33. C. Turchetti, G. Masetti and Y. Tsividis, "On the small-signal behaviour of the MOST in quasistatic operation", Solid State Electronics, vol. 26, pp. 941-949, 1983.

PART II

CIRCUIT BUILDING BLOCKS

INTRODUCTION

This part of the book deals with the detailed design of basic circuits used in telecommunications chips. CMOS operational amplifiers are discussed by Bedrich Hosticka, including detailed considerations of noise and power supply rejection. Eric Vittoz presents techniques for very low power operation of a variety of circuit functions, and follows this with a chapter on dynamic analog techniques, including a discussion of switch nonidealities and their compensation; both chapters focus on CMOS circuits. The subject of NMOS op amp design is taken up by Daniel Senderowicz in the following chapter. Paul Gray and David Hodges present analog-digital conversion techniques including oversampling, self-calibration, and ratio-independent schemes. Most appropriately, the next chapter deals with the design of digital filters; it is written by Walter Ulbrich. Digital filtering techniques in VLSI are developing at a rapid pace, but the monolithic filtering technique used most extensively sofar is that of switched capacitors; the next two chapters deal with this subject. Adel Sedra presents switched-capacitor filter synthesis techniques, including recent "exact design" methods that can be used for signal frequencies up to half the sampling rate. Paul Gray and Rinaldo Castello consider performance limitations in switched-capacitor filter, and make projections on what can be expected from scaled technologies. Recent techniques for integrated continuous-time filters are then discussed by Yannis Tsividis. This part of the book concludes with a presentation of several nonlinear analog MOS circuits, written by Bedrich Hosticka.

Typical use of many of the circuits presented in this part will be found in the representative systems described in Part III. The reader will also find there additonal circuit techniques.

3

CMOS OPERATIONAL AMPLIFIERS

B.J. HOSTICKA
Lehrstuhl Bauelemente der Elektrotechnik
University of Dortmund, D-4600 Dortmund 50
Federal Republic of Germany

1 INTRODUCTION

This tutorial presents basic design techniques for operational amplifiers implemented in CMOS technology. A number of topics will be addressed: small-signal MOSFET characteristics, basic analog circuits, operational amplifiers, power supply rejection ratio, noise, slew-rate, and output stages, and practical design examples will be given.

Operational amplifier has become one of the basic network components in design of analog circuits and have found wide use in range of applications in data acquisition, instrumentation, signal processing, etc. Its behavior is of essential importance and greatly affects performance of the circuits in which it is used. Hence thorough understanding of the operational amplifier is necessary and great attention has to be paid to its design. While the monolithic operational amplifier was originally devised as a stand-alone general-purpose component and was commercially available as a discrete device over number of years, the LSI operational amplifier is a somewhat different matter [1-3]. This follows from the fact that such an amplifier is usually assigned a specific permanent task on the chip that is known to the designer beforehand and the designer can tailor the amplifier according to the application. Thus the designer finds himself faced with a new set of design constraints and rules.

In recent years considerable interest has arisen in MOS analog circuits due to their compatibility with MOS digital hardware [4-6]. The design of MOS operational amplifiers, however, has been impeded by relatively low MOSFET transconductance and not negligible process variations. On the other hand, the design is often simplified by the fact that MOS amplifiers normally operate only into fixed on-chip capacitive loads. In this way, an MOS operational amplifier can be optimally designed into the environment in which it is supposed to operate. It appears to be very useful to establish a macrocell library containing a family of MOS/LSI operational amplifiers, at best in conjunction with efficient computer simulation package and embedded in mixed-mode environment.

2 SMALL-SIGNAL MOSFET CHARACTERISTICS

The operation of the MOS transistor relies on a field-controlled charge flow between two heavily doped regions (called source and drain) on a silicon surface (see Fig. 1a). The electric field is created between a conductor plate called gate and the lightly doped silicon substrate separated by a thin silicon dioxide layer that acts as a dielectric. An appropriate voltage applied to the gate induces an inversion charge on the silicon surface, which forms a conducting channel between the source and drain [7,8]. The input and output characteristics of a typical N-channel enhancement-mode MOSFET are shown in Figs. 2a and 2b, respectively. The MOSFET can be operated either as an analog switch or voltage-controlled current source (VCCS), this being subject to biasing conditions. If the applied gate-to-source voltage V_{GS} exceeds the threshold voltage V_T, the substrate surface is strongly inverted. In the neighborhood of and below V_T the surface is only weakly inverted and corresponding subthreshold drain current I_D is rather low. Since the input electrode, i.e. gate, is coupled only capacitively to the transistor, the input resistance of the MOSFET is infinite.

In many analog applications, e.g. in linear circuits, the variation of voltages and currents are small compared to the quiescent values, so that small-signal analysis can be employed. A simplified small-signal MOSFET circuit diagram is illustrated in Fig. 3 [9,10]. At low frequencies two small-signal parameters are of main interest: the transconductance $g_m = \partial I_D / \partial V_{GS}$ and the output resistance r_{DS} ($1/r_{DS} = \partial I_D / \partial V_{DS}$, where V_{DS} is the drain-to-source voltage).

The third parameter, the substrate transconductance $g_{mB} = \partial I_D / \partial V_{BS}$ is due to variation in the threshold voltage resulting from changes in the source-to-substrate voltage V_{BS}. This phenomenon is called "body- or substrate-effect". g_{mS} can be determined as $g_{mB} = K_1 g_m / \left[2\sqrt{(\phi_s - V_{BS})} \right]$ in strong inversion and it is usually much smaller than g_m. The body-effect improves common-mode rejection ratio of MOS differential stages because it lowers the common-mode voltage gain and it does not affect the differential-mode gain. The performance of MOS transistors in common-gate configuration improves too because the variations in V_{BS} have the same phase as those in V_{GS}. On the other hand, the common-drain (i.e. source-follower) performance suffers because the phase difference of these variations is 180°.

Let us now consider the low-frequency voltage gain A_{vo} and gain-bandwidth product GBW of an MOSFET as a function of a bias current $I_{D,BIAS}$. If the transistor operates in common-source configuration and the input gate is driven by an ideal voltage source, the gain is $A_{vo} = -g_m r_{DS}$ and the bandwidth is GBW = $g_m / (2\pi C_{GD})$ (in practice the capacitance at the drain would include all strays and capacitive loading). The dependence of these parameters in $I_{D,BIAS}$ can be expressed as $g_m \propto \sqrt{I_{D,BIAS}}$ and $r_{DS} \propto 1/I_{D,BIAS}$ for MOSFET channel in strong inversion, and $g_m \propto I_{D,BIAS}$ and

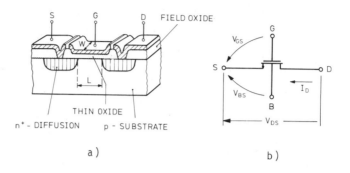

Fig. 1: a) Cross section of N-channel MOSFET,
b) electrical symbol

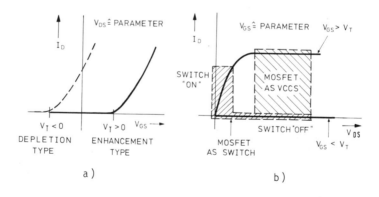

Fig. 2: a) Input MOSFET characteristic,
b) output characteristic

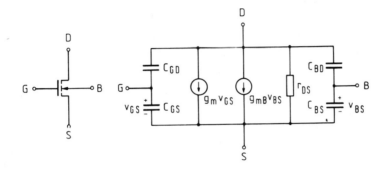

Fig. 3: Small-signal MOSFET equivalent circuit

$r_{DS} \propto 1/I_{D,BIAS}$ in weak inversion (i.e., when the transistor operates at subthreshold current) as depicted in Fig. 4a [11-13]. The subthreshold operation provides much better transconductance-to-current ratio than the strong inversion operation [14,15]. The values of A_{vo} and GBW are plotted in Fig. 4b. The designer obviously faces a dilemma: he can design either a broad-band amplifier (high GBW) or a high-gain amplifier (high A_{vo}).

So far no consideration was given to dependence of small-signal parameters on technology. Minimum available channel length and gate oxide thickness play eminent roles and as they get smaller the MOSFET transconductance increases. Since the MOS downscaling involves a number of parameters some advantages may be outweighed by undesired effects. But when applying the right scaling rules, the analog MOS circuits can benefit from shrinking of MOS dimensions [16].

3 BASIC ANALOG CMOS CIRCUITS

As mentioned in the previous section, the MOS transistor basically operates as VCCS. To convert the output current back into a voltage, a load device has to be provided. In the following, three simplest inverting voltage amplifiers will be reviewed to demonstrate the principle of analog CMOS design [17]. The basic CMOS push-pull inverter, well known from digital CMOS techniques, is shown in Fig. 5a [18]. It consists of two transistors, NMOS and PMOS, both operating as VCCS and load at the same time. Despite its appealing simplicity, it is not used too often because the quiescent current depends on the input voltage. In the second inverter (see Fig. 5b), the NMOS transistor serves as VCCS while the PMOS load device operates at a constant current owing to its fixed gate bias. The voltage gain is lower by 6 dB compared with the inverter in Fig. 5a, but the quiescent current is easily established. The third inverter, depicted in Fig. 5c, employs again a PMOS load device but this time connected as a diode. Such a configuration offers a rather low gain but it is quite broad-band, and it is often used in conjuction with current mirrors discussed below.

One of the principal attributes of the common-source configurations just discussed is the high-frequency feedback occuring from the output node back to the input through the gate-to-drain capacitance C_{GD}. Small excursions of the input voltage produce large output voltage variations of opposite polarity and thus large currents flow into C_{GD}. Hence C_{GD} appears at the input magnified by the factor voltage gain plus one (this effect is called Miller capacitance). This can cause a considerable bandwidth reduction if the common-source voltage amplifier is not driven by an input voltage source with low source resistance.

MOS current mirrors have become one of the standard analog configurations used time and again. The current mirror circuit serves either as a constant (or even time-varying) current source in biasing applications

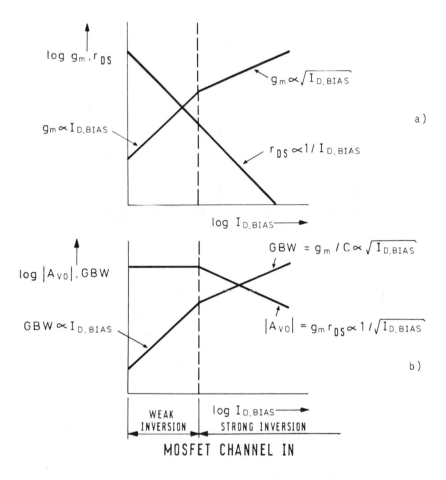

Fig. 4: a) Transconductance g_m and output resistance r_{DS}
as functions of the bias current $I_{D,BIAS}$,
b) low-frequency gain A_{vo} and gain-bandwidth product GBW
as functions of the bias current $I_{D,BIAS}$

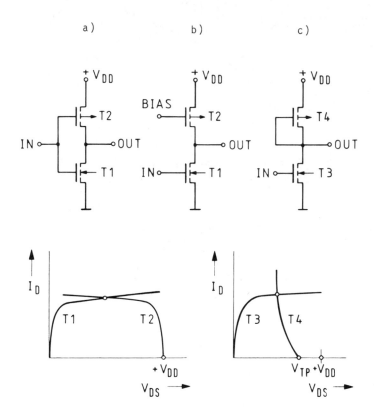

Fig. 5: a) CMOS push-pull inverter,
b) inverter with a current source load,
c) inverter with a diode load

or as an active load device for amplifier stages. The simplest form is illustrated in Fig. 6. Transistor T_2 has the same gate-to-source voltage V_{GS} as the diode-connected transistor T_1. Thus for the case of identical devices T_1 and T_2, the output and reference current are equal. If the transistor geometry ratios (W/L) are made different, the two drain currents, I_{REF} and I_{OUT}, will have a constant ratio given by the geometry. Hence any desired I_{OUT} can be derived from a given I_{REF} just by scaling the transistor geometry. The foregoing analysis assumed identical threshold voltages for both devices. Any mismatch of the threshold voltages of T_1 and T_2 (designated as V_{T1} and V_{T2}, respectively) will generate an error in the output current given by $\Delta I_{OUT}/I_{OUT} = 2\Delta V_T/(V_{GS}-V_{T1}) = (g_m/I_{OUT})\Delta V_T$, where $\Delta V_T = V_{T2}-V_{T1}$ [15]. This implies a rather poor current matching at subthreshold current levels.

Another basic analog circuit is the differential stage. This is a circuit designed to amplify the difference between two input signals. An example of an NMOS differential stage is shown in Fig. 7a. It consists of two source-coupled NMOS transistors, biased by a current source I_0. If both transistors are precisely matched, the stage is strictly symmetrical. Suppose two symmetrical (common-mode) signals are applied to the inputs, i.e. $V^+_{IN} = V^-_{IN}$, then the tail current I_0 will be equally split between the two legs of the differential stage. If the input signals are antisymmetrical (differential-mode), they cause an antisymmetrical current difference $\Delta I = g_m(V^+_{IN}-V^-_{IN})$, assuming small-signal disturbances. As far as the load devices are concerned, a PMOS current mirror or PMOS diode-connected transistors are typically used, but more sophisticated loads are feasible as well. The load choice is strongly influenced by the overall amplifier architecture.

If the two input devices are not precisely matched, a current error will occur. We usually try to refer this error to the input of the differential stage and define the so called input offset voltage ΔV_{os} which is the voltage that must be applied to the input in order to obtain zero current error at the output. Thus we require $I_0/2 = g_{m1}V_{GS1} = g_{m2}V_{GS2}$ (see Fig. 7b). $\Delta g_m = g_{m1}-g_{m2}$ is the transconductance mismatch and $\Delta V_{os} = V_{GS1}-V_{GS2}$ is the input offset voltage. Since $V_{GS2} \approx I_0/g_m$ for small-signal conditions, we can derive $\Delta V_{os} = (I_0/g_m) \cdot (\Delta g_m/g_m)$. This suggests low offsets in the subthreshold region. There are several possible sources of offset in the differential stages, such as transistor geometry mismatch, threshold voltage mismatch, etc. Since the differential stage usually serves as input stage of operational amplifiers, its offset contributes mostly to the total input offset of the amplifier because it is not reduced by any voltage gain.

An important feature of the differential stage is the rejection of common-mode signals. It is conveniently described by common-mode rejection ratio (CMRR), i.e. the ratio of the differential-mode gain to the common-mode gain. Since the differential stage is supposed to respond to the differential signals only, any response to the common-mode signals produces undesirable errors at the output. Hence CMRR should be as large as possible. A typical source of finite CMRR is the nonideal tail current source I_0.

65

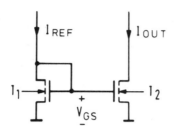

$$I_{REF} = K_1(V_{GS}-V_{T1})^2$$

$$I_{OUT} = K_2(V_{GS}-V_{T2})^2$$

FOR PERFECT MATCHING:

$$\frac{I_{OUT}}{I_{REF}} = \frac{(W/L)_2}{(W/L)_1} = \frac{G_{M2}}{G_{M1}}$$

MISMATCH: E.G. $V_{T1} \neq V_{T2} \Rightarrow V_{T2} = V_{T1} + \Delta V_T$

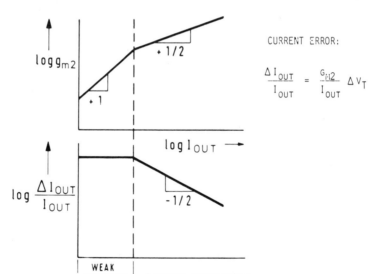

CURRENT ERROR:

$$\frac{\Delta I_{OUT}}{I_{OUT}} = \frac{G_{M2}}{I_{OUT}} \Delta V_T$$

Fig. 6: NMOS current mirror

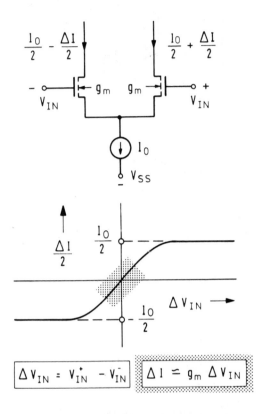

Fig. 7: a) NMOS differential stage,

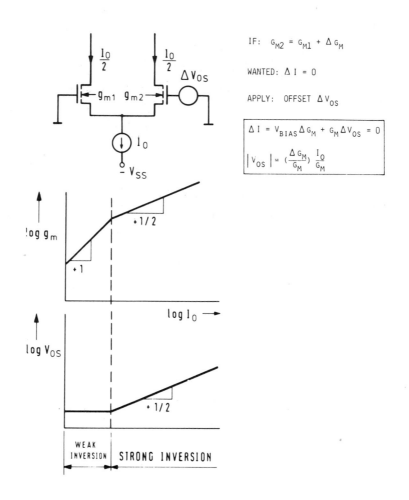

IF: $G_{M2} = G_{M1} + \Delta G_M$

WANTED: $\Delta I = 0$

APPLY: OFFSET ΔV_{OS}

$$\Delta I = V_{BIAS} \Delta G_M + G_M \Delta V_{OS} = 0$$

$$\left| V_{OS} \right| = \left(\frac{\Delta G_M}{G_M} \right) \frac{I_0}{G_M}$$

Fig. 7: b) Transistor mismatch in the differential stage

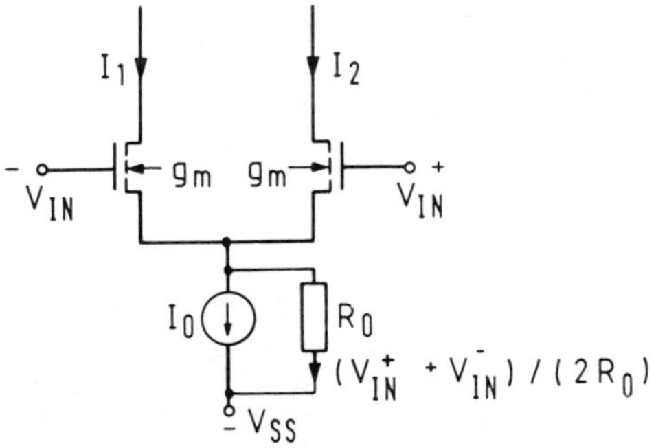

$$I_{COMMON-MODE} = \frac{I_1 + I_2}{2} = \frac{V_{IN}^+ + V_{IN}^-}{4R_0}$$

Recall : $\Delta I \simeq g_m (V_{IN}^+ - V_{IN}^-)$

$$CMRR = \frac{\dfrac{\Delta I}{V_{IN}^+ - V_{IN}^-}}{\dfrac{I_{CM}}{2(V_{IN}^+ + V_{IN}^-)}} \simeq 2 g_m R_0$$

Fig. 7: c) Common—mode rejection ratio

69

Assume it has an output resistance R_o, then the current flow through R_o corresponds to the common-mode current $I_{CM} = (I_1+I_2)/2 = [(V_{IN}^++V_{IN}^-)/2]/(2R_o)$, as shown in Fig. 7c, and so

$$CMRR = \frac{\Delta I/(V_{IN}^+-V_{IN}^-)}{I_{CM}/[(V_{IN}^++V_{IN}^-)/2]} = 2g_mR_o.$$

Thus a "good" current source is required with high output resistance to achieve a high CMRR. Current mirrors are frequently employed to bias differential stages.

The last basic analog circuit to be examined is the cascode stage in Fig. 8. While the device T1 operates as VCCS (common-source configuration) and regulates the output current, the common-gate transistor T2 increases the output resistance and decreases the input capacitance. Both effects can be easily explained. The high resistance seen at the output is caused by the output resistance r_{DS1} of the transistor T1 operating as a high source resistance for the device T2. An attempt to increase the output current of the cascode stage by a current source connected to the drain of T2 would increase the drain-to-source voltage V_{DS1} and thus lower the gate-to-source voltage V_{GS2}. As a consequence, the dynamic output resistance of the cascode stage increases to: $R_{OUT} = r_{DS2}[1+r_{DS1}/r_{DS2}+(g_{m2}+g_{mS2})r_{DS1}]$ [19]. If the input voltage changes by v_{in}, the drain current of T1 varies by $g_{m1}v_{in}$, and the gate-to-source voltage of T2 varies by $g_{m2}v_{GS2}$. The effective capacitance seen at the input is then: $[1+g_{m1}/(g_{m2}+g_{mS2})]C_{GD1}+C_{GS1}$. Thus Miller capacitance is reduced if g_{m2} is chosen sufficiently large.

4 OPERATIONAL AMPLIFIERS

An "ideal" operational amplifier is defined as a voltage-controlled voltage source (VCVS) which exhibits an infinite voltage gain, infinite input impedance, zero output impedance, and infinite bandwidth. By applying a well-defined passive feedback around the operational amplifier, a large number of network functions can be generated, such as precise amplification, integration, differentiation, etc.

Throughout the years, two types of operational amplifiers have evolved: one- and two-stage configurations [1,2]. The reference to one- and two-stage configuration is usually done on the understanding that only stages providing high voltage gain are counted. Such stages tend to contribute the dominant frequency poles and dictate the frequency compensation. Traditionally, the most commonly used approach has been the two-stage concept which was originally developed for bipolar technology. This configuration features good common-mode rejection ratio and range, high

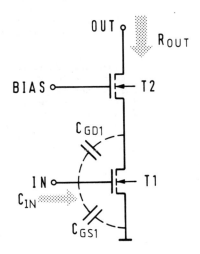

Fig. 8: Cascode stage

voltage gain and output swing. Nevertheless, the one-stage amplifier exhibits superior performance in some applications, as it will become apparent below.

Consider the two-stage CMOS operational amplifier shown in Fig. 9a [20-22]. It consists of a biasing network, a differential input stage, and an inverting output stage. It is interesting to note that the use of current mirror as a load for the differential stage provides conversion from the differential to single-ended signal. Since the output resistance is high, the amplifier can drive only capacitive loads. The amplifier is frequency-compensated by the Miller capacitor C_C but at the cost of introducing a zero in the right-half plane that can seriously degrade the phase margin. The usual remedy is to insert a nulling resistor R_Z in series with C_C.

For small signals, the two-stage amplifier can be modeled as illustrated in Fig. 9b. In this figure, the DC voltage gain of the i-th stage is approximated by $A_{vi} = g_{mi}R_i$, where g_{mi} is the transconductance and R_i is the output resistance. Similarly, C_i is the output capacitance. The capacitance at the output node of the 2-nd stage must include the capacitive load C_L. Let us assume the operational amplifier of Fig. 9b to be connected in unity-gain configuration (dashed line), which represents the worst case for the frequency stability. The voltage transfer function to the output is then given by [23]

$$\frac{v_{out}}{v_{in}} = \frac{A_v[1 - sC_C(1/g_{m2}-R_Z)]}{(1+A_v) + sa + s^2b + s^3c}$$

where

$$C_2' = C_2 + C_L$$

$$A_v = A_{v1}A_{v2}$$

$$a = C_C[A_v(1/g_{m1}-1/g_{m2}+R_Z)+R_1+R_2+R_Z]+(C_1R_1+C_2'R_2)$$

$$b = R_1R_2(C_1C_2'+C_1C_C+C_2'C_C)+R_ZC_C(C_1R_1+C_2'R_2)$$

$$c = C_1C_2'C_CR_1R_2R_Z$$

To simplify the analysis, we make several assumptions that are usually fulfilled in practice:

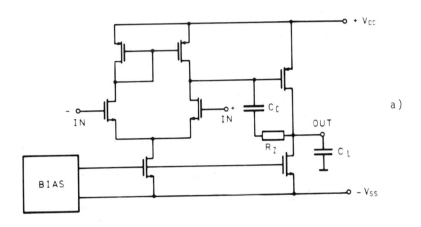

a)

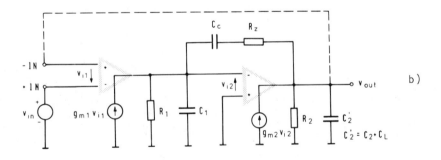

b)

Fig. 9: a) CMOS two-stage amplifier,
       b) small-signal equivalent circuit

73

$A_v \gg 1$, $R_Z \ll R_1$, $R_Z \ll R_2$,

$C_1 \ll C_C$, $C_1 \ll C_L$, and $C_2 \ll C_L$.

If we further assume that the poles are reasonably wide spaced ($|p_1| \ll |p_2| \ll |p_3|$), the pole and zero locations in the transfer function can be estimated as [3]:

$$p_1 \simeq -g_{m1}/C_C ; \qquad p_2 \simeq -g_{m2}/C_L ;$$

$$p_3 \simeq -1/(\dot{C}_1 R_Z) ; \qquad z_1 = [C_C(1/g_{m2} - R_Z)]^{-1} .$$

Note that z_1 vanishes when $R_Z = 1/g_{m2}$. The capacitive loading must fulfill the condition $|p_1| \ll |p_2|$, i.e. $C_L \ll C_C(g_{m2}/g_{m1})$ in order to ensure a sufficient phase margin. The most dominant pole of the open-loop operational amplifier ($p_{1,open-loop} \simeq -(g_{m2}R_1R_2C_C)^{-1}$) has been moved by the amount of A_v to the cross-over frequency of the open-loop amplifier. This frequency is usually called gain-bandwidth product (GBW) or unity-gain bandwidth and since it is given by GBW = $A_v p_{1,open-loop} = g_{m1}/C_C$ it is the same as p_1.

Let us now consider the one-stage CMOS operational amplifier in Fig. 10a [13,24]. This amplifier contains a biasing network, a differential input stage with diode-connected active loads, and a high gain cascode output stage. The conversion of the differential to single-ended signal is carried out in the output stage. As the input stage has a rather low gain and almost all the gain is obtained at the output, this can be regarded as a single-stage operational amplifier. The output node is the only high-resistance node in the entire amplifier. The amplifier is compensated by capacitive loads at this node. This saves the compensation capacitor and eliminates any problems connected with the right-half plane zero. A simplified small-signal equivalent circuit is depicted in Fig. 10b. The DC voltage gain is $A_{vs} = g_{ms}R_s$, the output capacitance is C_s, and the loading capacitance is C_L. If the amplifier is connected in unity-gain configuration, the transfer function is [23]

$$v_{out}/v_{in} \simeq [1 + s(R_s C_s'/A_{vs})]^{-1} = [1 + s(C_s'/g_{ms})]^{-1},$$

where $A_{vs} \gg 1$ and $C_s' = C_s + C_L$. This analysis neglects any poles inside the amplifier. A care must be taken to keep the loading capacitance at the output node sufficiently large so that this pole is really the most dominant and an adequate phase margin is ensured.

74

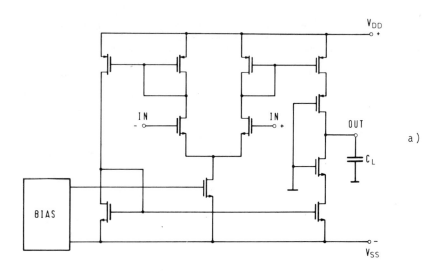

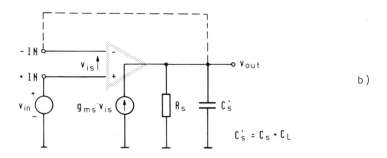

Fig. 10: a) CMOS one-stage amplifier,
        b) small-signal equivalent circuit

Power-supply rejection ratio (PSRR) is an important parameter, unfortunately often underestimated. The size of today LSI and VLSI chips implies that an operational amplifier must coexist with large amount of various hardware on the same chip. Even with extremely careful design it is nearly impossible to prevent coupling of extraneous signals into power supply rails of the amplifier. If such signals are not sufficiently rejected by the amplifier, they can be conveyed into the signal path and degrade the performance of the operational amplifier. PSRR of an amplifier is defined as ratio of the open-loop voltage gain from the input to the output to that from the power supply to the output.

There are different ways how the interference from the power supply rails can be coupled into the amplifier. In particular, various parasitics and certain amplifier characteristics can be responsible. Consider an example of a differential stage shown in Fig. 11a. The potential V_Y varies when V_{SS} changes: on the on hand, this is due to the body effect of the input transistors, on the other hand, it may be aided by the dependence of I_o on V_{SS} and thus

$$\frac{dV_Y}{dV_{SS}} = -\frac{1}{2(g_m + g_{mS})}\frac{dI_o}{dV_{SS}}.$$

Similarly, any dependence of I_o on V_{DD} produces variations in V_Y as well:

$$\frac{dV_Y}{dV_{DD}} = -\frac{1}{2(g_m + g_{mS})}\frac{dI_o}{dV_{DD}}.$$

Changes in V_Y generate a displacement current through C_{GS} and this produces potential changes at the input that is driven only capacitively. For example, the displacement current at the inverting input of SC integrators can reach the output through the feedback capacitor and decrease PSRR. This type of PSRR degradation can be considerably reduced if I_o is made supply independent and if the input devices of the differential stage are placed into their own isolated well.

At high frequencies, coupling of interference from power supplies may occur due to architecture of the operational amplifier. This is typical for two-stage amplifiers where the signals can reach the output through the compensation capacitor because the second stage uses the power supply rail as a reference (see Fig. 11b). Since this effect is closely related to noise

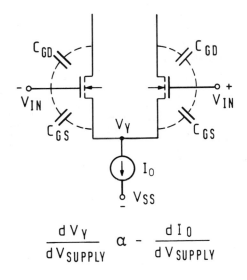

$$\frac{dV_Y}{dV_{SUPPLY}} \quad \alpha \quad - \quad \frac{dI_0}{dV_{SUPPLY}}$$

Fig. 11: a) Power-supply rejection ratio of a differential stage

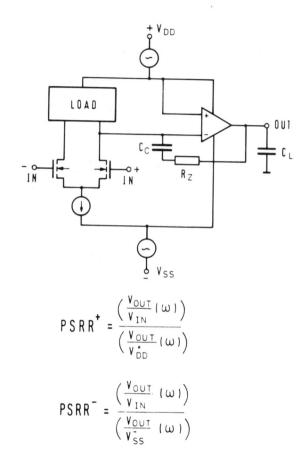

$$PSRR^+ = \frac{\left(\dfrac{V_{OUT}}{V_{IN}}(\omega)\right)}{\left(\dfrac{V_{OUT}}{V_{DD}^+}(\omega)\right)}$$

$$PSRR^- = \frac{\left(\dfrac{V_{OUT}}{V_{IN}}(\omega)\right)}{\left(\dfrac{V_{OUT}}{V_{SS}^-}(\omega)\right)}$$

Fig. 11: b) Power-supply rejection ratio of the two-stage amplifier

performance it will be analyzed together with noise in the section below.

6 NOISE PERFORMANCE

In this section we shall discuss the noise problems of CMOS operational amplifiers. Although the noise performance of an amplifier depends on a particular configuration, a rough general estimate is possible.

The major components of the gate-referred noise spectrum of a typical MOS transistor are 1/f-noise and thermal noise [25] (see Fig. 12). The 1/f-noise decreases with frequency and is independent of bias. This noise is process dependent and can be reduced either by using devices with inherently lower 1/f-noise, such as MOS transistors with large geometries or nonstandard MOS-compatible JFET transistors, or by applying special circuit techniques, e.g. correlated double sampling or chopper stabilization [26,27].

The thermal noise of MOS transistors is given by $\overline{v_n^2}/\Delta f = 8kT/3g_m$, where g_m is the transconductance. Approximating the equivalent noise bandwidth by $g_m/4C$, we obtain $\overline{v_n^2} = 2kT/3C$, where C is the bandwidth limiting capacitor.

To find the equivalent input noise of an entire gain stage, we have to model all transistors with their equivalent noise sources and calculate the total equivalent input noise [28]. While the equivalent input noise of a gain stage driver appears fully at the input of the stage, the equivalent input noise of an active load can be referred to the stage input as divided by $[g_{m,driver}/g_{m,load}]^2$ independently of load configuration (see Fig. 13). This discussion can be extended to the noise evaluation of entire amplifiers where the noise tends to be accumulated due to the contribution of individual devices but the total noise will again be inversely proportional to the capacitor that limits the amplifier bandwidth [23].

Apart from antialiasing filters and unity-gain buffers with static biasing, most analog MOS circuits for low-frequency applications tend to use switched-capacitor (SC) techniques [6]. In SC networks the time constant given by the capacitor and the ON switch resistance must be much smaller than the sampling period in order to ensure the most complete charge transfer. The same consideration applies to the time constant, i.e. bandwidth, of the operational amplifier. As a result, all the white noise is greatly undersampled and folded back into the baseband below the Nyquist frequency [29,30]. This foldover effect usually does not substantially increase the flicker noise component due to its 1/f roll-off and, hence, the folded white noise becomes easily dominant in SC networks [31]. It will be shown that this is exactly where the architecture of the operational amplifier comes in.

79

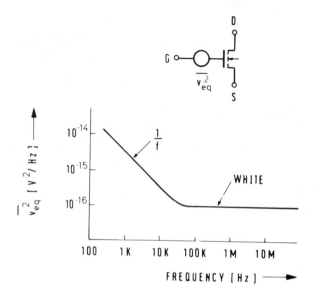

Fig. 12: Noise spectrum of a typical MOSFET

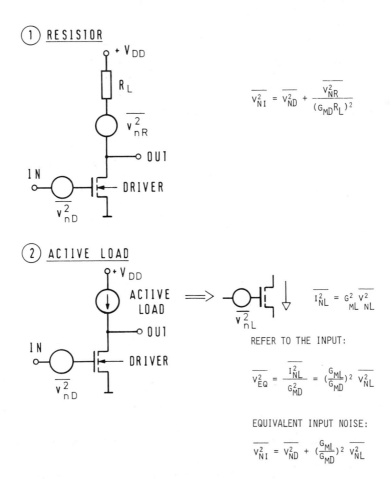

① **RESISTOR**

$$\overline{v_{NI}^2} = \overline{v_{ND}^2} + \frac{\overline{v_{NR}^2}}{(G_{MD}R_L)^2}$$

② **ACTIVE LOAD**

$$\overline{I_{NL}^2} = G_{ML}^2\,\overline{v_{NL}^2}$$

REFER TO THE INPUT:

$$\overline{v_{EQ}^2} = \frac{\overline{I_{NL}^2}}{G_{MD}^2} = (\frac{G_{ML}}{G_{MD}})^2\,\overline{v_{NL}^2}$$

EQUIVALENT INPUT NOISE:

$$\overline{v_{NI}^2} = \overline{v_{ND}^2} + (\frac{G_{ML}}{G_{MD}})^2\,\overline{v_{NL}^2}$$

Fig. 13: Input equivalent noise of amplifier stages

Consider once more the small-signal model of the two-stage CMOS operational amplifier of Fig. 9a shown in Fig. 14a. If v_{ni} is the equivalent input noise voltage of the i-th stage, the transfer functions to the output are

$$\frac{v_{out1}}{v_{n1}} = \frac{A_v\left[1 - sC_C(1/g_{m2}-R_Z)\right]}{(1+A_v) + sa + s^2b + s^3c}$$

$$\frac{v_{out2}}{v_{n2}} = A_{v2}\frac{\left\{1 + s\left[C_C(R_1+R_Z) + C_1R_1\right] + s^2C_1C_CR_1R_Z\right\}}{(1+A_v) + sa + s^2b + s^3c}$$

where C_2', A_v, a, b, and c are defined as in the section on operational amplifiers. If we apply the same assumptions , i.e.:

$$A_v \gg 1, \quad R_Z \ll R_1, \quad R_Z \ll R_2,$$
$$C_1 \ll C_C, \quad C_1 \ll C_L, \quad \text{and} \quad C_2 \ll C_L.$$

we obtain for wide spaced poles the following pole and zero locations:

$$p_1 \simeq -g_{m1}/C_C \; ; \qquad p_2 \simeq -g_{m2}/C_L \; ;$$
$$p_3 \simeq -1/(C_1R_Z) \; ; \qquad z_1 = \left[C_C(1/g_{m2} - R_Z)\right]^{-1} \; ;$$
$$z_2 \simeq -1/(C_CR_1) \; ; \qquad z_3 \simeq -1/(C_1R_Z) \to -\infty.$$

The poles are identical for both functions, while z_1 appears only in the function v_{out1}/v_{n1} and vanishes when $R_Z=1/g_{m2}$. The zeros z_2 and z_3 appear in v_{out2}/v_{n2} only.

Figs. 14b and 14c depict the Bode diagrams of the transfer functions v_{out1}/v_{n1} and v_{out2}/v_{n2}, respectively, for $R_Z = 1/g_{m2}$. The frequency components of v_{n2} that lie in the frequency band $p_1 \ll \omega \ll p_2$ are conveyed directly to the amplifier output node [23]. Actually, this also applies to any signal appearing on the positive power supply rail because the second stage uses this rail as a reference [32]. Most of the two-stage operational amplifiers use one of the power supply rails as a reference for the second stage. Hence the corresponding PSRR will be degraded above the closed loop

82

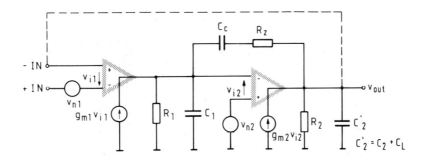

Fig. 14: a) Small-signal equivalent circuit of the two-stage amplifier

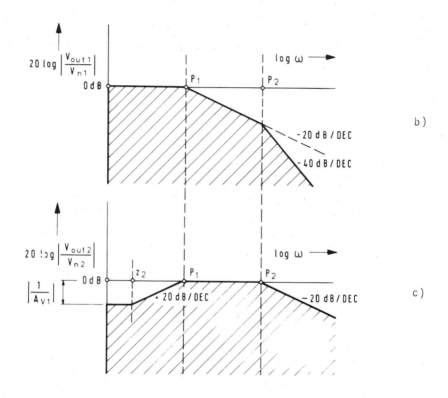

Fig. 14: b) Bode diagram of the transfer function v_{out1}/v_{n1},
          c) Bode diagram of the transfer function v_{out2}/v_{n2}

bandwidth.

From Figs. 14b and 14c, the equivalent noise bandwidths can be estimated as $f_{B1} = g_{m1}/(4C_C)$ and $f_{B2} \simeq (g_{m2}/C_L-g_{m1}/C_C)/4 \simeq g_{m2}/(4C_L)$ [33]. White noise behaviour of MOS transistors can be characterized by an equivalent input noise resistance $R_N = \overline{v_n^2}/4kT\Delta f$, which is independent of frequency and approximately equal to the inverse of transistor transconductance [34,35]. $\overline{v_n^2}/\Delta f$ is the spectral density of the equivalent input noise.

Considering that all the noise sources are uncorrelated, the total noise of one stage can be determined as

$$\overline{v_{ni}^2} = 4kT\gamma_i f_{Bi}/g_{mi},$$

if we include the relative contribution of all transistors of the i-th gain stage in the factor γ_i. In practice, γ is in the range between 1 and 4 [34]. The total equivalent input noise of the amplifier is then given by

$$\overline{v_{ntot}^2} \simeq kT (\gamma_1/C_C + \gamma_2/C_L).$$

This implies that the noise given by $kT\gamma_1/C_C$ is the minimum value for large capacitive loads.

Let us now consider the one-stage CMOS operational amplifier in Fig. 10a and its small-signal model shown in Fig. 15a. If the amplifier is connected in unity-gain configuration, the transfer function is

$$v_{outs}/v_{ns} \simeq [1 + s(R_s C_s'/A_{vs})]^{-1} = [1 + s(C_s'/g_{ms})]^{-1},$$

where v_{ns} is the equivalent input noise voltage (see Fig. 14b) and $A_{vs} \gg 1$. The equivalent noise bandwidth is

$$f_{Bs} = g_{ms}/(4C_s') \simeq g_{ms}/(4C_L)$$

85

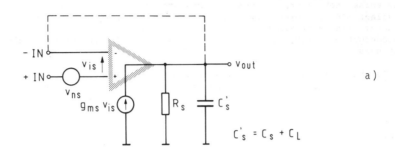

a)

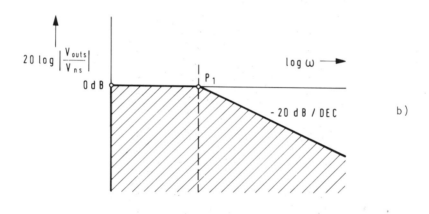

b)

Fig. 15: a) Small-signal equivalent circuit of the one-stage amplifier,
b) Bode diagram of the transfer function v_{outs}/v_{ns}

86

for large capacitive loads, and the total equivalent input noise of the amplifier is

$$\overline{v_{ntot}^2} = kT\gamma_s/C_L.$$

The advantage of the one-stage amplifier is that, unlike in the two-stage amplifier, the voltage gain of any transfer function to the output node rolls off at the same frequency. Since the output resistance of the one-stage amplifier we have just investigated is high, this amplifier is again useful for driving capacitive loads only. Any resistive load R_L would decrease the DC voltage gain. For such cases we have to add a stage with a low output resistance (R_o), such as source follower. Consider a simple model of the one-stage amplifier with a unity-gain output stage as shown in Fig. 16a. The Bode diagrams of the corresponding transfer functions are shown in Fig. 16b and 16c. If wide spaced, the poles can be estimated as

$$p_1 \simeq - g_{ms}/C_s' \simeq - g_{ms}/C_C$$

$$p_2 \simeq - 1/(R_oC_o') \simeq - 1/(R_oC_L)$$

$$z_1 = - 1/(R_sC_s') \simeq - 1/(R_sC_C)$$

for $R_o \lll R_L$ and $R_o \lll R_s$. A frequency-compensation capacitor C_C has been connected to the output of the first stage to maintain the most dominant pole there. The compensation at the output would require an oversized capacitor due to low R_o. Low capacitive loading of this amplifier would produce very high white noise. Clearly, the addition of the low-resistance output stage degrades the good noise performance of single-stage amplifiers.

7 SLEW RATE

So far only small-signal properties of CMOS amplifiers have been considered. Nevertheless, it may happen in some cases that the input stage of the operational amplifier is driven into nonlinear region of operation due to large input signals. This becomes obvious when considering the transfer characteristic of the NMOS differential stage depicted in Fig. 9a. The maximum current this stage can sink or supply under large-signal conditions is given by the tail current I_o, and this limits the large-signal

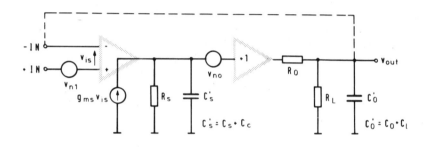

Fig. 16: a) Small-signal equivalent circuit of a one-stage amplifier
with a unity-gain low resistance output stage

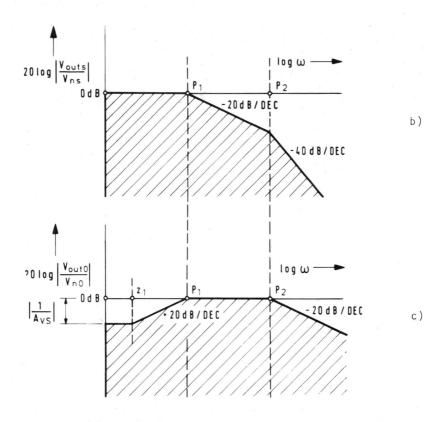

Fig. 16: b) Bode diagram of the transfer function v_{outs}/v_{ns},
c) Bode diagram of the transfer function v_{outo}/v_{no}

bandwidth of any amplifier that uses a constant current tail source. A classical example where this applies is the two-stage CMOS operational amplifier in Fig. 9a. The maximum rate at which the output can change (called slew-rate) is given by the tail current I_0 and the compensation capacitor C_C: SR = I_0/C_C. Recall now the calculation of the pole and zero locations associated with the amplifier. It followed there that GBW = g_{m1}/C_C and based on this it can be estimated that the slew-rate is SR = GBW$\cdot(I_0/g_{m1})$. The ratio I_0/g_{m1} can be made high for MOS in strong inversion and MOS amplifiers are thus capable of displaying high slew-rates. This discussion also applies to the one-stage amplifier of Fig. 10a, where the loading capacitance at the output can be charged or discharged by a limited current – this time defined by the tail current source of the input stage and multiplied by the ratios of the current mirrors that drive the output gain stage.

8 OUTPUT STAGES

It has already been pointed out that MOS/LSI amplifiers drive mostly fixed on-chip capacitive loads. For these applications, a low-resistance output stage is not only unnecessary but it can even degrade the noise performance as discussed above. Driving off-chip loads, either resistive ones or very large capacitances, requires addition of an output stage to the basic amplifier. This stage must possess a low output resistance so that it remains broad-band for large capacitive loads and it can supply power into resistive loads without performance degradation.

MOS source followers have one distinct disadvantage: it is difficult to make their output resistance sufficiently low due to inherently low MOSFET transconductance. As this transconductance is proportional to the square-root of the current in strong inversion, it is not economical to design a low output resistance just by increasing the quiescent current.

The simplest output stage is the common-drain configuration (see Fig. 17a) [3]. It is a class-A circuit with a quite good current-sourcing capability but its current sinking-capability is limited by the bias current of the tail current source. Another serious drawback of this circuit is that the positive output swing is limited by the gate-to-source voltage of the MOS transistor. A better current-sourcing and -sinking capability can be gained when the class-AB push-pull output stage of Fig. 17b is used [3]. This type of stage is frequently used in bipolar operational amplifiers. The quiescent current is set by the diode-connected transistors incorporated into the preceding gain stage. The output swing is again limited by the necessary gate-to-source voltage of both output transistors.

The output stage design may be decidedly eased if bipolar or low-threshold MOS devices are available. As a conclusion, it can be generally stated that there is no standard CMOS low-resistance output stage

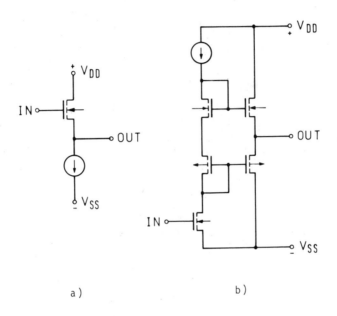

Fig. 17: a) class-A source follower,
        b) class-AB source follower

and its design still presents a difficult task.

9 DIFFERENTIAL OUTPUT AMPLIFIERS

Operational amplifier is basically a high-gain controlled source, commonly a voltage amplifier. The input is usually differential, i.e. ideally totally voltage-floating, and the output port is earthed. While it is usually impossible to realize voltage-floating outputs, it appears easier to design an operational amplifier with voltage-symmetric output, which is differential but not floating. Such a configuration makes it possible to design circuits with balanced signal paths. These circuits inherently reject all comon-mode interference signals and thus exhibit high PSRR [27].

An example of a CMOS differential output operational amplifier is depicted in Fig. 18 [4]. An important design feature is the common-mode feedback (CMFB) loop that maintains stable bias. The CMFB circuit can be either of continuous- or discrete-time type.

10 DESIGN EXAMPLES

In this section two different designs of operational amplifiers will be presented and experimental results demonstrated. Both amplifiers have been integrated in CMOS silicon-gate technology with 40 nm gate oxide and 1.8×10^{14} cm^{-3} p-substrate. The doping of the n-wells was 1.8×10^{16} cm^{-3} [36].

The first device is a high-gain two-stage operational amplifier with low-resistance output stage (see Fig. 19a). All the bias currents in the amplifier are set by the transistor T1 and mirrored into the amplifier by T2. The transistor pair T3, T4 provides biasing for PMOS current sources, while NMOS current sources are biased directly by T2. The architecture of the amplifier strongly resembles the one of Fig. 9a. The input differential stage consists of transistors T8-T12, the gain inverting stage of T13, T14, and the output class-A source follower of T15, T16. The amplifier is again compensated by the Miller capacitor C_C but the nulling resistor has been substituted by the transistor T17. The value of r_{DS} of T17 was chosen so that the zero z_1 cancelled the second pole p_2. It can be readily found that the condition for this is $r_{DS} = (1+C_2/C_C)/g_{m2}$ [22]. It suffices to fulfill $|p_1|\lll|p_3|$ to maintain stability and thus C_C can be made much smaller. r_{DS} of T17 should track any changes in g_{m2} to allow for process, temperature and power supply variations. This is accomplished by the tracking circuit T5-T7. It can be shown that if $K = (W/L)_{T13}/(W/L)_{T5} = (W/L)_{T14}/(W/L)_{T7}$, then $(W/L)_{T17} = [K(W/L)_{T6}(W/L)_{T14}]^{1/2} \cdot C_C/(C_C+C_2)$ guarantees good tracking.

92

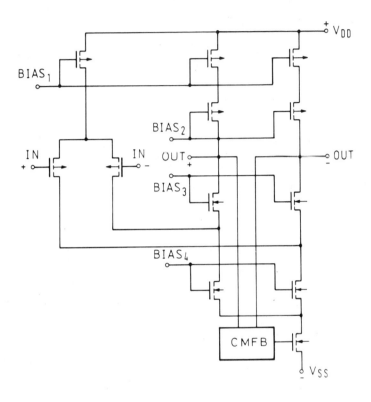

Fig. 18: CMOS differential output operational amplifier

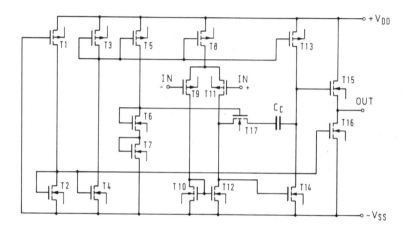

Fig. 19: a) CMOS two-stage operational amplifier

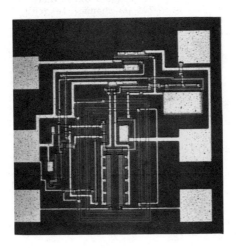

Fig. 19: b) Chip photomicrograph of the two-stage amplifier

TABLE I

Performance of the Two-Stage
CMOS-Operational Amplifier

Technology   Si-Gate CMOS with
                              40 nm gate oxide and
                              7.5 µm channel length

Power supply voltage   ±3.75 V

Power dissipation   900 µW

Open-loop gain   ≥80 dB

Unity-gain bandwidth   1 MHz at 12.5 pF load

Phase margin   57° at 12.5 pF load

Noise spectral density   0.8 µV/√Hz at 80 Hz
                               60 nV/√Hz at 100 kHz

Slew-rate   ≥2 V/µsec at 12.5 pF load

PSRR   60 dB

Common-mode input range   +2.5 to −3.7 V

The transistor channel length was chosen 7.5 μm. This degraded the bandwidth but helped obtain open-loop voltage gain in excess of 80 dB. The measured data are summarized in Table I. The input capacitance is below 2 pF and the output open-loop resistance is about 4 kΩ. The chip photomicrograph can be seen in Fig. 19b. The chip area is 0.2 mm^2 without bonding pads.

The second device is a high-speed operational amplifier shown in Fig. 20a [37]. It is basically a one-stage architecture but the output current of the input stage (T9-T13) is fed directly into the output stage (T7, T8, T15, T16) through the cascode devices (T6, T14). The bias is provided by T1, T2, and it is mirrored by T3-T5 to bias the amplifier.

The transistor channel length was deliberately chosen smaller (3.5 μm) to achieve higher bandwidth. The experimental results are given in Table II and the chip photomicrograph is shown in Fig. 20b. Varying the reference current through T1, T2 changed not only power dissipation but the bandwidth and open-loop gain too. At 1 mW power consumption, the amplifier displayed GBW = 4 MHz and A_v = 71 dB with a load of 10 pF. An increase of power dissipation to 4 mW raised GBW to 9.2 MHz but A_v sunk to 63.5 dB. The chip area was 0.1 mm^2 without bonding pads.

11 CONCLUSION

In this work design techniques for CMOS operational amplifiers have been presented. It has been shown that the design is greatly affected by intended application as there is no ingenious CMOS amplifier that would fulfill any given task. The selected architecture has a profound impact on frequency compensation, noise, and PSRR performance. Three following design guidelines for design of CMOS operational amplifiers can be stated:
1. Use low-resistance output stages only for resistive loads.
2. Use one-stage amplifiers without low resistance output stage for low-noise switched capacitor applications.
3. Use one-stage amplifiers or amplifiers with differential output for good PSRR at higher frequencies.

12 ACKNOWLEDGEMENT

The author wishes to thank all colleagues at Lehrstuhl Bauelemente der Elektrotechnik at the University of Dortmund for useful discussions and support. Special thanks go to J. Fichtel and M. Wrede who were responsible for the design of the operational amplifiers. A careful proofreading of this manuscript by W. Brockherde and K.-G. Dalßas is greatfully acknowledged.

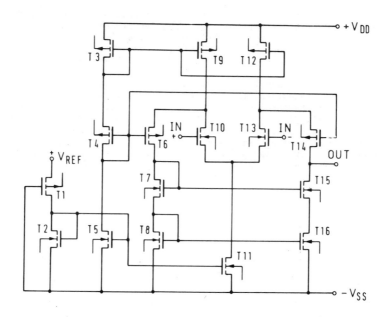

Fig. 20: a) CMOS one-stage operational amplifier

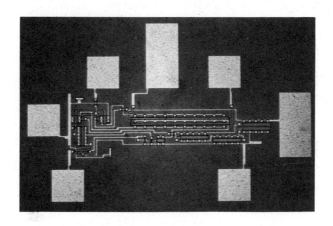

Fig. 20: b) Chip photomicrograph of the one-stage amplifier

TABLE II

Performance of the Single-Stage
CMOS-Operational Amplifier

Technology	Si-Gate CMOS with 40 nm gate oxide and 3.5 μm channel length
Power supply voltage	±5 V
Power dissipation	2 mW
Open-loop gain	68 dB
f_{3dB}	3.1 kHz
Unity-gain bandwidth	6.13 MHz at 10 pF load
Phase margin	54° at 10 pF load
Noise spectral density	0.42 $\mu V/\sqrt{Hz}$ at 80 Hz 19 $nV/\sqrt{Hz}$ at 1 MHz
Slew-rate	>4 V/μsec at 10 pF load

13 REFERENCES

[1] P.R. GRAY and R.G. MEYER, "Recent advances in monolithic operational amplifier design," IEEE Trans. Circuits Syst., vol. CAS-21, pp. 317-327, May 1974.

[2] J.E. SOLOMON,"The monolithic op amp - A tutorial study," IEEE J. Solid-State Circuits, vol. SC-9, pp. 314-332, Dec. 1974.

[3] P.R. GRAY, "Basic MOS operational amplifier design" in Analog MOS Integrated Circuits (Ed. P.R. Gray, D.A. Hodges, and R.W. Brodersen), IEEE Press: New York, 1980.

[4] P.R. GRAY and R.G. MEYER,"MOS operational amplifier design - A tutorial study," IEEE J. Solid-State Circuits, vol. SC-17, pp. 969-982, Dec. 1982.

[5] D.J. ALLSTOT and W.C. BLACK,"Technological design considerations for monolithic MOS switched-capacitor filtering systems," Proc. IEEE, vol. 71, pp. 967-986, Aug. 1983.

[6] P.R. GRAY, D.A. HODGES, and R.W. BRODERSEN: Analog MOS Integrated Circuits, IEEE Press: New York, 1980.

[7] A.S. GROVE: Physics and Technology of Semiconductor Devices, Wiley: New York, 1967.

[8] R.S. MULLER and T. KAMINS: Device Electronics for Integrated Circuits, Wiley: New York, 1977.

[9] H. SCHICHMANN and D.A. HODGES,"Modeling and simulation of insulated-gate field-effect transistors," IEEE J. Solid-State Circuits, vol. SC-3, pp. 285-289, Sept. 1968.

[10] Y. TSIVIDIS,"Design considerations in single-channel MOS analog integrated circuits - A tutorial," IEEE J. Solid-State Circuits, vol. SC-13, pp. 383-391, June 1978.

[11] S. LIU and L.W. NAGEL,"Small-signal MOSFET models for analog circuit design," IEEE J. Solid-State Circuits, vol. SC-17, pp. 983-998, Dec. 1982.

[12] W. STEINHAGEN and W.L. ENGL,"Design of integrated analog CMOS circuits - A multichannel telemetry transmitter," IEEE J. Solid-State Circuits, vol. SC-13, pp. 799-805, Dec. 1978.

[13] B.J. HOSTICKA,"Dynamic CMOS amplifiers," IEEE J. Solid-State Circuits, vol. SC-15, pp. 887-894, Oct. 1980.

[14] E. VITTOZ and J. FELLRATH,"CMOS analog integrated circuits based on weak inversion operation," IEEE J. Solid-State Circuits, vol. SC-12, pp. 224-231, June 1977.

[15] O.H. SCHADE, Jr.,"BiMOS micropower IC`s", IEEE J. Solid-State Circuits, vol. SC-13, pp. 791-798, Dec. 1978.

[16] T. ENOMOTO, T. ISHIKAWA, M.-A. YASUMOTO, and T. AIZAWA,"Design, fabrication, and performance of scaled analog IC´s," IEEE J. Solid-State Circuits, vol. SC-18, pp. 395-402, Aug. 1983.

[17] W.N. CARR and J.P. MIZE: MOS/LSI Design and Applications, McGraw-Hill: New York, 1972.

[18] M.I. Elmasry: Digital MOS Integrated Circuits, IEEE Press: New York, 1981.

[19] B.J. HOSTICKA,"Improvement of the gain of MOS amplifiers," IEEE J. Solid-State Circuits, vol. SC-14, pp. 1111-1114, Dec. 1979.

[20] F.H. MUSA and R.C. HUNTINGTON,"A CMOS monolithic $3^1/2$ digit A/D converter," in Dig. Tech. Papers, 1976 Int. Solid-State Circuits Conf., Philadelphia, PA, pp. 144-145, Febr. 1976.

[21] R. GREGORIAN and W.E. NICHOLSON, Jr.,"CMOS switched-capacitor filters for a PCM voice CODEC," IEEE J. Solid-State Circuits, vol. SC-14, pp. 970-980, Dec. 1979.

[22] W.C. BLACK, D.J. ALLSTOT, and R.A. REED,"A high performance low power CMOS channel filter," IEEE J. Solid-State Circuits, vol. SC-15, pp. 929-938, Dec. 1980.

[23] B.J. HOSTICKA, W. BROCKHERDE, and M. WREDE,"Effects of the architecture on noise performance of CMOS operational amplifiers," in Proc. 1983 European Conf. Circuit Theory and Design, Stuttgart, pp. 238-241, Sept. 1983.

[24] F. KRUMMENACHER,"High voltage-gain CMOS OTA for micropower switched capacitor filters," Electron. Lett., vol. 17, no. 4, pp. 160-164, Febr. 1981.

[25] R.S.C. COBBOLD: Theory and Applications of Field-Effect Transistors, Wiley-Interscience, 1970.

[26] M.B. DAS and J.M. MOORE,"Measurement and interpretation of low frequency noise in FET´s," IEEE Trans. Electron Devices, vol. ED-21, pp. 247-257, Apr. 1974.

[27] K.-C. HSIEH, P.R. GRAY, D. SENDEROWICZ, and D.A. MESSERSCHMITT, "A low-noise chopper-stabilized differential switched-capacitor filtering technique," IEEE J. Solid-State Circuits, vol. SC-16, pp. 708-715,

Dec. 1981.

[28] J.C. BERTAILS,"Low frequency noise considerations for MOS amplifier design," IEEE J. Solid-State Circuits, vol. SC-14, pp. 773-776, Aug. 1979.

[29] C.-A. GOBET and A. KNOB, "Noise analysis of switched capacitors," ISCAS Proc., April 1981, pp. 856-859.

[30] B. FURRER and W. GUGGENBUEHL, "Noise analysis of sampled-data circuits," ISCAS Proc., April 1981, pp. 860-863.

[31] J.H. FISCHER, "Noise sources and calculations techniques for switched capacitor filters," IEEE Journal of Solid-State Circuits, vol. SC-17, pp. 742-752, Aug. 1982.

[32] V.G. RUEDIGER and B.J. HOSTICKA, "The response of 741 op amps to very short pulses," IEEE J. of Solid-State Circuits, vol. SC-15, pp. 910-913, Oct. 1980.

[33] A.B. CARLSON, Communication Systems, 2nd Edition, McGraw-Hill, 1975.

[34] E.A. VITTOZ, "Micropower IC," Digest of Techn. Papers ESSCIRC '80, pp. 174-189, Sept.1980

[35] G. REINBOLD and P. GENTIL,"White noise of MOS transistors operating in weak inversion," IEEE Trans. Electron Devices, vol. ED-29, no. 11, pp. 1722-1725, Nov. 1982.

[36] G. ZIMMER, H. FIEDLER, B. HOEFFLINGER, E. NEUBERT, and H. VOGT, "Performance of a scaled Si gate n-well CMOS technology," Electron. Lett., vol. 17, no. 18, pp. 666-667, Sept. 1981.

[37] R. HOFMANN, G. LUTZ, B.J. HOSTICKA, M. WREDE, G. ZIMMER, and J. KEMMER,"Development of readout electronics for monolithic integration with diode strip detectors," to be published.

4

MICROPOWER TECHNIQUES

Eric A. Vittoz

Centre Suisse d'Electronique et de Microtechnique S.A.,
CSEM (Formerly CEH), Maladière 71, 2000 Neuchâtel
and
Swiss Federal Institute of Technology (EPFL), Lausanne
Switzerland

1   INTRODUCTION

Micropower integrated circuit techniques and technologies have originally been developed for applications in electronic watches [1],[2]. Modern
watch circuits have a complexity ranging from a few thousands to a few tens
of thousands of transistors, combine processor-type architectures with some
critical analog functions, and are routinely produced by tens of millions per
year with power consumption below 0.5 µW at 1.5 V. The need for very low
power consumption has been extended to other kinds of battery operated systems such as pocket calculators, pace-makers, hearing aids, paging receivers
and various types of very small size portable instruments. Another interesting
niche is that of data acquisition systems powered by optical fibers to avoid
galvanic coupling.

Most of these applications require low voltage operation (1 to 3 volts)
and current consumption typically below 50 µA, which is a very severe limitation specially for analog subcircuits. Some trade-off with dynamic range must
and can generally be accepted, but specific techniques had to be developed.

It is believed that some of these techniques can be adapted to normal
telecommunication circuits to reduce the power per function and allow an increase of complexity, while limiting the power dissipated on chip.

Furthermore, some approaches developed for micropower prove to be efficient in high frequency design.

After mentioning a few particular points related to technology, this
paper will discuss devices and circuits aspects of micropower integrated
circuits. Some examples of analog and digital subcircuits will then be presented.

2   TECHNOLOGY

CMOS is unquestionably the best technology for micropower, and it has been used for that purpose almost exclusively for more than 10 years. It combines an unexcelled capability for low power digital with a high flexibility for analog, and the density problem has long been solved by adequate circuit techniques.

The combination of dynamic and static power P consumed by a CMOS gate is approximately given by the well known simple formula

$$P = f\ C\ V_B^2 + V_B I_o \qquad\qquad 2.1$$

where f is the output frequency, V_B the supply voltage, C the output capacitance and I_o the average leakage current. Capacitance C is kept low by keeping substrate and well doping low (1 to 5 $.10^{15} cm^{-3}$) and by using shallow diffusions and small dimensions. Gate oxide must not be too thin (70-100 nm) in order to limit gate capacitance.

It can be shown that the smallest possible supply voltage V_B for digital operation is about 200 mV [3]. Meanwhile, if the threshold voltage V_T is too low, the residual channel current at zero gate voltage becomes a dominant part of I_o, and static power consumption is increased. Optimum values of V_T lie around 0.5 V. An accuracy of $\pm$ 0.1 V can be obtained by ion implantation of both substrate and wells, and by carefully optimized annealing.

Strict control of the whole process is needed to minimize junction reverse currents. These can be kept below 10 nA/mm^2 at 1.5 V and ambient temperature.

Si-gate is preferred for its self-alignment properties. The polysilicon layer can be used extensively for interconnections since its resistivity is negligible at low current. It has also been used to form capacitors together with the Al layer [4]. Some micropower technologies allow p or n doping of the same polysilicon layer, which offers interesting devices and circuit techniques as will be shown later. Recent technologies provide two layers of polysilicon to implement non-volatile memories and denser capacitors [5].

No reason has been found to change the original choice of p-type wells.

3   DEVICES

3.1 MOS transistors

When the drain current of a MOS transistor is decreased by reducing the gate voltage, the density of mobile current carriers eventually becomes negligible with respect to that of fixed charges in the channel. The device has entered the weak inversion mode of operation where the usual parabolic

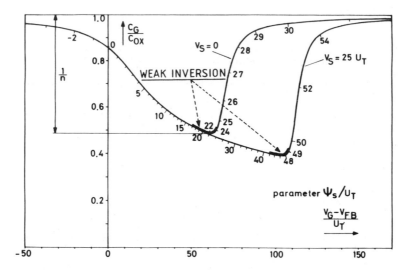

Fig. 3.1   Weak inversion on a C-V curve (Copyright © 1977 IEEE).

transfer characteristics for strong inversion are no longer valid. The MOS structure operates then at the bottom of the dip of the C-V curve as shown in figure 3.1.

For long channel transistors with a negligible density of fast surface states and negligible leakage current to substrate, the drain current I_D in weak inversion may be expressed as [6]:

$$I_D = I_{DO} \, e^{V_G/nU_T} \, (e^{-V_S/U_T} - e^{-V_D/U_T})$$ 3.1

where V_G, V_S and V_D are the gate, source and drain potentials with respect to local substrate, and $U_T = kT/q$.

The slope factor n (usually between 1.3 and 2) results from the capacitive voltage divider which relates the variation of gate voltage V_G to that of surface potential ψ_s. It is a fairly controllable parameter which can be found on the C-V plot of figure 3.1. Practically, it depends slightly on source voltage V_S, but may be considered constant within a range of ψ_s wider than 7 to 8 U_T, which corresponds to at least 3 orders of magnitude of I_D.

On the contrary,

$$I_{DO} \sim \frac{W}{L} \, e^{\frac{-V_T}{nU_T}}$$ 3.2

is a highly unpredictable parameter, which plays a role similar to saturation current I_S of bipolar transistors. As a consequence, MOS transistors in weak inversion should always be biased at a fixed drain current I_D and not at a fixed gate voltage V_G.

Matching of similar devices on the same chip is related to that of V_T (5 mV $\hat{=}$ $\sim$10 %) and current can be weighted by effective width to length ratio W/L.

Equation 3.1 yields the characteristics shown in figure 3.2

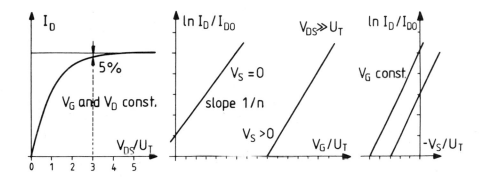

Fig. 3.2   Characteristics in weak inversion.

Output characteristics saturate as soon as $V_{DS} = V_D-V_S$ exceeds a few U_T. This very low saturation voltage (< 100 mV) is a first advantageous feature of weak inversion, which allows to maximize the dynamic range for a given supply voltage. Transfer characteristics from gate are exponential with slope factor n. Transfer characteristics from source are identical to those of bipolar transistors, and can be shifted by changing gate voltage V_G. This identity is more than casual since both are barrier-controlled devices with diffusion as the dominant charge transport mechanism. The most important difference lies in the fact that carriers diffusing between source and drain at the surface of a MOS transistor in weak inversion are effectively majority carriers with respect to their immediate environment. They can thus diffuse much further than their diffusion length in the bulk.

The maximum saturation current in weak inversion is roughly given by

$$I_D = \beta U_T^2 \qquad\qquad 3.3$$

where $\beta = \mu\, C_{ox}\, \dfrac{W}{L}$ is the usual transfer parameter in strong inversion.

The maximum value of β achievable with a transistor of reasonable size (0.02 mm^2) is about 10 mA/V^2 which corresponds to a maximum possible operating current in weak inversion of a few microamperes. For a typical, minimum-sized transistor, this limit is between 10 and 100 nA.

The minimum operating current is limited by carrier generation in the drain and channel depletion layers. It is therefore roughly proportional to the overall area of the transistor. Minimum-sized transistors may be used with a drain current as low as 100 pA at 50°C.

In order to keep symmetry between drain and source, the following model will be used for strong inversion [7]:

$$I_D = \beta (V_D - V_S) [V_G - V_T - \frac{n}{2} (V_D + V_S)] \quad \text{for } V_D \leqslant \frac{V_G - V_T}{n}$$

$$I_D = \frac{\beta}{2n} (V_G - V_T - nV_S)^2 \qquad \qquad \text{for } V_D \geqslant \frac{V_G - V_T}{n}$$

3.4

where V_T is gate threshold voltage for $V_S = 0$ and n is the same parameter as in weak inversion, which takes into account bulk modulation effects. This model can be obtained by second order series expansion of the classical model [8] around $V_D = 0$ and $V_S = 0$.

Models 3.1 and 3.4 are design models which sacrify some accuracy to simplicity. For computer simulation, more accurate models that include the medium current range between weak and strong inversion are being used [9],[10].

Small signal gate to drain transconductances in saturation are easily derived as

$$g_m = \frac{\partial I_D}{\partial V_G} = \begin{cases} \dfrac{I_D}{nU_T} & \text{(weak inversion) 3.5} \\[2em] \dfrac{2I_D}{V_G - V_T - nV_S} = \sqrt{\dfrac{2 \beta I_D}{n}} = \dfrac{\beta}{n} (V_G - V_T - nV_S) & \text{(strong inversion)} \\ & \qquad\qquad\qquad 3.6 \end{cases}$$

For a given transistor, it is thus proportional to I_D in weak inversion and to $I_D^{1/2}$ in strong inversion.

The source to drain transconductance

$$g_{ms} = \frac{\partial I_D}{\partial (-V_S)} = n g_m$$

3.7

is equal in weak inversion to that of a bipolar transistor operated at the same current.

The maximum voltage amplification achievable with a transistor is limited by the non-zero output conductance g_o of the device, which is due to channel shortening. In both modes of operation, it is approximately proportional to drain current and can be characterized by an extrapolated voltage V_E roughly proportional to channel length L:

$$g_o = I_D/V_E \qquad\qquad 3.8$$

As shown in figure 3.3, the amplification factor $A_{omax} = g_m/g_o$ has a maximum value

$$A_{omax} = V_E/nU_T \qquad\qquad 3.9$$

in weak inversion, but it decreases like $I_D^{-1/2}$ in strong inversion. This maximum possible gain in weak inversion reflects the maximum of the transconductance-to-current ratio. It may be further increased by increasing the channel length of the transistor.

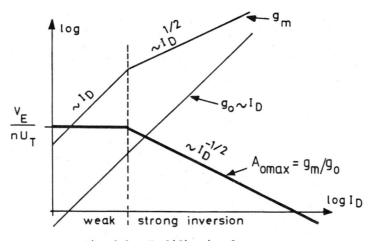

Fig. 3.3   Amplification factor.

The noise behaviour of a MOS transistor must be modelled by at least 2 independent noise sources as shown in figure 3.4.

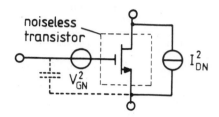

noiseless transistor

V_{GN}^2

I_{DN}^2

<u>Fig. 3.4</u> Noise model of a MOS transistor.

In strong inversion, the noise current source is the thermal noise of the non-homogeneous resistive channel [11], with a spectral density in saturation:

$$\frac{dI_{DN}^2}{df} = \frac{8}{3} kT\ g_m\ = 2q\ I_D\ \cdot\ \frac{8U_T}{3(V_G - V_T - nV_S)} \qquad 3.10$$

Since the flow of carriers in weak inversion is controlled by a potential barrier, it is reasonable to assume an associated shot noise in saturation with spectral density:

$$\frac{dI_{DN}^2}{df} = 2q\ I_D \qquad 3.11$$

which has been verified experimentally [12],[13].

Noise voltage V_{GN}^2 is the noise related to the Si-SiO$_2$ interface. Although it has been shown to depend slightly on gate voltage [14], one may assume for sake of simplicity in the design phase

$$\frac{dV_{GN}^2}{df} = 4\ kT\ \frac{\wp}{f.WL} \qquad 3.12$$

Where $\wp$ is a technology dependent parameter, usually larger for n than for p-channel transistors. Typical values measured in a low-threshold p-well technology are $\wp_n = 2\ \Omega m^2/s$ and $\wp_p = 0.05\ \Omega m^2/s$.

It is usually convenient to define a total equivalent noise voltage at the gate

$$dV_N^2 = dV_{GN}^2 + d\ \frac{I_{DN}^2}{g_m^2} \qquad 3.13$$

which can be represented by frequency dependent equivalent noise resistance

$$R_N = \frac{1}{4\,kT} \cdot \frac{dV_N^2}{df}$$

3.14

Combination of 3.10 to 3.14 yields

$$R_N = \frac{\mathcal{G}}{f\,W.L} + \begin{cases} \dfrac{2}{3g_m} & \text{strong inversion} \\[2em] \dfrac{n}{2g_m} & \text{weak inversion} \end{cases}$$

3.15

At high frequencies, R_N is approximately equal to $1/g_m$. It therefore decreases with increasing I_D. But if I_D is fixed (limited by power consumption), R_N is reduced by increasing W/L until it reaches a minimum in weak inversion, as shown in figure 3.5. This is another advantage of the maximum value of g_m/I_D in weak inversion.

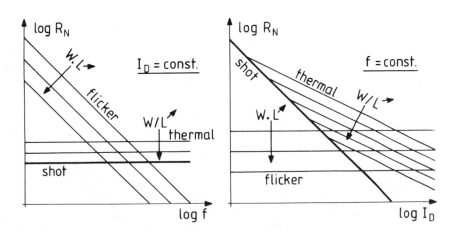

Fig. 3.5   Variations of noise resistance.

At low frequencies, 1/f flicker noise dominates and R_N is independent of I_D. It can only be reduced by increasing W.L.

Much like noise, mismatch between two identical transistors must be characterized by 2 independent statistical parameters: threshold mismatch ΔV_T and β-mismatch $\quad \mathcal{E}_\beta = \Delta\beta/\beta$.

Drain current mismatch of 2 transistors with the same gate voltage V_G (and same V_S)

$$\frac{\Delta I_D}{I_D} = \varepsilon_\beta - \frac{g_m}{I_D} \Delta V_T \qquad\qquad 3.16$$

is thus improved by decreasing g_m/I_D (strong inversion), whereas gate voltage difference of 2 transistors at the same current I_D

$$\Delta V_G = \Delta V_T - \frac{I_D}{g_m} \varepsilon_\beta \qquad\qquad 3.17$$

is improved by increasing g_m/I_D and reaches a minimum in weak inversion.

3.2 CMOS compatible bipolar transistors [15]

Fig. 3.6 shows the source-to-drain transfer characteristics in saturation, measured on a normal n-channel enhancement mode MOS transistor realized in a low-voltage p-well CMOS technology.

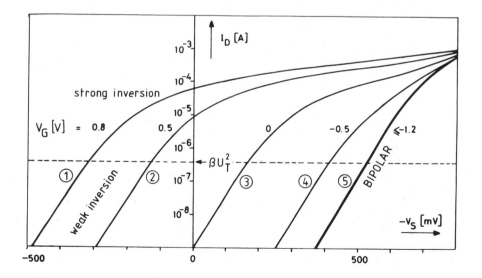

Fig. 3.6   Source-to-drain transfer characteristics.
(Copyright © 1983 IEEE)

In normal MOS operation, drain current flows for positive gate voltage V_G (curves 1 and 2). As predicted by the model (equation 3.1 and fig. 3.2), I_D is approximately proportional to

$$e^{-V_S/U_T}$$

below the limit βU_T^2 of weak inversion. If V_G is reduced to zero (curve 3), the same law typical of MOS operation is still obtained, but for negative values of V_S (source junction slightly conducting). However, if a large negative gate voltage is applied (in this case $V_G \leqslant -1.2$ V), all curves merge into curve 5. The I_D-V_S relationship becomes independent of V_G, stays exponential to much higher currents and has a slope exactly equal to U_T. Transport of carriers from source to drain has been pushed away below the surface of the device and is achieved by pure bipolar operation.

Figure 3.7 shows the major flows of current carriers in this mode of operation.

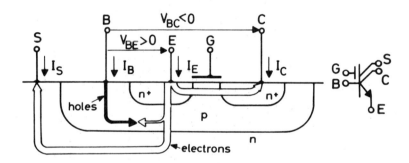

<u>Fig. 3.7</u>   Major flows of carriers in lateral bipolar operation.
(Copyright © 1983 IEEE)

Because of the absence of a p^+ buried layer which would be needed to prevent injection of electrons toward the substrate, this lateral npn bipolar is combined with a vertical npn. Emitter current I_E is thus split unto a base current I_B, a lateral collector current I_C and a substrate collector current I_S. Therefore, neither of the common base current gains

$$\alpha = \frac{-I_C}{I_E} \text{ and } \alpha_S = \frac{-I_S}{I_E} \qquad\qquad 3.18$$

can be close to 1. However, due to the very small rate of recombination inside the p-well and the high emitter efficiency, both common emitter current
gains

$$\beta = \frac{I_C}{I_B} \quad \text{and} \quad \beta_S = \frac{I_S}{I_B} \qquad\qquad 3.19$$

can be large.

To favor the lateral device, I_C/I_S can be maximized by minimizing the
emitter area and the lateral base width, and by having the emitter surrounded
by the collector. Such an optimized layout in 6 μm Si-gate technology yields
typical gain values of $\beta = 100$ at $I_C = 1$ μA, with I_S approximately equal to
I_C.

Besides its true bipolar characteristics, this device offers interesting matching and low-frequency noise properties, both thought to be linked
to the removal of surface effect dependence. Measurements show an improvement in input offset voltage by a factor 10 and a reduction of more than
40 dB of 1/f noise between MOS operation and lateral bipolar operation of
the same devices.

High current performance is limited by strong injection which occurs
somewhat below 100 μA for minimum size devices.

Although no data are available yet, high frequency performance is certainly limited by the large amount of minority carriers in the base.

The lateral collector can be removed in common collector configurations,
but keeping the whole structure and connecting the lateral collector to substrate improves the total β gain.

The negative gate voltage required to prevent surface conduction in low
V_T technologies can easily be produced on chip. For technologies with large
V_T, the gate can simply be connected to emitter. The correct criterion for
maximum gate voltage is to ensure a negligible value of dI_C/dV_G.

3.3 Passive devices

No special requirements are imposed on capacitors for the implementation
of micropower circuits. Scaling down values of functional capacitors to help
reducing power consumption is bounded by acceptable kT/C noise limits.

On the contrary, the realization of good resistors is made difficult
by the need for high values. The highest value per unit area is usually obtained by p-well diffusions, but applications of this kind of resistor are
limited by their high intrinsic time constant and by their sensitivity to
substrate modulation.

Doped polysilicon provides excellent resistors, but their values are
not compatible with submicroampere current consumption. Lightly doped poly

resistors available in some technologies do not achieve the accuracy required in most analog circuits.

As mentioned in section 2, p and n doping of the same polysilicon layer has been maintained in some micropower technologies. Without really reducing the density of digital circuits, this feature allows the realization of lateral pn diodes as shown in figure 3.8 [16].

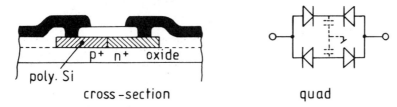

cross-section   quad

Fig. 3.8   Lateral polysilicon diode.

Although its characteristics are far from being ideal, this device provides the basic advantage of having no leakage to substrate. It finds application in the realization of very high impedance on-chip voltage multipliers required for non-volatile memories [5]. It may also be used around the origin of its I-V characteristics as a very high impedance equivalent resistor. Values of many giga-ohms can thus be obtained with negligible area. When symmetrical characteristics are required, the quad configuration has been found to be the best solution to equilibrate rectification of AC signals by the (very low) parasitic capacitance at the common point of back-to-back connected diodes.

3.4 Parasitic effects

Parasitic capacitors and noise impose a trade-off between bandwidth (or speed) and current drain. However, an absolute limit of operation current is given by leakage currents. Each branch of an analog circuit must operate at an average current level sufficiently higher than the local leakage current. This is true for dynamic digital circuits as well, but not for static digital circuits which can withstand leakage currents much larger than their nominal current drain.

It must be remembered that leakage is mostly due to generation of electron-hole pairs within space charge regions (with a tendency to increase at the Si-SiO$_2$ interface), and is therefore approximately proportional to the volume of such regions. As a consequence, it is for example impossible to operate a transistor in strong inversion ($V_{GS} \gg V_T$) at very low currents: this would require a very long channel, thus a very large space charge region below this channel, the leakage of which would become comparable to the drain current. If minimum size, and thus weak inversion, is acceptable, leakage can be kept negligible up to 80°C for drain currents as low as 1 nA. If threshold V_T is low (0.3 to 0.6 V), residual channel current at V_{GS} = 0 may be an important contribution to leakage when a transistor is used as a switch.

Impact ionization causes a component of output conductance of the transistor, which cannot be eliminated by a cascode configuration, since it appears between drain and local substrate. Micropower technologies have high drain breakdown voltages (> 20 V) because of their low doping of substrate and well. Impact ionization is therefore negligible for 3 Volt operation.

The otherwise difficult problem of latch-up has long been solved in a very simple and safe way in micropower. Normal layout procedures ensure a value of holding current above 1 mA (p-well resistivity of about 2 kΩ/square, necessary voltage drop of about 0.5 V). Latch-up can thus be prevented by limiting the current available from the power source below this value, by means of a series resistor of a few Kilo-ohms. This resistor causes negligible voltage drop at nominal currents of a few microamperes.

4   BASIC CIRCUIT TECHNIQUES

4.1 Current mirrors

As a result of relation 3.16, inaccuracy of a current mirror (fig. 4.1) is given by

$$\frac{\Delta I}{I} = \mathcal{E}_\beta - \frac{g_m}{I} \Delta V_T \qquad\qquad 4.1$$

and is worst in weak inversion where g_m/I reaches the maximum value

$$\frac{g_m}{I} = \frac{1}{nU_T} \qquad\qquad 4.2$$

Thus, for n = 1.9, a threshold offset ΔV_T = 5 mV results in a 10 % contribution to inaccuracy, which is much larger than usual values of $\mathcal{E}_\beta$. The only way to improve accuracy (besides using optimum layout and non-minimum size) is to reduce g_m/I by going into strong inversion where for V_S = 0

$$\frac{g_m}{I} = \frac{2}{V_G - V_T} \qquad\qquad 4.3$$

Accuracy is then improved by a factor $(V_G-V_T)/2nU_T$ which is approximately 10 for V_G-V_T = 1 V. The same is true for the contribution of 1/f flicker noise to output noise current.

According to relations 3.10 and 3.11, output white noise current is also reduced in strong inversion, but only by a factor $\sqrt{3(V_G-V_T)/8U_T}$.

The amount by which V_G can be increased above V_T is obviously limited by the total value of supply voltage V_B. However, a more severe limit is given by the decrease of saturation voltage which may reduce the dynamic range of analog circuits.

As shown in figure 4.1, this saturation voltage which is a few U_T in weak inversion is equal to $(V_G-V_T)/n$ in strong inversion. Therefore for a low supply voltage, accuracy and noise must often be traded for maximum possible voltage amplitude.

Furthermore, it has been shown in section 3 that the presence of leakage allows no other choice but weak inversion when currents are very low.

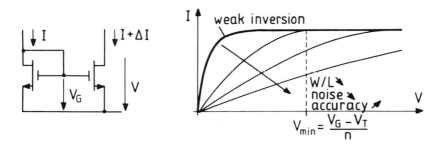

Fig. 4.1   Current mirror and output characteristics.

Compatible bipolars offer the possibility of improving accuracy by one order of magnitude [15] while maintaining a reasonably low value of minimum output voltage.

4.2 Elementary amplifiers

Common source amplifiers may be differenciated by the way the load transistor is connected and by the mode of operation of both active and load devices.

As shown in figure 4.2, the load transistor T_2 can be connected as a passive dipole, driven by active transistor T_1.

117

Voltage gain is then:

$$
-A_o = \frac{g_{m1}}{g_{m2}} = \begin{cases} \sqrt{\dfrac{n_2\,\beta_1}{n_1\,\beta_2}} & \text{①} \quad T_1 \text{ and } T_2 \text{ in strong inversion} \\[2ex] \dfrac{1}{n_1 U_T}\cdot\sqrt{\dfrac{n_2 I_o}{2\,\beta_2}} & \text{②} \quad \begin{array}{l} T_1 \text{ in weak inversion} \\ T_2^1 \text{ in strong inversion} \end{array} \\[2ex] \dfrac{n_2}{n_1} & \text{③} \quad T_1 \text{ and } T_2 \text{ in weak inversion} \end{cases} \qquad 4.4
$$

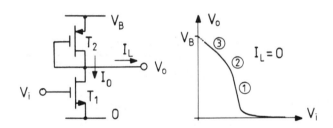

Fig. 4.2 Passive load.

The gain is independent of current I_o when both transistors are in strong inversion, decreases with $\sqrt{I_o}$ when the active transistor operates in weak inversion to reach a value close to 1 when the load also reaches weak inversion. Thus no voltage gain can be obtained at very low current.

Equivalent noise resistor for $|A_o| \gg 1$ is equal to that of T_1.

If load transistor T_2 is connected as a current source (output of a mirror), the DC voltage gain is only limited by the paralleled output conductances of the 2 transistors in saturation. Thus

$$
-A_o = \frac{g_{m1}}{g_{o1}+g_{o2}} = \begin{cases} V_E\sqrt{\dfrac{2\,\beta_1}{n_1 I_o}} & \text{for } T_1 \text{ in strong inversion} \\[2ex] \dfrac{V_E}{n_1 U_T} & \text{for } T_1 \text{ in weak inversion} \end{cases} \qquad 4.5
$$

where

$$\frac{1}{V_E} = \frac{1}{V_{E1}} + \frac{1}{V_{E2}}$$ 4.6

Total equivalent input noise resistance is

$$R_{Nt} = R_{N1} + (\frac{g_{m2}}{g_{m1}})^2 R_{N2} \xrightarrow{T_1 \equiv T_2} 2R_N$$ 4.7

It can be reduced to that of T_1 if T_2 has a large value of V_G-V_T, which limits the maximum output amplitude.

The most efficient amplifier is the complementary inverter of figure 4.3, in which both transistors are active.

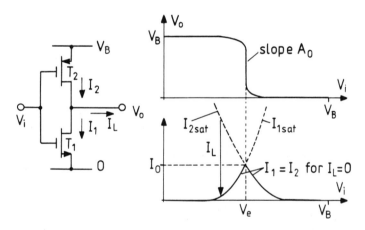

Fig. 4.3   Complementary inverter.

Around equilibrium voltage $V_i = V_e$, total transconductance is

$$g_{mt} = \frac{\partial I_L}{\partial V_i} = g_{m1} + g_{m2}$$ 4.8

119

and reaches in weak inversion the maximum value achievable in a single stage with current I_o:

$$g_{mt} = \frac{I_o}{U_T} \left(\frac{1}{n_1} + \frac{1}{n_2}\right) = \frac{2I_o}{nU_T} \qquad \text{for } T_1 \equiv T_2 \qquad 4.9$$

Voltage gain

$$-A_o = \frac{g_{m1} + g_{m2}}{g_{o1} + g_{o2}} = A_{omax} \qquad \text{for } T_1 \equiv T_2 \qquad 4.10$$

is comprised between amplification factors of T_1 and T_2.

Total equivalent input noise resistance is

$$R_{Nt} = \frac{R_{N1}\, g_{m1}^{\,2} + R_{N2}\, g_{m2}^{\,2}}{(g_{m1} + g_{m2})^2} = \frac{R_N}{2} = \frac{n^2 U_T}{4\, I_o} \qquad 4.11$$
$$\text{for } T_1 \equiv T_2 \quad \text{weak inversion} \atop \text{white noise}$$

Thus, if both transistors have same R_N and g_m, equivalent input noise is 3 dB below that of a single transistor, and 6 dB better then when T_2 is connected as a current source (relation 4.7).

In order to use this configuration, a solution must be found to control bias current I_o independently of V_B or threshold values.

The differential amplifier shown in figure 4.4 is a widely used configuration.

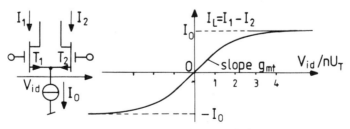

Fig. 4.4   Differential amplifier.

120

The contribution of the mismatch of T_1 and T_2 to input offset voltage is the difference ΔV_G of their gate voltages for $I_1 = I_2$. According to 3.17, this difference is minimum in weak inversion. Thus, except when some linearity of transconductance is required, this circuit is normally biased in weak inversion.

Its large signal transfer characteristics are obtained from model 3.1

$$I_{1(2)} = \frac{I_o}{1 + \exp\left[-(+)\dfrac{V_{id}}{nU_T}\right]} \qquad 4.12$$

and the current available in the load is (after mirroring I_1 and/or I_2)

$$I_L = I_1 - I_2 = I_o \tanh\frac{V_{id}}{2nU_T} \qquad 4.13$$

Small signal total transconductance

$$g_{mt} = \frac{\partial I_L}{\partial V_{id}} = g_{m1(2)} = \frac{I_o}{2nU_T} \qquad 4.14$$

is 4-times smaller than that of the complementary inverter biased at the same current.

Total equivalent input noise resistance is

$$R_{Nt} = 2R_N \qquad 4.15$$

Since only half of the current I_o is available for each transistor, the white noise component of R_{Nt} in weak inversion is increased by a factor 8 with respect to the complementary inverter operated at the same current, which corresponds to 9 dB increase of equivalent input noise voltage.

Table 4.1 summarizes values of g_{mt} and R_{Nt} (white noise) normalized to those of the complementary inverter operated at the same current and in weak inversion.

If some input current can be accepted, active MOS transistors can be replaced by compatible bipolars. Performances approximately equivalent to weak inversion ($\alpha < 1$ compensates $n > 1$) are obtained even at higher currents and 1/f noise is reduced [15].

121

		g_{mt}	R_{Nt}	total noise voltage [dB]
weak inversion	Complementary inverter	1	1	0
	Differential pair	1/4	8	9
	Current load	1/2	2 to 4	3 to 6
	Passive load	1/2	2	3
Strong inversion		divide by $\dfrac{V_G-V_T}{2nU_T}$	multiply by $\dfrac{V_G-V_T}{3nU_T}$	add $10 \log \dfrac{V_G-V_T}{3nU_T}$

<u>TABLE 4.1</u> Normalized transconductance g_{mt} and noise resistance R_{Nt}.

4.3 Low voltage cascoding

Addition of a cascode transistor (common gate configuration) decreases output conductance by a factor g_{ms}/g_o. No significant noise is added. Single stage cascode amplifiers may thus achieve a DC voltage gain A_o higher than 2-stage non-cascode amplifiers. Figure 4.5 shows a cascode pair biased for minimum saturation voltage, in order to ensure sufficient output amplitude even at low supply voltage.

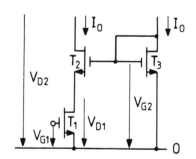

T_1 : common emitter

T_2 : common base

T_3 : bias

<u>Fig. 4.5</u> Low-voltage cascode.

<u>Strong inversion:</u> [29]

$$\frac{W_1}{L_1} = \frac{W_2}{L_2} = 4 \frac{W_3}{L_3}$$

4.16

Thus, according to 3.4 (all transistors in saturation):

$$V_{G1} - V_T = V_{G2} - V_T - nV_{D1} = \frac{1}{2}(V_{G2} - V_T)$$

elimination of V_{G2} yields

$$V_{D1} = \frac{V_{G1} - V_T}{n}$$

Drain voltage of T_1 is just at the limit of saturation, which is accept-able since voltage variations at this node are very small. The minimum out-put voltage V_{D2} that ensures saturation is

$$V_{D2min} = 2 \frac{V_{G1} - V_T}{n} = 2 \sqrt{\frac{2I_o}{\beta_1 n}} \qquad 4.17$$

Weak inversion

$$\frac{W_1}{L_1} = \frac{W_2}{L_2} = K \frac{W_3}{L_3} \qquad 4.18$$

Application of 3.1 yields:

$$V_{D1} = U_T \ln K \qquad 4.19$$

To ensure saturation, V_{D1} must be equal to

$$\Delta V = 4 \text{ to } 6 \ U_T \qquad 4.20$$

therefore K = 60 to 400, which corresponds to a much smaller current density in T_1 and T_2 than in T_3. In most practical situations, T_1 and T_2 operate at the upper limit of weak inversion for which

$$V_{G1} \cong V_T \quad \text{and} \quad V_{G2} \cong V_T + nV_{D1}$$

Saturation of T_1 is ensured if

$$V_{D1} = \frac{V_{G2} - V_T}{n} = \Delta V$$

T_3 is therefore operated in strong inversion, giving

$$V_{G2} - V_T = \sqrt{\frac{2nI_o}{\beta_3}}$$

Thus

$$\beta_3 = \frac{2\,I_o}{n\,\Delta V^2} \qquad\qquad 4.21$$

The minimum output voltage that ensures saturation is

$$V_{D2min} = 2\,\Delta V = 200 \text{ to } 300 \text{ mV} \qquad\qquad 4.22$$

4.4 Current references

Low power generation of the reference current required to bias analog circuits is complicated by the lack of good resistors of high value and by the small difference between V_T and supply voltage V_B.

Figure 4.6 shows the solution which is generally adopted [6].

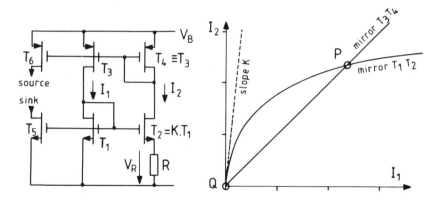

Fig. 4.6   Low power current reference.

Transistors T_3 and T_4 are identical and operate preferably in strong inversion to improve mirror accuracy.

Mirror T_1-T_2 has a gain $K = (W_2/L_2)/(W_1/L_1)$ larger than 1, but it is degenerated by resistor R. Equilibrium is reached at point P when K is compensated by voltage drop V_R across R.

For T_1 and T_2 in weak inversion, and located on the same substrate $(V_{G1} = V_{G2})$, relation 3.1 yields

$$R \ I_2 = V_R = U_T \ \ln K \qquad\qquad 4.23$$

Low currents can be obtained with reasonably low values of R, for example:

For $K = 10$ $(V_R \cong 60 \text{ mV})$, $I_1 = I_2 = 1 \ \mu A$ requires $R = 60 \ K\Omega$.

Current is extracted by sink T_5 and/or source T_6.

The second solution at point Q is unstable, so no start-up circuitry is usually necessary.

The dependence of output current on U_T usually turns out to be an advantage since it compensates the temperature dependence of transconductance in weak inversion. Sensitivity to variations of V_B may be reduced by cascoding.

4.5 Voltage reference

Micropower voltage references have been proposed in which the variation of base-emitter voltage V_{BE} of the vertical bipolar transistor to substrate was compensated by a voltage proportional to absolute temperature (PTAT) that was obtained from MOS transistors in weak inversion [17],[18]. Accuracy was limited by threshold mismatch ΔV_T. The same limitation appears when the PTAT voltage is obtained from two vertical bipolars biased at different current densities. Independence from ΔV_T and excellent results have been obtained by using compatible lateral bipolars [15], but minimum power is limited by the lack of high value resistors. Weighting and summing V_{BE} and ΔV_{BE} can be obtained by switched capacitor circuits [19],[20]. Accuracy is then limited by charge injection, and dynamic power consumption must be high enough to achieve independence from leakage currents.

Figure 4.7 shows the principle of a static voltage reference specially well suited to very low power [21].

It is based on the availability of p^+ and n^+ doping of the same poly-silicon layer.

Transistor T_1 is a normal n-channel transistor with n^+-doped gate and threshold V_T. Transistor T_2 is also n-channel, but has a p^+-doped gate, which shifts its threshold voltage by an amount close to bandgap voltage V_{Gap}.

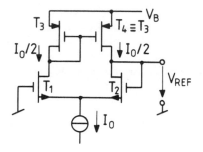

Fig. 4.7 Principle of
very low power
voltage reference.

Both transistors are operated in weak inversion. Then, since drain currents are equal:

$$\frac{W_1}{L_1} \exp \frac{V_{G1}-V_T}{nU_T} = \frac{W_2}{L_2} \exp \frac{V_{G2}-V_T-V_{Gap}}{nU_T}$$

which yields

$$V_{Ref} = V_{G2}-V_{G1} = V_{Gap} + nU_T \ln \frac{W_1/L_1}{W_2/L_2} \qquad 4.24$$

After compensation of the variation of V_{Gap} with temperature by the second term, a temperature stability of $\pm$ 30 ppm/°C is obtained. Total current can be as low as 10 nA. Accuracy is about $\pm$ 50 mV without adjustment.

4.6 Analog switches

The lowest voltage at which a technology can be used for analog circuits is limited by the implementation of analog switches. Switches are realized by means of n and/or p channel transistors. They are switched on and off by connecting their gates to the most positive or to the most negative potential available, namely V_B or 0 (fig. 4.8).

The on-conductance

$$g = \frac{\partial I_D}{\partial (V_D-V_S)} \text{ around } V_D = V_S$$

is a strong function of voltage V_F at which the switch is floating. Examination of 3.4 shows that as V_F increases from zero, the effective gate voltage of n-channel transistors decreases n times faster. The same is true for p-channel when V_F is decreased from V_B.

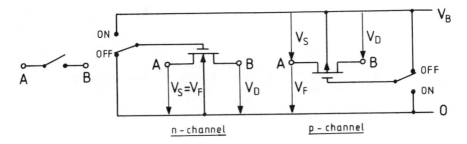

Fig. 4.8   Transistors used as switches.

The conductances g_n and g_p are represented in figure 4.9 as functions of the relative value of V_F.

If V_B is larger than a critical value V_{Bcrit}, a parallel connection of p and n switches ensures conduction for any value of V_F.

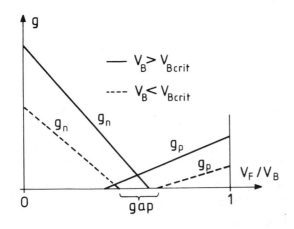

Fig. 4.9   Conductance of p and n switch.

If V_B is smaller than V_{Bcrit}, a gap appears in mid-range in which neither switch is conducting. Calculation of V_{Bcrit} from model 3.4 yields

$$V_{Bcrit} = \frac{V_{Tp} n_n + V_{Tn} n_p}{n_p + n_n - n_p n_n} \qquad 4.26$$

This formula, based on the implicit assumption of linear bulk modulation in model 3.4, is somewhat pessimistic.

In practice, a typical technology with maximum values V_{Tn} = 0.7 V and V_{Tp} = 0.6 V has V_{Bcrit} = 2.3 V [4]. Thus although digital operation is possible below 1 V, analog circuits with switches need at least 2.3 V. This important limitation may be circumvented by means of on-chip clock voltage multiplication [22].

4.7 Dynamic CMOS gates

Minimization of the dynamic power consumption of digital CMOS circuits can be obtained by reducing the number of transistors required to implement a given function. This is often possible by using dynamic gates which are an extension of the general CMOS gate shown in figure 4.10 [23].

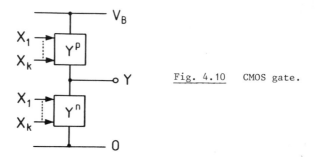

Fig. 4.10   CMOS gate.

A CMOS gate delivering an output variable Y is a series combination of a block of n-type transistors and a block of p-type transistors, each driven by input variables X_1 to X_k. The blocks can be characterized by their respective conduction functions Y^n and Y^P, which cover respectively 0 and 1 squares of the Karnaugh map of Y. In normal static CMOS gates:

$$Y^P \cdot Y^n = 0 \quad \text{to avoid direct paths from } V_B \text{ to } 0$$

$$Y^P + Y^n = 1 \quad \text{to ensure low output impedance.}$$

In a dynamic gate, $Y^P + Y^n = 0$ is allowed. It will be shown, in section 5.4, that dynamic gates can be advantageously generalized beyond clocked CMOS [24] and pull-up transistor logics.

5   EXAMPLE OF BUILDING BLOCKS

5.1 Single stage operational transconductance amplifiers (OTA)

    The major problems in the realization of micropower operational am-
plifiers is that of achieving reasonable speed and acceptable dynamic range.
Battery operation allows more relaxed requirements on power supply rejection,
since all power noise is generated on the chip and may be more easily filtered out at very low current.

    A key factor in reducing power is to avoid any other compensation capac-
itor than the load, which is only possible if the major part of voltage
amplification is achieved at the output node, that is by using single stage
OTA's.

    A simple OTA with differential input is shown in figure 5.1.

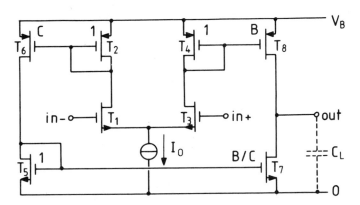

Fig. 5.1   Simple OTA with differential input.

Assuming a ratio B of mirror T_4-T_8, the total transconductance is

$$g_{mt} = B\, g_{m1(3)}$$
                                                                    5.1

    The circuit, loaded by C_L, behaves essentially as an integrator with
time constant

$$\tau_u = C_L / g_{mt}$$
                                                                    5.2

129

which is the inverse of the unity gain angular frequency ω_u. This corresponds to a dominant pole at a very low frequency that depends on DC gain A_o, as shown on figure 5.2.

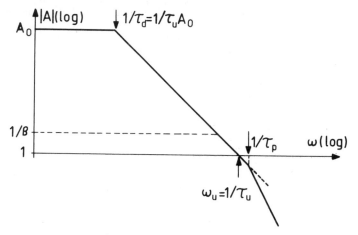

Fig. 5.2   Frequency behaviour of gain A.

Let us assume a single non-dominant parasitic pole with time constant τ_p. This time constant may be that of the output nodes of the differential pair T_1-T_3. It may also represent the effect of all parasitic poles in any amplifier, which can be shown to be equivalent to a single pole for [25]

$$\omega \leqslant \frac{1}{2\tau_p}$$

The open loop high frequency gain may thus be expressed as

$$A = \frac{1}{s\,\tau_u(1 + s\,\tau_p)} \qquad 5.3$$

An opamp is usually used with an amount of voltage feedback $1 \geqslant \beta \gg 1/A_o$.

The settling time T_S necessary to reach equilibrium with a residual error ε after application of a small unit step may be calculated with the gain given by 5.3. Approximative result is [4],[26]

$$T_S \cong (2\,\tau_p + \frac{\tau_u}{\beta})\,\ln\,\varepsilon^{-1} \qquad 5.4$$

130

For small bias currents ($I_o \cong 0.5$ µA), 1/f noise usually still dominates in the audio frequency range, even for large-sized input transistors T_1-T_3. However, in most practical applications, noise bandwidth is larger than 100 kHz and white noise predominates in the total noise power. This is specially true in SC circuits clocked at frequency f_c, for which the whole noise power of the amplifier is transposed below $f_c/2$ by undersampling.

The input referred equivalent noise bandwidth in closed loop is

$$\Delta f = \int_0^\infty \frac{1}{(1 + \frac{1}{\beta A})} \, df \qquad 5.5$$

which yields, with gain A given by 5.3

$$\Delta f = \frac{\beta}{4 \, \tau_u} \qquad 5.6$$

The noise bandwidth is independent of parasitic time constant τ_p. This can be explained qualitatively by the fact that any reduction of noise at high frequencies due to an increase of τ_p is compensated by some peaking at lower frequencies.

Total equivalent input noise is then

$$v_N^2 = 4 \, kT \, R_{Nt} \, \Delta f = \frac{\beta \, kT \, R_{Nt}}{\tau_u} \qquad 5.7$$

Now, by inspection of 3.15, R_{Nt} may be expressed as a function of total transconductance g_{mt} given by 5.1

$$R_{Nt} = \frac{\gamma}{g_{mt}} \qquad 5.8$$

where γ is a factor ranging from nB when noise of pair T_1-T_2 predominates (all other transistors have much smaller g_m/I_D) to n.(2B + 1 + B/C) at very low current (all transistors in weak inversion, see fig. 5.1 for definition of ratios B and C). Combination of 5.8, 5.7, 5.2 and 5.1 yields

$$v_N^2 = \frac{\gamma \beta kT}{C_L} \qquad 5.9$$

The amount of noise introduced by a single-stage transconductance amplifier is thus inversely proportional to load capacitance C_L and independent of current (except for small possible variations of γ). This result has been verified experimentally.

131

Noise can also be reduced by increasing C_L/B, which will increase settling time T_S. Examination of 5.4 shows that

$$\frac{\tau_u}{\beta} \leqslant \frac{T_S}{\ln \varepsilon^{-1}}$$

which can be introduced in 5.7 to give:

$$V_N^2 \geqslant \frac{kTR_{Nt}}{T_S} \ln \varepsilon^{-1} \qquad\qquad 5.10$$

The minimum noise for a given settling time T_S depends on R_{Nt} and is thus increased at low current. If current is fixed, according to 5.8, noise is minimum when the input pair is in weak inversion.

To achieve high value of voltage gain A_o, cascode transistors must be added to output pair $T_7 - T_8$.

Table 5.1 gives experimental results obtained with this circuit for mirror ratios $B = 2$ and $C = 1$[27].

It may be noted that this circuit consumes 13 times more current (CEF = 13) [28] and has 10 dB more noise (NEF = 10 dB) than $1/f$-noise-free complementary inverter operated in weak inversion with same load C_L and same unity gain frequency f_u.

Total supply voltage	V_B	= 3 V
Total current	I_{tot}	= 2.5 µA (I_o = 0.5 µA)
DC gain	A_o	= 97 dB
Unity gain frequency	f_u	= 135 kHz
Slew rate	S.R.	= 0.1 V/µs
Total noise above 10 Hz		60 µV
Noise corner frequency		30 kHz
Output swing		2.2 V
Dynamic range		82 dB
Input capacitance		0.7 pF
Current excess factor [28]	CEF	= 13
Noise excess factor	NEF	= 10 dB

$C_L = 10$ pF
$\beta = 1$

TABLE 5.1   Experimental results with cascode version of figure 5.1 [27].

132

Contribution to parasitic time constant τ_p (which does not affect noise but can increase settling time for low values of load C_L) of mirrors T_4-T_8 and/or T_2-T_6 can be eliminated by using the folded cascode scheme [29].

If low frequency noise is cumbersome, all noisy n-channel transistors can be replaced by compatible bipolars. An equivalent input noise density below 0.1 $\mu V/\sqrt{Hz}$ for frequencies as low as 1 Hz has been obtained with tail current I_O = 2 μA and minimum-sized devices [15].

Thanks to weak inversion operation of its input stage, amplifier of figure 5.1 is able to settle within a short period of time inspite of its very low power consumption. This is only true for small input steps. As soon as differential input voltage exceeds about 3 nU_T (see figure 4.4), output current saturates and slew rate is limited by the very small bias current. Calculation of settling time for large input step ΔV_i and $\tau_p \ll \tau_u/\beta$ yields [25]:

$$T_S \simeq \frac{\tau_u}{\beta} \ \ln \frac{\sinh(\ \Delta V_i/2nU_T)}{\sinh(\ \varepsilon\ \Delta V_i/2nU_T)} \qquad\qquad 5.11$$

This relation is plotted on figure 5.3. Settling time is increased by a factor 2 to 3 for $\Delta V_i \cong 1$ V.

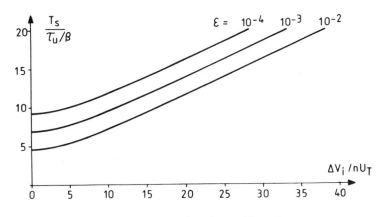

Fig. 5.3   Large signal settling time.

One way to circumvent this fundamental limitation in switched capacitor circuits is to increase tail current at the beginning of each clock cycle, that is when input steps are expected [30]. Power consumption is then systematically increased, even in standby or with small signals. Better power

133

efficiency is obtained by means of amplifiers that operate in class AB, and are therefore able to supply additional current only when it is required.

As a first possibility, one may use adaptive biasing schemes such as the one shown in figure 5.4 [31].

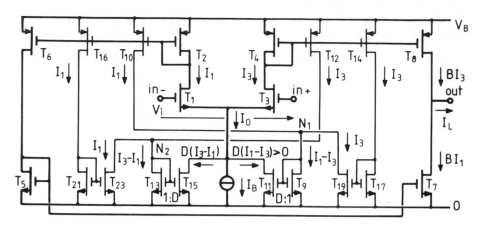

Fig. 5.4   Amplifier with adaptive bias.

The core T_1 to T_8 (shown in thick kines) is identical to amplifier of figure 5.1. At equilibrium, tail current I_0 is equal to bias current I_B.

If a differential input signal is applied and $I_1 > I_3$, difference I_1-I_3 is obtained at node N_1, and is added to I_B after multiplication by factor D in mirror T_9-T_{11}.

If $I_3 > I_1$, no current is available from node N_1, but I_3-I_1 is obtained at node N_2, and is added to I_B after multiplication by factor D in mirror $T_{13}-T_{15}$.

If input pair T_1-T_3 is in weak inversion, application of 4.13 yields:

$$-I_L = B\ (I_1-I_3) = BI_B \cdot \frac{\tanh(V_i/2nU_T)}{1-D|\tanh(V_i/2nU_T)|} \qquad 5.12$$

This relation is represented in figure 5.5 which shows that for $D = 0$ (normal amplifier), $|I_L|$ never exceeds BI_B whereas for $D = 2$ (for example), $-I_L \gg BI_B$ even for $V_i \cong nU_T$.

Slew rate limitations are thus eliminated at the cost of increased complexity, power and noise.

134

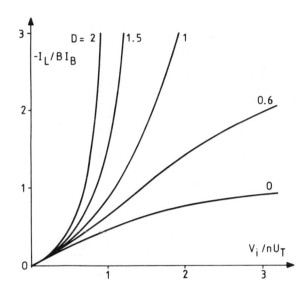

Fig. 5.5 Output current of adaptive bias amplifier.

If a non-differential input can be accepted, a perfect solution is to use just a complementary inverter [32]. As shown in figure 4.3, output current I_L is zero for an equilibrium value V_e of input voltage V_i, which plays the role of virtual ground voltage, with equilibrium current drain I_o. Maximum possible transconductance and minimum possible noise resistance with current I_o are achieved if both transistors are in weak inversion. Weak inversion also provides maximum DC gain which can be easily boosted up by simple or multiple cascoding.

For large variations of V_i around V_e, output current I_L may widely exceed bias current I_o, which eliminates any slew rate limitation.

Some circuitry must be added to control bias current I_o and to attenuate the effect of power supply variation.

A first possibility is depicted in figure 5.6. The supply voltage V_B of all amplifiers is maintained at the required value by means of an identical inverter T_{no}-T_{po} biased with current I_o. The ground potential V_e is available from this reference inverter. Additional grounds may be obtained by short-circuiting input and output of some of the amplifiers. The whole circuit must be fed at a voltage higher than V_B to allow correct operation of the follower-amplifier A.

The value of V_B for small signal bias in weak inversion is approximately equal to the sum of p and n-channel thresholds. The link between threshold voltages and supply voltage is eliminated in the circuit represented in figure 5.7. Bias current I_o is imposed independently of V_B by current mirror T_b-T_p. The behaviour of this circuit is similar to that of an inverter for

135

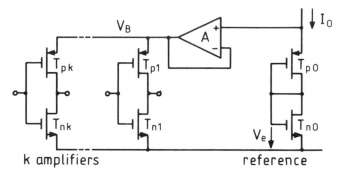

Fig. 5.6   Regulation of V_B.

k amplifiers   reference

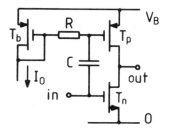

Fig. 5.7   AC coupled CMOS inverter-
amplifier.

frequencies higher than 1/RC. The high
value resistance R may be implemented by
means of polysilicon lateral diodes.
This amplifier has been used to implement
a very low-power SC-filter [33]. Its only, but very cumbersome, drawback is a
poor PSRR of about 6 dB which is not usually acceptable even for battery operation. This situation can be considerably improved by using the differentiating properties of dynamic bias [32],[34].

5.2   Quartz crystal oscillators

Micropower crystal oscillators have been given much consideration for
watch applications where accuracy is of prime importance. Virtually all such
circuits are based on the 3-point oscillator shown in figure 5.8.

A very powerful way of analyzing any crystal oscillator is to split it
as shown in the same figure [35]. Z_m is the linear frequency dependent
motional impedance of the resonator. Because of the high quality factor Q,
current   i   through Z_m may be considered purely sinusoïdal. The rest of the
circuit is nonlinear, but can be considered independent of frequency in the
vicinity of oscillation frequency. It can be characterized by its impedance
at fundamental frequency $Z_{c(1)}$ which is reduced to normal linear impedance
Z_c for small amplitudes. Sustained oscillation corresponds to

$$Z_m + Z_{c(1)} = 0 \qquad\qquad 5.13$$

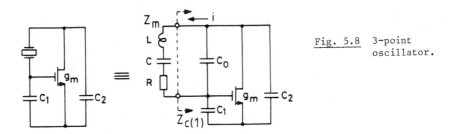

Fig. 5.8 3-point
oscillator.

from which amplitude and frequency can be computed. Replacing $Z_{c(1)}$ by Z_c in
5.13 yields the critical transconductance for oscillation (for
$QC/C_o \gg 2 + C_o/C_1$, and all losses neglected except R of resonator)

$$g_{mc} = \frac{\omega}{QC} (C_1 + 2C_o)^2 \qquad\qquad 5.14$$

and the small signal frequency pulling with respect to resonant frequency
of Z_m

$$\frac{\Delta\omega}{\omega} = \frac{C}{C_1 + 2C_o} \qquad\qquad 5.15$$

where capacitors C_1 and C_2 of figure 5.8 have been supposed equal for simplicity.

Critical transconductance g_{mc} may be reduced by reducing C_1 (C_o is always
as small as possible), at the cost of an increase of $\Delta\omega/\omega$ and thus a
degradation of frequency stability.

The transistor is operated in weak inversion to achieve minimum critical value I_{DC} of drain current I_D.

The first nonlinearity which usually intervenes when $I_D > I_{DC}$ and amplitude increases is that of the exponential transfer characteristics of the
transistor. Fourier analysis yields amplitude ΔV_G of oscillation at the
gate as a function of DC drain current I_D shown in figure 5.9.

For $I_D \gg I_{DC}$:

$$\Delta V_G \cong \frac{2I_D}{I_{DC}} nU_T = \frac{2I_D}{g_{mc}} \qquad\qquad 5.16$$

Frequency is affected by this nonlinearity only in a negligible way.

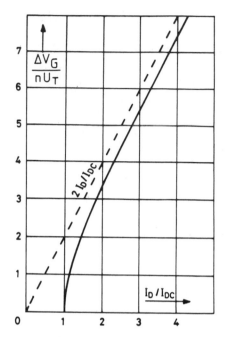

Fig. 5.9   Amplitude of oscillation.

In order to limit power consumption, I_D must be adjusted slightly above I_{DC} by means of an amplitude regulator.

Figure 5.10 shows, as an example, the complete diagram of an oscillator that is extensively used for low and medium frequency applications [35].

Active transistor T_1 is biased by current source T_2 and diode D_1. C_6 insures DC decoupling of gate from pin Q_1.

The regulator is based on weak inversion operation of transistor T_3 and T_5 and provides a sharp drop of I_D when the amplitude of oscillation reaches a few nU_T [6]. It can be applied in other kinds of oscillators as well [36]. In absence of oscillation, the regulator is equivalent to current reference of figure 4.6 and provides the start-up value of I_D.

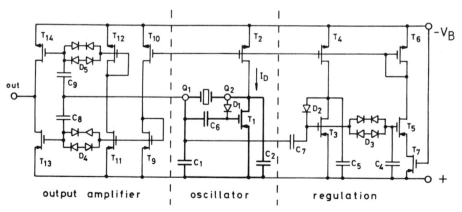

output amplifier   oscillator   regulation

Fig. 5.10   Complete diagram of a grounded-source oscillator with amplitude regulation.

138

The small amplitude of oscillation is efficiently increased to logic level by the AC-coupled CMOS inverter-amplifier. All diodes are lateral polycrystalline diodes. Floating capacitors have values below 2 pF. This circuit may consume as few as 20 nA at 32 kHz with a start-up current of many microamperes.

Active transistor T_1 could be replaced by an AC coupled CMOS inverter to reduce power, but this would allow direct injection of power supply noise into the heart of the oscillator. It has also been proposed to replace the non-critical low-pass filter D_3-C_4 by an SC filter [37].

5.4 Sequential logic building blocks

In order to limit the dynamic power consumption, the average number of nodes which transit during any period of time must be kept as low as possible. High frequency synchronous circuits must therefore be avoided and replaced when possible by asynchronous cells. Each cell should have a minimum number of nodes changing state to achieve a given function. It should furthermore contain a minimum number of minimum sized transistors in order to obtain a low total parasitic capacitance. Logic structures with critical races must therefore be discarded in favour of race-free circuits which work independently of their delays [38]. Dynamic CMOS gates must be used whenever possible.

As an example, let us consider the asynchronous divide-by-2 cell defined by the following set of equations [39]

$$A = \overline{IC}$$
$$B = \overline{(I + D)A}$$
$$C = \overline{IE + AB}$$
$$D = \overline{B}$$
$$E = \overline{C}$$

5.17

I is the input variable, and each internal variable A to E is produced by a CMOS gate (D and E by simple inverters). The gates are interconnected according to the set of equations and the whole circuit comprises 22 transistors (2 per variable on the right hand side of equations). The graph of transitions of this structure represented on figure 5.11 shows that no more than one variable tends to transit in any

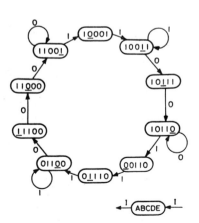

Fig. 5.11   Graph of transitions of divider-by-2.

139

given state, hence there can be no race between variables; thus the logic function does not depend on the relative speed of the various gates which can be optimized independently to minimize power. Each internal variable transits at half the frequency of the input and may thus be used as the output of the divider. A method for synthesizing a race-free implementation of any simple sequential function has been developed [40].

According to section 4.7, the p and n-block conduction functions corresponding to equations 5.17 are

$$
\begin{array}{ll}
A^n = \underline{IC} & A^p = \overline{I} + \overline{C} \\
B^n = \underline{IA} + DA & B^p = \overline{A} + \overline{DI} \\
C^n = IE + \underline{AB} & C^p = \overline{IA} + \overline{IB} + \overline{EA} + \overline{EB} \\
D^n = \underline{B} & D^p = \overline{B} \\
E^n = \underline{C} & E^p = \overline{C}
\end{array} \qquad 5.18
$$

Each term in the above equations corresponds to a series connection of transistors to output node. Now in graph of figure 5.11, the variable which had to transit to reach each state has been identified and underlined. The term responsible for each transition has been in turn identified in equations 5.18 and underlined. All remaining terms are used only to maintain established states against leakage currents and may thus be dropped for high frequency circuits. This yields the reduced set of equations

$$
\begin{array}{ll}
A^n = IC & A^p = \overline{I} \\
B^n = IA & B^p = \overline{A} \\
C^n = AB & C^p = \overline{IB}
\end{array} \qquad 5.19
$$

which can be implemented by 3 dynamic gates and a total of 10 transistors only. Dynamic power consumption is reduced by a factor 2 to 3 with respect to the static version.

This technique amounts to keeping only the transistors necessary to cause transitions, plus eventually some transistors needed to maintain long-lasting states [23],[39]. It has been efficiently applied to all kinds of logic cells, including some low-frequency circuits driven by short logic pulses.

6   CONCLUSION

Microwatt and submicrowatt operation of digital and analog CMOS integrated circuits requires some trade-off with other performances. Low supply voltage affects the maximum amplitude of signals and degrades the symmetry of current mirrors. The minimum possible voltage is limited by the need to realize analog switches.

Low current requirements limit the use of resistors, reduces speed and bandwidth, and increases noise voltage spectral densities. However, the possibility of operating transistors in weak inversion helps counter these limitations. The very low value of saturation voltage permits peak-to-peak amplitudes close to supply voltage. The maximum value of g_m/I_D provides a large DC gain per stage, as well as minimum noise and maximum bandwidth for a given current. Well controlled exponential transfer characteristics permit interesting analog subcircuits.

Other favourable factors are the absence of any thermal gradient on chip, an easy elimination of latch-up and reduced PSRR requirements. Special design techniques include single stage cascode transconductance amplifiers, elimination of slew rate limitations, complementary dynamic logic circuits, and the use of special devices such as the lateral diode in the polysilicon gate layer and compatible bipolars.

ACKNOWLEDGMENT

The author would like to thank M.G. Degrauwe and F. Krummenacher for very helpful discussions.

REFERENCES

[1]   F. LEUENBERGER, "Solid-State devices in watches", Solid-State Devices, 1973, Conference Series 19, The Institute of Physics, London and Bristol, pp. 31-50 (Invited paper ESSDERC'73).

[2]   E. VITTOZ, "LSI in Watches", Solid-State Circuits 1976, Editions du Journal de Physique, Paris, pp. 7-27 (Invited paper ESSCIRC'76).

[3]   R.M. SWANSON and J.D. MEINDL, "Ion-implanted complementary MOS transistors in low-voltage circuits", IEEE J. Solid-State Circuits, Vol. SC-7, pp. 146-153, April 1972.

[4]   E. VITTOZ and F. KRUMMENACHER, "Micropower SC Filters in Si-Gate Technology",Proceedings ECCTD'80, Warshaw, pp. 71-72, 1980.

[5]   B. GERBER et al, "A 1.5 V Single-Supply One-Transistor CMOS EEPROM",
      IEEE J. Solid-State Circuits, Vol. SC-16, pp. 195-199, June 1981.

[6]   E. VITTOZ and J. FELLRATH, "CMOS analog integrated circuits based on
      weak inversion operation", IEEE J. Solid-State Circuits, Vol. SC-12,
      pp. 224-231, June 1977.

[7]   J.D. CHATELAIN, "Dispositifs à semiconducteurs", Traité d'Electricité,
      Vol. 7, (EPF-Lausanne), Georgi 1979, pp. 225-235.

[8]   H.K.J. IHANTOLA and J.L. MOLL, "Design theory of a Surface Field-Effect
      Transistor", Solid-State Electronics, Vol. 7, pp. 423-430, 1964.

[9]   P. ANTOGNETTI et al, "CAD Model for Threshold and Subthreshold Conduc-
      tion in MOSFET's", IEEE J. Solid-State Circuits, Vol. SC-17, pp. 454-
      458, June 1982.

[10] H.J. OGUEY and S. CSERVENY, "MOS Modelling at Low Current Density",
      Summer Course on Process and Devices Modelling, KU-Leuven, 1983.

[11] F.M. KLAASSEN and J. PRINS, "Thermal noise of MOS Transistors", Philips
      Research Reports, Vol. 22, pp. 505-514, 1967.

[12] J. FELLRATH and E. VITTOZ, "Small signal model of MOS transistors in
      weak inversion", Proc. Journées d'Electronique 1977, EPF-Lausanne,
      pp. 315-324.

[13] J. FELLRATH, "Shot noise behaviour of subthreshold MOS transistors",
      Revue de Physique Appliquée, Vol. 13, pp. 719-723, Dec. 1978.

[14] M. AOKI et al, "Low 1/f Noise Design of Hi-CMOS Devices", IEEE Trans.
      on Electron Devices, Vol. ED-29, pp. 296-299, Febr. 1982.

[15] E. VITTOZ, "MOS Transistors Operated in the Lateral Bipolar Mode and
      their Application in CMOS Technology", IEEE J. Solid-State Circuits,
      Vol. SC-18, pp. 273-279, June 1983.

[16] M. DUTOIT and F. SOLLBERGER, "Lateral polysilicon p-n diodes", J.
      Electrochem. Soc., Vol. 125, pp. 1648-1651, Oct. 1978.

[17] G. TZANATEAS, C.A.T. SALAMA and Y.P. TSIVIDIS, "A CMOS bandgap voltage
      reference", IEEE J. Solid-State Circuits, Vol. SC-14, pp. 655-657,
      June 1979.

[18] E. VITTOZ and O. NEYROUD, "A low-voltage CMOS bandgap reference", IEEE
      J. Solid-State Circuits, Vol. SC-14, pp. 573-577, June 1979.

[19] O. LEUTHOLD, "Integrierte Spannungsüberwachungsschaltung", Meeting of
      IEEE Swiss Chapter on Solid-State Devices and Circuits, Bern, Oct. 1981.

[20] B.S. SONG and P.R. GRAY, "A Precision Curvature-Compensated CMOS Band-gap Reference", Digest of ISSCC 83, pp. 240-241.

[21] H.J. OGUEY and B. GERBER, "MOS voltage reference based on polysilicon gate work function difference", IEEE J. Solid-State Circuits, Vol. SC-15, pp. 264-269, June 1980.

[22] F. KRUMMENACHER, H. PINIER and A. GUILLAUME, "Higher Sampling Rates in SC circuits by On-Chip Clock-Voltage Multiplication", Digest of ESSCIRC' 83, Lausanne, pp. 123-126.

[23] E. VITTOZ and H. OGUEY, "Complementary dynamic MOS logic circuits", Electronics Letters, Vol. 9, Febr. 22, 1973.

[24] Y. SUZUKI, K. ODAGAWA and T. ABE, "Clocked CMOS calculator circuitry", IEEE J. Solid-State Circuits, Vol. SC-8, pp. 462-469, Dec. 1973.

[25] F. KRUMMENACHER, "Etude et dimensionnement d'un OTA cascode CMOS", Rprt. LEG 82.04, Swiss Federal Institute of Technology, Lausanne (EPFL), 1982.

[26] E. VITTOZ, "Microwatt Switched Capacitor Circuit Design", Electro-component Science and Technology, Vol. 9, pp. 263-273, 1982 (Summer Course on SC circuits, KU-Leuven, 1981).

[27] F. KRUMMENACHER, "High voltage gain CMOS OTA for micropower SC filters", Electronics Letters, Vol. 17, pp. 160-162, 1981.

[28] M.G. DEGRAUWE and W.M.C. SANSEN, "Current Efficiency of MOS Trans-conductance Amplifiers", IEEE J. Solid-State Circuits, Vol. SC-19, pp. 349-359, June 1984.

[29] T.C. CHOI et al, "High Frequency CMOS Switched-Capacitor Filters for Communications Application", IEEE J. Solid-State Circuits, Vol. SC-18, pp. 652-664, Dec. 1983.

[30] B.J. HOSTICKA, "Dynamic CMOS Amplifiers", IEEE J. Solid-State Circuits, Vol. SC-15, pp. 887-894, Oct. 1980.

[31] M.G. DEGRAUWE et al, "Adaptive Biasing CMOS Amplifier", IEEE J. Solid-State Circuits, Vol. SC-17, pp. 522-528, June 1982.

[32] F. KRUMMENACHER, E. VITTOZ and M. DEGRAUWE, "Class AB CMOS Amplifiers for Micropower SC Filters", Electronics Letters, Vol. 17, pp. 433-435, 25th June 1981.

[33] F. KRUMMENACHER, "Micropower Switched Capacitor Biquadratic Cell", IEEE J. Solid-State Circuits, Vol. SC-17, pp. 507-512, June 1982.

[34] E. VITTOZ, "Dynamic Analog Techniques", Summer Course on Design of MOS-VLSI Circuits for Telecommunications, SSGRR-L'Aquila, Italy, June 18-29, 1984.

[35] E. VITTOZ, "Quartz oscillators for watches", Proc. Int. Congress of Chronometry, Geneva 1979, pp. 131-140.

[36] E. VITTOZ, "Micropower Switched-Capacitor Oscillator", IEEE J. Solid-State Circuits, Vol. SC-14, pp. 622-624, June 1979.

[37] L.L. LEWYN et al, "CMOS Oscillator Subsystem Optimized for minimum Power Consumption", Proc. ISCAS 83.

[38] E. VITTOZ, C. PIGUET and W. HAMMER, "Model of the Logic Gate", Proc. Journées d'Electronique 1977, EPF-Lausanne, pp. 455-467.

[39] E. VITTOZ, B. GERBER and F. LEUENBERGER, "Silicon-gate CMOS frequency divider for the electronic wrist watch", IEEE J. Solid-State Circuits, Vol. SC-7, pp. 100-104, April 1972.

[40] C. PIGUET, "Synthèse de systèmes logiques asynchrones à l'aide des propriétés des fonctions logiques décroissantes", Ph. D. Thesis No 395, Swiss Federal Institute of Technology, Lausanne (EPFL), 1981.

5

DYNAMIC ANALOG TECHNIQUES

Eric A. Vittoz

Centre Suisse d'Electronique et de Microtechnique S.A.,
CSEM (Formerly CEH), Maladière 71, 2000 Neuchâtel
and
Swiss Federal Institute of Technology (EPFL), Lausanne
Switzerland

1   INTRODUCTION

The absence of any control current is one of the most significant advantages offered by MOS transistors. It allows storage of the gate control voltage in a capacitor, for a period of time limited by leakage current due to unavoidable pn junctions. This property is used to simplify digital circuits by temporarily maintaining established logic states as voltages across capacitors. This kind of circuits requires that these states do not last for too long, hence the appellation of dynamic logic circuits. Dynamic analog circuits will be defined on this analogy. They differ from charge exchange circuits (which include SC circuits) by their insensitivity to exact values and non linearities of the capacitors.

2   SAMPLE AND HOLD

2.1 Principle

Dynamic analog circuits are all based on the elementary sample and hold circuit represented in figure 2.1.

R is the total resistance of the circuit. Sampling duration is assumed to be long enough to permit full sampling:

$$T_s \; \gg \; \tau = RC \qquad\qquad 2.1$$

When the switch opens, after instant t_n, voltage v_c across C has a constant value

$$v_{cn} = v_i'(t_n) \qquad\qquad 2.2$$

145

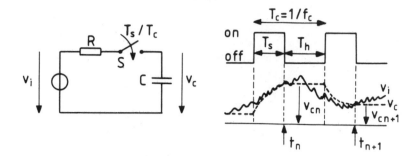

Fig. 2.1 Sample and hold.

where v_i' is the input signal v_i band-limited by transfer function

$$G(\omega) = \frac{1}{1 + j\omega\tau}$$

2.3

Switch S is implemented by single or complementary MOS transistors [1].

2.2 Transfer function

The circuit of figure 2.1 may be analyzed on the basis of the equivalent block diagram shown in figure 2.2, where the value of v_c during sampling time

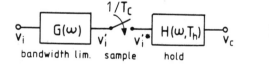

Fig. 2.2 Block diagram of sample and hold.

T_S is not considered. $G(\omega)$ is given by 2.3.

The Fourier transform of sampled input v_i' is [2]

$$V_i'^*(\omega) = \frac{1}{T_c} \sum_{n = -\infty}^{+\infty} G \cdot V_i(\omega + \frac{2\pi n}{T_c})$$

2.4

146

and the transfer function of the hold device is [2]

$$H(\omega, T_h) = T_h \frac{\sin \frac{\omega T_h}{2}}{\frac{\omega T_h}{2}} e^{-j\frac{\omega T_h}{2}} \qquad 2.5$$

Thus

$$V_c(\omega) = H(\omega, T_h) \cdot V_i'^*(\omega) \qquad 2.6$$

One may define a transfer function for the fundamental signal (n = 0)

$$W_{co}(\omega) = \frac{T_h}{T_c} \cdot \frac{\sin(\frac{\omega T_h}{2})}{\frac{\omega T_h}{2}} \cdot \frac{e^{-j\frac{\omega T_h}{2}}}{1 + j\omega \tau} \qquad 2.7$$

2.3 Sampling noise

When the switch is closed, the noise of the circuit may be represented by an equivalent noise component of input voltage v_i, of spectral density

$$\frac{dV_{iN}^2}{df} = 4 \, kTR_N \qquad 2.8$$

where

$$R_N = \gamma R \qquad 2.9$$

is the equivalent noise resistance of the circuit [1]. In the ideal case, when noise is only due to thermal noise of R, $R_N = R$ thus $\gamma = 1$.

Total noise voltage v_{cN} across C has mean square value

$$V_{cN}^2 = \int_0^\infty \frac{dV_{iN}^2}{df} \cdot |G(\omega)|^2 \, df \qquad 2.10$$

By introducing 2.3, 2.8 and 2.9

$$V_{cN}^2 = \gamma \frac{kT}{C} \qquad 2.11$$

147

When the switch opens at instant t_n, the instantaneous value of noise voltage v_{cN} is stored in capacitance C. A noise sample Δv_{cn} is thus added to signal sample v_{cn} (see figure 2.3).

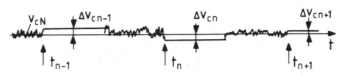

Fig. 2.3 Switching noise.

Successive noise samples are not correlated. They have a variance equal to V_{cN}^2.

Figure 2.4 shows the power spectral densities of continuous noise ($T_s = T_c$, switch always closed) and sampled noise ($T_h = T_c$, very short sampling periods) calculated from relation 2.6.

Most of the noise is thus transposed to low frequencies by the sample and hold process. When neither T_h nor T_s is negligible, continuous and sampled spectra are approximately weighted by T_s/T_c and T_h/T_c respectively.

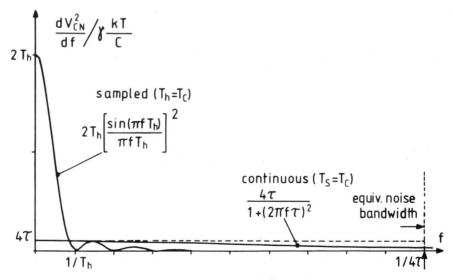

Fig. 2.4 Power spectral densities.

2.4   Charge injection by the switch

Another limitation to the accuracy of a sample and hold circuit is the disturbance of the sampled voltage v_c when the transistor-switch turns off. Consider the circuit of figure 2.5 in which R_i and C_i have been added to the basic sample-hold.

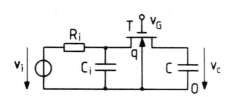

Fig. 2.5   Circuit for analysis of charge injection.

When the gate voltage is reduced from V_{G1} to below V_T during turn off, the charge q which leaves the channel may go to source, drain and/or substrate. The fraction q_c of q, which is released into capacitance C, causes an error of v_c

$$\Delta v_c = \frac{q_c}{C} \qquad\qquad 2.12$$

Let us assume that the fraction of q going to substrate is negligible, and that time constant $R_i C_i$ is much larger than the switch-off time. Charge q is thus shared between C_i and C. If the transistor is symmetrical and $C_i = C$, half of the charge will be released in each capacitor. But if $C_i \neq C$, a voltage appears across the transistor during the descent of gate voltage, which may result in a different sharing of charge q.

This problem can be analyzed by the model of figure 2.6.

V_{Te} is the effective gate voltage at which the channel is blocked:

$$V_{Te} \cong V_T + nv_{cn} \quad [1] \qquad\qquad 2.13$$

where V_T is the threshold for zero source voltage, v_{cn} is the particular value of v_c at the sampling period being considered, and n the effect of bulk modulation.

149

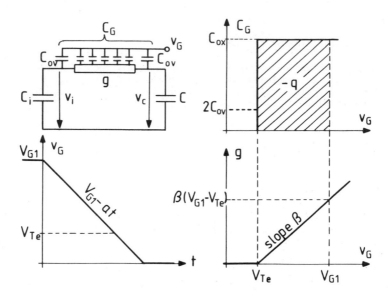

Fig. 2.6   Model for analysis of charge injection.

For $v_G > V_{Te}$, channel conductance is

$$g = \beta (v_G - V_{Te}) \qquad\qquad 2.14$$

and gate capacitance C_G has value C_{ox}, sum of the distributed capacitance to (homogeneous) channel and of two overlap capacitances C_{ov}.

For $v_G < V_{Te}$, $g = 0$ (weak inversion neglected) and C_G is reduced to $2\,C_{ov}$.

Gate voltage v_G is assumed to decrease linearly with time from its on-value V_{G1}

$$v_G = V_{G1} - at \qquad\qquad 2.15$$

Now $\Delta v_c \ll V_{G1} - V_{Te}$, thus the potential at each end of g may be considered constant. Let us assume that this is true as well for any point of the channel, hence that the current flowing through the distributed capacitor during the descent of v_G causes negligible voltage drop even in the middle of the channel.

150

This can be easily translated into the following condition:

$$a \ll \frac{\beta (V_{G1} - V_{Te})^2}{C_G}$$ 2.16

If this condition is fulfilled, the linear decrease of v_G across $C_G = C_{ox}$ is equivalent to a distributed current source, of total value $-aC_{ox}$, which flows symmetrically towards both ends of the channel, which yields the final model of figure 2.7 [3]

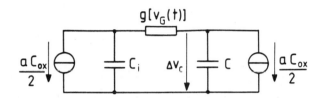

Fig. 2.7   Final model of charge injection.

Introducing 2.14 and 2.15 in this model yields the following normalized differential equation:

$$\frac{dV_c}{dT} = (T - B) \left[(1 + \frac{C}{C_i}) V_c + 2 \frac{C}{C_i} T \right] - 1$$ 2.17

where

$$V_c = \Delta v_c / \left[\frac{C_{ox}}{2} \cdot \sqrt{\frac{a}{\beta C}} \right]$$ 2.18

$$T = t / \sqrt{\frac{C}{a \beta}}$$ 2.19

$$B = (V_{G1} - V_{Te}) \cdot \sqrt{\frac{\beta}{aC}}$$ 2.20

$V_c = 0$ for $T = 0$; numerical integration for $T = 0$ to B yields the value reached by V_c when the transistor is off ($v_G = V_{Te}$). Results are reported on figure 2.8 for various values of C_i/C.

Measurements obtained with a very broad transistor (W/L = 10 mm/22 µm) confirm these results [4]. Some easily calculable additional charge is due to the effect of capacitive divider C_{ov}/C when v_G decreases further below V_{Te}. Results of numerical and breadboard simulation of the effect of R_i have been reported [5] as well as an analytic solution for $C_i/C = \infty$ [6].

Figure 2.8 suggests various possible strategies to reduce charge injection.

151

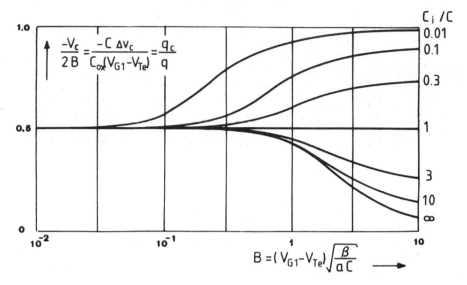

$$\frac{-V_c}{2B} = \frac{-C \, \Delta v_c}{C_{ox}(V_{G1}-V_{Te})} = \frac{q_c}{q}$$

$$B = (V_{G1}-V_{Te})\sqrt{\frac{\beta}{aC}}$$

Fig. 2.8   Charge injection in C (for $v_G = V_{Te}$) [3]

One is to choose $C_i \gg$ C and B large (> 10), in order to let all the charge injected into C flow-back into the voltage source, This can be obtained by reducing the slope a of the gate voltage decay, which reduces the speed of operation. Residual effect due to C_{ov} after blocking will remain.

Another strategy is to choose C_i = C [7]. For any value of B, half of charge q flows into C and can be compensated by half-sized dummy switches [8] as shown in figure 2.9.

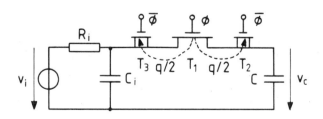

Fig. 2.9   Compensation by dummy switches.

If T_2 and T_3 are exactly identical to T_1, but have half its channel area, their switching on, just after T_1 has been switched off, absorbs charge q and no charge flows into C. The practical amount of compensation depends not only on matching of transistors, but also on that of C_i and C. It is not always feasible to have C_i = C and a resistor R_i of high enough value in series with the voltage source to be sampled.

A third strategy is to choose a small value of B to permit compensation by dummy switch T_2 even for $C_i \neq$ C and/or R_i = 0. Dummy switch T_3 is usually not necessary if C_i/C is not too small or R not too large. Care must be taken not to violate 2.16 which can be expressed as

$$B^2 \gg \frac{C_G}{C} \qquad\qquad 2.21$$

In case of violation, the above analysis is not valid any more; channel potential might drop during switch-off which causes a leakage of charge to substrate.

When the switch is made up of complementary transistors, they partially compensate each other's charge q (matching is poor). In addition, all strategies may be combined with a differential approach [9] which further reduces the effect of residual charge injection.

2.5 Differentiation by sample and hold

The circuit of figure 2.1 may be redrawn as shown in figure 2.10.

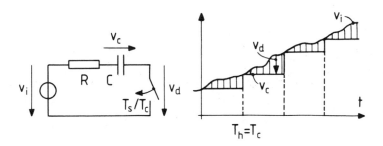

Fig. 2.10   Differentiation by sample and hold.

Assuming $\tau \ll T_s \ll T_c$, the evolution of voltages v_i, v_c and v_d is shown on the same figure.

153

Since $v_d = v_i - v_c$, the transfer function for the <u>fundamental</u> signal is

$$W_{do}(\omega) = 1 - W_{co}(\omega) \qquad\qquad 2.22$$

where $W_{co}(\omega)$ is given by relation 2.7 with $T_c = T_h$ and τ negligible. Simple calculations yield

$$\left| W_{do}(\omega) \right| = \sqrt{(1 - \frac{\sin(\omega T_h)}{\omega T_h})^2 + (\frac{1 - \cos(\omega T_h)}{\omega T_h})^2} \qquad\qquad 2.23$$

This result is represented on figure 2.11.

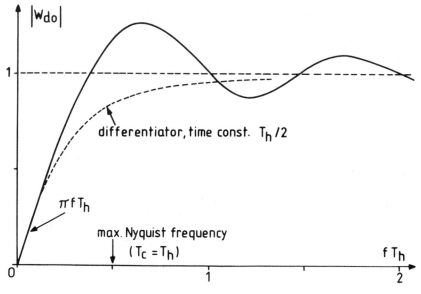

<u>Fig. 2.11</u>   Differentiation transfer function for fundamental signal.

If $T_h < T_c$, the whole transfer function is multiplied by T_h/T_c. Owing to sampling, any signal of frequency above $f_c/2 = 1/2T_c$ will be transposed below $f_c/2$, after possible attenuation by $G(\omega)$ (see relation 2.4).

For $\omega T_h \ll 2$: $\qquad\qquad W_{do}(\omega) \cong \dfrac{j\,\omega\,T_h}{2}$ $\qquad\qquad$ 2.24

Low frequencies are attenuated by a factor $\pi f T_h$ (30 dB for $f T_h = 10^{-2}$).

3 DYNAMIC COMPARATOR

3.1 Basic circuit

The basic dynamic comparator [10] is shown in figure 3.1.

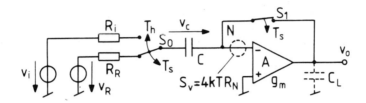

Fig. 3.1 Basic dynamic comparator.

A is a transconductance amplifier (transconductance g_m) with a dominant pole due to output node.

In the initialization phase of duration T_s shown in the figure, the circuit reaches equilibrium with time constant

$$\tau_u = (C + C_L)/g_m \quad (\text{if } R_R g_m \ll 1) \qquad 3.1$$

At equilibrium, voltage v_c across capacitor C is exactly the difference between reference V_R and equilibrium voltage at node N (virtual ground of the amplifier). When S_1 is then opened, no change occurs except for noise and charge injection. When S_0 is then switched to input v_i, any difference $v_i - V_R$ is transmitted to node N, and output voltage v_o starts slewing at rate

$$\frac{dv_o}{dt} = -\frac{g_m}{C_L}(v_i - V_R) \qquad 3.2$$

Effective gain A_e therefore depends on time T_h alotted for comparison

$$A_e = \frac{g_m}{C_L} T_h \qquad 3.3$$

It is ultimately limited by DC gain A_o of the amplifier. Accuracy is limited by various independent phenomena.

3.1.1. Noise – Noise of the comparator is due to two distinct mechanisms.

Each time switch S_1 is opened, a noise sample is stored in capacitance C. It will correspond during comparison to a stochastic input error Δv_{i1} of variance V_{i1}^2. If $g_m R_R \ll 1$, relation 2.11 gives

$$V_{i1}^2 = \frac{\gamma kT}{C + C_L} \tag{3.4}$$

where

$$\gamma = R_N g_m \tag{3.5}$$

characterizes the noise of the amplifier. If $\gamma \gg 1$, V_{i1}^2 can be reduced by increasing R_R at the expense of an increase in settling time constant τ_u.

During each comparison phase, output noise current of spectral density

$$S_I = 4kT \, R_N g_m^2 = 4 \, \gamma \, kT g_m \tag{3.6}$$

is integrated in load capacitance C_L. It can be shown that the variance of output noise voltage obtained after time T_h is

$$V_{oN}^2 = \frac{T_h}{2 C_L^2} \, S_I \tag{3.7}$$

Combination of 3.7, 3.6 and 3.3 gives the variance of the resulting equivalent input error Δv_{i2}

$$V_{i2}^2 = \frac{2 \, \gamma \, kT}{g_m T_h} \tag{3.8}$$

Comparison of the two noise contributions

$$\frac{V_{i2}^2}{V_{i1}^2} = \frac{2(C+C_L)}{g_m T_h} = \frac{2 \, \tau_u}{T_h} \tag{3.9}$$

shows that sampling noise Δv_{i1} usually dominates.

Continuous noise Δv_{i2} would be increased by a large value of resistance R_i. Additional contributions of noise may come from the power supply if PSRR is not sufficient. White power supply noise corresponds to an increased value of γ .

156

3.1.2. <u>Charge injection</u> – Charge q_c released in capacitance C when S_1 is switched off causes an additional input error

$$\Delta v_{i3} = \frac{-q_c}{C} \qquad\qquad 3.10$$

It can be shown that the analysis of section 2.4 is not invalidated by the presence of the amplifier A, provided its transconductance g_m is smaller than the on-conductance of switch S_1. Since load capacitance C_L must be kept small, the best strategy for reducing Δv_{i3} is to compensate by a dummy switch at node N while ensuring equipartition of charge q by a correct choice of B (Fig. 2.8), and/or to use fully-differential circuits [9].

3.1.3. <u>Leak of charge</u> – Any leak of charge from node N results in a contribution to input error. In particular, leakage current I_1 of junctions associated with switch S_1 causes an error

$$\Delta v_{i4} = \frac{I_1\, T_h}{C} \qquad\qquad 3.11$$

which limits the maximum value of comparison time length T_h.

A particularly deleterious effect may occur if many comparisons are carried out after the same initialization phase, as is the case in some A/D converters [9]. A large value of v_i-V_R may bootstrap potential of node N beyond that of substrate (or well), and forward bias a junction of the switch. An important loss of charge will result and alter subsequent comparisons.

3.2 Multistage comparator

Multistage comparators are usually required to improve sensitivity and speed beyond those achievable by the basic circuit of figure 3.1. It turns out that clever multistage design also improves accuracy by reducing effects of noise and charge injection. Figure 3.2 shows a possible implementation represented in its initialization phase.

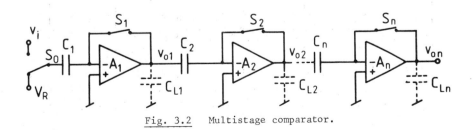

Fig. 3.2   Multistage comparator.

157

Improved speed and sensitivity result from n^{th} order integration of $v_i - V_R$ in the n successive stages, which gives an effective gain for n identical stages

$$A_{en} = \frac{1}{n!} \ (\frac{g_m \ T_h}{C_L})^n \hspace{3cm} 3.12$$

It is ultimately limited by the product of the DC gains of the n amplifiers.

If switch S_1 is opened first, after the initialization phase, charge injection and sampled noise cause an error voltage across C_1. At equilibrium, it results in an output error Δv_{o1} which is stored in capacitance C_2. Perturbations caused by switching of S_1 are thus compensated, provided amplifier A_1 does not saturate. The same is true when switches S_2 to S_n are then opened sequentially. The only residual error at the input is that due to switch S_n divided by the gain of the n-1 first stages [11],[12]. This very efficient scheme can virtually eliminate offset. Values as low as 5 µV have been reported for a differential implementation [13]. Errors accumulated during comparison phase T_h (noise Δv_{i2} and charge leak Δv_{i4}) are not compensated.

The gain per stage must be limited to prevent saturation. If even the DC offset does not cause saturation, another equivalent implementation can be used as shown in figure 3.3.

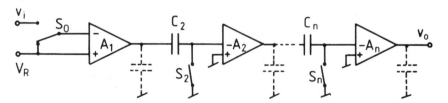

Fig. 3.3   Alternative multistage comparator.

It is the scheme that was proposed originally [11], in which capacitor C_1 and switch S_1 are not needed.

In practical realizations, the gain of each amplifier is defined by using a well controlled load device which can be either a diode-connected transistor [12],[13] or a resistor [9]. Differential input is required in the original scheme of figure 3.3 (low offset) but not in that of figure 3.2 (any offset acceptable). Fully symmetrical implementations (differential input and output) are preferred to further reduce charge injection, to partially compensate charge leak and to improve PSRR [9].

158

4.1 First order compensation

The basic dynamic comparator of figure 3.1 may be seen as a scheme that compensates any offset voltage of amplifier A. In order to preserve differential input or fully differential structure [14], the circuit may be slightly modified as shown in figure 4.1.

The amplifier is only available during the amplification phase shown in the figure. Unavailability during compensation (or auto-zeroing) phase can often be accepted, or avoided by two time shared amplifiers.

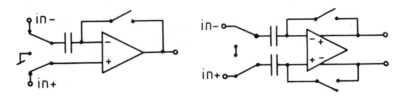

Fig. 4.1   First order compensation of offset voltage.

This scheme actually differentiates any internally generated signal, which in addition to offset cancellation provides attenuation of low frequency components according to figure 2.11, where T_h is the duration of each amplification phase. Low frequency 1/f noise and low frequency components of power supply variations are thus strongly attenuated. High frequency noise power is transposed to low frequencies by the sampling process, as shown by figure 2.4. This increases the noise in the bandwidth of operation. Furthermore, offset compensation is limited by the charge injected by the compensation switch.

As opposed to a comparator, an operational amplifier must be stable in unity gain configuration. This prevents the elimination of sampling noise and charge injection by a multistage scheme, except if a drastic reduction of bandwidth can be accepted.

4.2 Compensation by low sensitivity auxiliary input

4.2.2. Principle – A new scheme for accurate offset compensation is depicted in figure 4.2.

A is an ordinary operational amplifier with DC gain A_i and offset ε_i; ε_i may also be the residual offset after first order compensation according to figure 4.1.

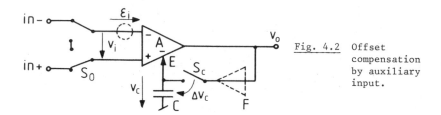

Fig. 4.2 Offset compensation by auxiliary input.

A compensation input E of reduced sensitivity is added to the main amplifier. It is controlled by compensation voltage v_c stored in capacitor C, so that output voltage v_o is given by

$$v_o = -A_i (v_i - \varepsilon_i) - f_c(v_c) \qquad 4.1$$

During compensation, the main input is short-circuited by switch S_o and the compensation loop is closed by switch S_c. Output voltage v_o and compensation voltage v_c take the common value v_{co}. If $f(v_{co}) \gg v_{co}$, v_{co} is given by

$$f_c(v_{co}) = A_i \varepsilon_i \qquad 4.2$$

as shown in figure 4.3.

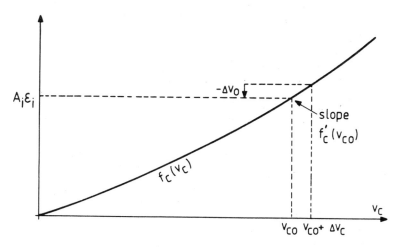

Fig. 4.3   Effect of Δv_c.

160

When switch S_c is opened, a perturbation Δv_c of compensation voltage is produced by charge injection and sampled noise, which causes a variation Δv_0 of output voltage all the smaller as the compensation gain $f_c'(v_{co})$ is smaller.

4.2.2. <u>Linear compensation</u> - For a linear compensation with gain A_c and off-set $\mathcal{E}_c$, output voltage is given by

$$v_0 = -A_i (v_i - \mathcal{E}_i) - A_c (v_c - \mathcal{E}_c) \qquad 4.3$$

If $A_i \gg A_c \gg 1$ and $\mathcal{E}_c$ has the same order of magnitude as $\mathcal{E}_i$, then the compensation phase brings output and compensation voltages to common value

$$v_{co} = \mathcal{E}_i \frac{A_i}{A_c} \qquad 4.4$$

The variation Δv_c of compensation voltage, produced when switch S_c opens, results in a modified value of output voltage

$$v_0 = v_{co} - A_c \Delta v_c \qquad 4.5$$

This value corresponds to an equivalent residual offset (input voltage v_i required to obtain $v_0 = 0$)

$$\mathcal{E}_{iR} = \frac{v_{co} - A_c \Delta v_c}{A_i} = \frac{\mathcal{E}_i}{A_c} - \frac{A_c}{A_i} \Delta v_c \qquad 4.6$$

Worst case occurs when $\mathcal{E}_i$ and Δv_c have both maximum absolute values $\mathcal{E}_{im}$ and Δv_{cm} but opposite signs, giving:

$$\mathcal{E}_{iR} = \frac{\mathcal{E}_{im}}{A_c} + \frac{A_c}{A_i} \Delta v_{cm} \qquad 4.7$$

A large value of A_c reduces the effect of the original offset but increases that of the perturbation by switch S_c. Optimum value is

$$A_{copt} = (\frac{\mathcal{E}_{im} A_i}{\Delta v_{cm}})^{1/2} \qquad 4.8$$

which gives a minimum residual offset in the worst case

$$\mathcal{E}_{iR} = 2. \left(\frac{\mathcal{E}_{im} \, \Delta v_{cm}}{A_i} \right)^{1/2} \qquad\qquad 4.9$$

The corresponding variation of $\mathcal{E}_{iR}$ for any values of $\mathcal{E}_i$ and Δv_c is shown in figure 4.4

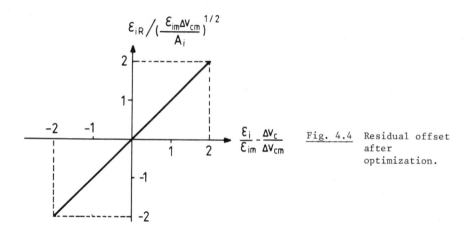

Fig. 4.4 Residual offset after optimization.

The following numerical example illustrates the effectiveness of such a scheme:

Maximum offset to be compensated	$\mathcal{E}_{im}$	= 10 mV
Maximum effect of charge injection	Δv_{cm}	= 10 mV
Gain of main amplifier	A_i	= 10^4
Optimum gain of compensation	A_{copt}	= 10^2
Residual offset	$|\mathcal{E}_{iR}|$	$\leqslant$ 0,2 mV

Figure 4.5 is an example of practical implementation.

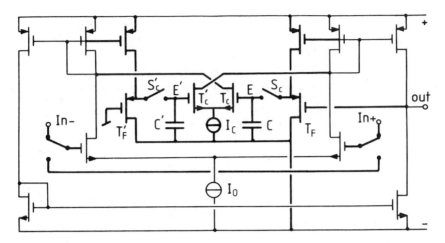

Fig. 4.5   Implementation of an accurate offset compensation.

It is a standard transconductance single stage operational amplifier to which a compensation circuitry shown in thicker lines has been added.

An additional differential pair T_c-T_c' connected in parallel with the main pair has a much smaller transconductance to achieve $A_c/A_i \ll 1$. This is obtained by choosing tail currents ratio I_c/I_0 as small as possible while still ensuring compensation of maximum current dissymmetry, and by operating T_c/T_c' at a large value of gate to source voltage (ultimately limited by supply voltage). In order to speed up compensation, source follower T_F has been added to drive storage capacitor C. Symmetrical implementation of C', S_c' and T_F' allows partial compensation of charge injection effects, thus reducing the effective value of Δv_{cm}.

Another way of realizing compensation is to adjust the ratio of current mirrors of the main amplifier, as illustrated by figure 4.6.

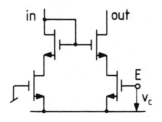

Fig. 4.6   Adjustable current mirror.

This principle has been applied to a low voltage low current instrumentation amplifier, with the result of a residual offset of about 200 μV[15]. Better results are obtainable for larger supply voltages.

4.2.3. Quadratic compensation - Gain A_c of the linear compensation must be optimized for the maximum offset $\mathcal{E}_{im}$ to be compensated. Therefore, as shown by figure 4.4, the effect of charge injection is constant and remains even for samples with very small initial offset ($\mathcal{E}_i \cong 0$). It will be shown that this can be drastically improved by using a quadratic compensation with

$$f_c(v_c) = Kv_c^2 \qquad\qquad 4.10$$

K must have the sign of $\mathcal{E}_i$. The following derivation will be carried out with $\mathcal{E}_i > 0$; the result for $\mathcal{E}_i < 0$ is obtained by simple symmetry.

For $A_i \mathcal{E}_i \gg v_{co}$, relation 4.2 gives

$$v_{co} = (\frac{A_i \mathcal{E}_i}{K})^{1/2} \qquad\qquad 4.11$$

As shown by figure 4.3, variation Δv_c produced when S_c is opened results in the modified value of output voltage

$$v_o = v_{co} - f'(v_{co}) \, \Delta v_c$$

or

$$v_o = v_{co}(1 - 2K \, \Delta v_c) \qquad\qquad 4.12$$

which corresponds to a residual offset

$$\mathcal{E}_{iR} = (\frac{\mathcal{E}_i}{A_i})^{1/2} (K^{-1/2} - 2K^{1/2} \, \Delta v_c) \qquad\qquad 4.13$$

Worst case occurs when Δv_c is negative with maximum absolute value Δv_{cm}. Optimization of K gives

$$K_{opt} = 1/2 \, \Delta v_{cm} \qquad\qquad 4.14$$

and corresponds to the minimum possible residual offset for worst case of

$$\mathcal{E}_{iR} = 2 \, (2 \, \frac{\mathcal{E}_i \, \Delta v_{cm}}{A_i})^{1/2} \qquad\qquad 4.15$$

164

The variation of $\mathcal{E}_{iR}$ for any value of Δv_c is shown in figure 4.7.

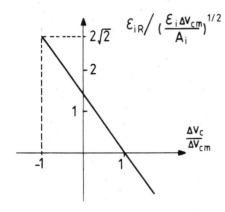

Fig. 4.7 Residual offset after optimization (quadratic).

Residual offset is always proportional to the square root of the initial offset.

Numerical example:

Δv_{cm} = 10 mV   and A_i = 10^4 requires K_{opt} = 50 v^{-1}

Then, for $\mathcal{E}_i$ = 10 mV   1 mV   0,1 mV

$\mathcal{E}_{iR}$ =282 µV   89 µV   28 µV

Worst case residual offset is $\sqrt{2}$ larger than with linear compensation but decreases to zero with the initial offset. Furthermore, as shown by figure 4.7, if Δv_c can be maintained positive with a value close to Δv_{cm} residual offset becomes very small.

This has been realized in the example of figure 4.8 in which quadratic compensation is applied to a standard cascode transconductance amplifier. The n-channel transistor T_{cn} produces quadratic compensation for $\mathcal{E}_i > 0$. Switch S_{cp} is a p-channel transistor and causes positive Δv_c that can be compensated according to relation 4.13. Symmetrical effect is obtained by T_{cp} and S_{cn} for $\mathcal{E}_i < 0$. Source followers F_n and F_p should be added to speed up compensation.

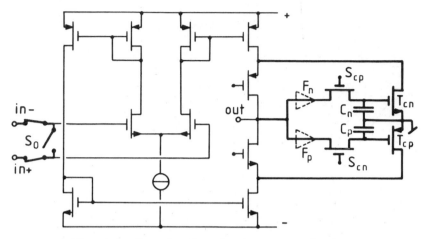

Fig. 4.8   Example of quadratic compensation.

5   DYNAMIC BIASING OF CMOS INVERTER-AMPLIFIERS

The standard CMOS inverter can provide excellent performances as an amplifier [1]. It offers maximum possible transconductance and minimum possible noise for a given current. Its inherent capability of operating in class AB eliminates all slew rate limitations. The only problem consists in finding a biasing scheme that is insensitive to threshold and power supply voltages, and that provides acceptable PSRR.

A good solution is to use dynamic bias as shown in figure 5.1 [16].

During biasing phase, all switches are in position opposite to that represented in the figure. Internal input is thus grounded by S_2 and output is disconnected by S_3. Transistor T_p is a current source that reflects bias current I_o into diode-connected transistor T_n.

During the amplification phase represented in the figure, T_n and T_p behave as a CMOS inverter with equilibrium value v_e of input voltage v_i equal to zero, except for a small offset due to charge injection and sampled noise.

Low frequency 1/f noise and power supply voltage variations are attenuated by differentiation, according to figure 2.11.

Sampled noise and charge injection can be reduced by first switching off S_4 at the end of the biasing phase. Disturbance of v_{c1} will be compensated

166

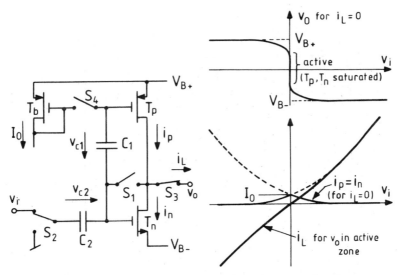

Fig. 5.1   CMOS inverter-amplifier with dynamic bias.

by a small difference in v_{c2} when equilibrium is reached. Total sampled noise referred to input is then

$$v_{iN}^2 = \gamma \frac{kT}{C_2} \qquad\qquad 5.1$$

where γ can be as low as 2/3 [1]. DC gain may be boosted up by cascoding T_p and T_n.

This amplifier can be easily transformed into a SC integrator by connecting an integrating capacitor between output node and gate of T_n. C_2 then becomes part of the SC network. It has been applied in an 8-channel multiplexed bandpass audio filter which achieves more than 60 dB of dynamic range with a power consumption of 72 μW at a total supply voltage of 3 V, and at 192 kHz clock frequency [17].

If compensation of offset and 1/f noise is not required, a slightly different scheme of dynamic bias, shown in figure 5.2 [18], can be adopted.

Voltages required across capacitors C_1 and C_2 to achieve activity of T_{p1} and T_{n1} for $v_i \cong V_R$ in amplifying phase are obtained during biasing phase by means of T_{p2} and T_{n2}. Bias current I_0 is controlled by resistor R. Mismatch of pairs $T_{p1}-T_{p2}$ and $T_{n1}-T_{n2}$ results in an input offset. Differentiation reduces low frequency components of power supply noise, but not 1/f noise generated in transistors (no correlation between noise of each transistor of pairs $T_{p1}-T_{p2}$ and $T_{n1}-T_{n2}$).

167

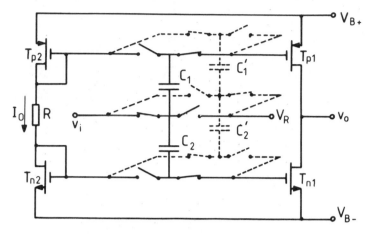

Fig. 5.2   Sampled differential amplifier [18].

As an advantage of this solution, the output voltage is not disturbed during the biasing phase. Quasi-continuous operation can be achieved by duplicating all switches and capacitors as shown in dotted line. The additional set of switches is activated in counterphase so that complementary inverter T_{p1}-T_{n1} amplifies in both phases , and input signal is not sampled. The signal frequency may thus exceed that of clock.

This amplifier can achieve a unity gain frequency of 130 MHz with a power dissipation of about 2 mW in 1.5 μm technology. It has been applied to a fifth order elliptic low-pass filter with a cut-off frequency of 3.58 MHz and a clock rate of 28.6 MHz [18].

6   CONCLUSION

Taking advantage of an easy implementation of the sample and hold function in MOS technologies, dynamic techniques help improve the performance of analog circuits. Temporary storage of voltages in capacitors provides accurate comparators, offset cancellation in operational amplifiers and biasing means for low noise, power efficient and fast CMOS inverter-amplifiers. Limitations due to charge injection and noise undersampling can be overcome by adequate implementation of sampling switches and storage capacitors. Charge injection effects may be further reduced by local compensation and/or by using symmetrical structures.

A basic property of the sample and hold process is to provide differentiation, which attenuates low frequency components of noise and of power supply fluctuations.

ACKNOWLEDGMENT

The author would like to thank F. Krummenacher, C.A. Gobet and Ch. Enz for very helpful discussions.

Part of the original work presented has been supported by the "Fonds National Suisse de la Recherche Scientifique, PN13".

7   REFERENCES

[1]   E. VITTOZ, "Micropower Techniques", Summer Course on Design of MOS-VLSI Circuits for Telecommunications, SSGRR - L'Aquila, Italy, June 18-29, 1984.

[2]   J.T. TOU, "Digital and Sampled-data Control Systems", McGraw-Hill, 1959, New York.

[3]   E. VITTOZ, "Microwatt Switched Capacitor Circuit Design", Summer Course on Switched Capacitor Circuits, KU-Leuven, Belgium, June 9-12, 1981, Republished in Electrocomponent Science and Technology, Vol. 9, 1982, pp. 263-273.

[4]   F. SALCHLI, "Injection d'horloge", CEH Technical Report No 276, April 1983.

[5]   R.C.T. YEN, "High Precision Analog Circuits Using MOS VLSI Technology", Memorandum No UCB/ERL M83/63, UC-Berkeley, 1983.

[6]   B.J. SHEU and C.M. HU, "Modeling the Switch-Induced Error Voltage on a Switched-Capacitor", IEEE Trans. on Circuits Syst., Vol. CAS-30, pp. 911-913, Dec. 1983.

[7]   L. BIENSTMAN and H.J. DE MAN, "An eight-channel 8-bit microprocessor compatible NMOS converter with programmable scaling", IEEE J. Solid-State Circuits, Vol. SC-15, pp. 1051-1059, Dec. 80.

[8]   E. SUAREZ, P.R. GRAY and D.A. HODGES, "All-MOS charge redistribution analog-to-digital conversion techniques - part II", IEEE J. Solid-State Circuits, Vol. SC-10, pp. 379-385, Dec. 75.

[9]   D.J. ALSTOTT, "A Precision Variable-Supply CMOS Comparator", IEEE J. Solid-State Circuits, Vol. SC-17, pp. 1080-1087, Dec. 82.

[10] Y.S. LEE et al, "A 1 mV MOS Comparator", IEEE J. Solid-State Circuits, Vol. SC-13, pp. 294-297, June 78.

[11] R. POUJOIS et al, "Low-level MOS transistor amplifier using storage techniques", ISSCC Digest Technical Papers, 1973, pp. 152-153.

[12] A.R. HAMADE, "A Single Chip All-MOS 8-Bit A/D Converter", IEEE J. Solid-State Circuits, Vol. SC-14, pp. 785-791, Dec. 78.

[13] R. POUJOIS and J. BOREL, "A low Drift Fully Integrated MOSFET Operational Amplifier", IEEE J. Solid-State Circuits, Vol. SC-13, pp. 499-503, August 78.

[14] T.C. CHOI and R.T. KANESHIRO, "High-Frequency CMOS Switched-Capacitor Filters for Communications Application", IEEE J. Solid-State Circuits, Vol. SC-18, pp. 652-664, Dec. 83.

[15] M. DEGRAUWE, E. VITTOZ and I. VERBAUWHEDE, "A Micropower CMOS Instrumentation Amplifier", Digest of Technical Papers, ESSCIRC'84, Edinburgh.

[16] F. KRUMMENACHER, E. VITTOZ and M. DEGRAUWE, "Class AB CMOS Amplifiers for Micropower SC Filters", Electronics Letters, Vol. 17, pp. 433-435, 25 June 1981.

[17] M. DEGRAUWE and F. SALCHLI, "A Multipurpose Micropower SC-filter", IEEE J. Solid-State Circuits, Vol. SC-19, pp. 343-348, June 1984.

[18] S. MASUDA et al, "CMOS Sampled Differential Push-Pull Cascode Operational Amplifier", Proc. ISCAS 84, p. 1211, Montreal, May 1984.

6

NMOS OPERATIONAL AMPLIFIERS

DANIEL SENDEROWICZ
MOS DIVISION
SGS-ATES
VIA OLIVETTI 2
20041 AGRATE BRIANZA, (MI)
ITALIA

1. INTRODUCTION

The idea of the operational amplifier (OPAMP) has proven to be a great blessing for the majority of analog system designers because it allows the implementation of a large family of functions such as summers, integrators, limiters and so forth without requiring that the engineer really understand neither how to bias a transistor (or a tube), nor how to prevent troublesome second order effects generally found in the design of discrete circuits (at the device level).

A dramatic increase in the usage of OPAMPS became more evident when integrated circuit (IC) technology made them easily available and economical. In bipolar analog IC's, fully integrated functions other than OPAMPS are implemented using the fewest possible number of active devices because of the area limitations imposed by the technology. For example a subsystem such as a preamplifier within a fully integrated audio processor would be implemented with just two or three transistors and some resistors and capacitors, since making a full-blown OPAMP for such a function would reduce the versatility of the device.

The rapid development of MOS technology for digital applications such as memory devices and microprocessors did not initially make a significant impact in the analog area. MOS transistors were considered unsuitable due to their low gain, large amount of $\frac{1}{f}$ (or Flicker) noise and long-term parameter unstability. It should be noted as an additional limitation that in the case of NMOS only one polarity of devices was available. On the other hand, one of the main advantages of the MOS transistor, that is its large input impedance, was not fully utilized until circuits using charge-redistribution techniques came about [1,2]. These techniques applied to the design of A/D and D/A converters, as well as switched-capacitors filters [3], and the need of support circuitry for charge-coupled devices (CCDs), another emerging technology, all created the need for the development of design techniques suitable for analog MOS.

171

In an analog MOS LSI the main active block that can be identified is the OPAMP. In the early stages of the development of these devices, the system engineer would normally specify the need for an OPAMP, so that the OPAMP designer had to produce a general purpose circuit equivalent to its discrete bipolar counterparts, that is, a circuit having: an input common mode range comparable to the total supply voltage, functionality over a wide range of supply voltages, good output driving performance, stability under any feedback configuration, etc. These initial criteria were the challenges that drove circuit designers to learn how to achieve such required specifications without the flexibility of bipolar technology, and with the addition of new constraints such as reduced power consumption and die area, to mention just a few.

This learning process brought up an issue similar to that faced by analog-bipolar designers: why design a general purpose OPAMP for a function of which the operating conditions are already known? It seems much more efficient to tailor the design accordingly which in turn optimizes the performance for a given amount of power consumption and die area. The main drawback of this approach is that there will be as many OPAMP designs as there are different functions present in the device. This represents such problems as excessive amount of design work, extensive turn-around time, and in some cases where changes are prescribed from the original design, loss in flexibility. It should also be mentioned here that the more current concepts in VLSI design emphasize the importance of modularity, that is, of having a limited library of standard cells that can be used for different functions. An answer to this can be to classify the applications based on functional aspects first, which being of qualitative nature define the general topology of the OPAMP, and then to further extend this to an evaluation of the range of operating conditions (quantitative), for example the variation of the loading conditions or the maximum allowed settling-time.

2. APPLICATIONS

A reduced in size, yet comprehensive classification of the different types of MOS OPAMPS that corresponds to the most important aspects of their circuit design can be defined by these two main families:

(1) OPAMPS for switched-capacitor (SC) integrators: Fig. 1 shows a typical integrator consisting of an input switched-capacitor C_1, a continuous feedback capacitor C_2, and a switched-capacitor C_3 corresponding to a loss. The important characteristic about this configuration is that it has no DC path from the output to the input, neither implemented by a resistor nor by a switch that is turned on and off. The main reason for this distinction, as it will become clear later, is that the bias potential of the input can be chosen in such way as to wave the need for a level-shifting stage. Furthermore, most of the SC integrators operate in the inverting mode, in other words the input potential remains fixed, so that simpler circuit configurations can also be yielded.

(2) OPAMPS for circuits with arbitrary feedback: this category includes all those applications that require a DC connection between the output and input. For example, let us look at two different types of gain-setting amplifiers, one implemented with capacitors and a reset switch (Fig. 2a), the other a non-inverting configuration using resistors (Fig. 2b). In the first case the output is reset to the same potential of the positive input (generally ground) every time that the switch is turned on, while in the second connection the input

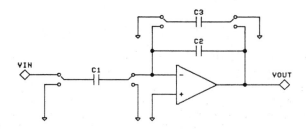

Figure 1. Switched-capacitor integrator

potential follows the input signal, a situation that generally requires a precisely designed level-shifting stage.

Within each of the two applications just mentioned which in turn define a given type of core cell, we can further distinguish other subgroups such as the following:

(a) Amplifiers that must drive resistive loads. These involve merely combining a main core cell with an output buffer. It should be noted that some connections such as the gain-setting amplifiers of Fig. 2b present a resistive loading to the output even though there are no resistors connected from the output to ground.

(b) Amplifiers for variable or fixed supply voltages.

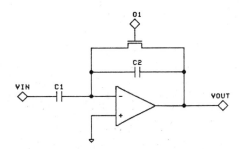

Figure 2a. Resettable gain set

173

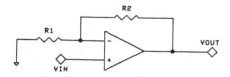

Figure 2b. Continuous gain set

(c) Single-ended and differential amplifiers. The former are the more conventional ones while the latter have their main application in the design of high-performance LSI circuits in which immunity to different sources of noise is a primary target [4].

(d) Specialized circuits such as chopper-stabilized OPAMPS for improved noise performance at low frequencies [5], dynamic amplifiers for reduced power consumption [6], amplifiers with prescribed phase response for high selectivity filters [7], etc.

3. GENERAL REQUIREMENTS

It would be very difficult to achieve a high-quality performance in a LSI chip without a top-down design strategy, that is, without forseeing the limitations of the real circuits when the building blocks are layed-out on the initial system drafts. This implies the need for a close interaction between system and circuit designers, analogous to a *closed-loop* system in which the initial *input* (the system diagram) is passed through a circuit design stage that might further modify the configuration until the final *output* becomes optimum.

We can divide the specifications of an OPAMP into two main groups. One group corresponds to the *inner* characteristics that can be extracted from observing the functional relation between input and output. For example:

- Voltage gain (A_o),
- Frequency/phase response (F_o),
- Settling time (T_d),
- Output noise (N_o) or input-referred (N_i). Only includes random effects internally generated such as thermal and $\frac{1}{f}$ noise,
- Input offset voltage (V_{os}),
- Input/output impedance (Z_i, Z_o),
- Harmonic distortion (T_{hd}).

The other group of specifications pertains to the immunity of the output to signals present in other physical nodes of the device such as the positive supply rail (V_{DD}), the negative supply rail (V_{SS}), the substrate potential, etc. For

174

example:

- Common-mode rejection-ratio (CMRR).
- Power supply rejection-ratio (PSRR).
- Substrate coupling (effect generally merged with the PSRR).

Of course the power consumption and die area are the key parameters for calculating the amount of functions that can be included in the device. Furthermore, they also serve as deciding factors in evaluating two configurations in which all other parameters are comparable.

What kind of considerations relate typical systems to OPAMP specifications? The answer to this question obviously varies depending on the type of system desired, but based on typical systems implemented in analog MOS, let us proceed to outline at least a set of general guidelines to follow, beginning with switched-capacitor filters as our subject.

3.1. Switched-capacitor Filters

This is by far the most extensive application of OPAMPs in an analog MOS chip. Most of the conventional SC filters use one OPAMP per integrator, as shown in Fig. 1. When the capacitor that implements the loss (C_3) has zero value (loss-less integrator) there is still a residual loss which is a function of the DC gain of the amplifier (assuming that the settling-time of the amplifier is less than the semi-period of the sampling frequency). This effect is analogous to the dissipative effects found in inductors and capacitors, and since one way of implementing SC filters is by analog simulation of a reactive ladder network (Fig. 3a, 3b) the loss is often indicated as the maximum Q (factor of merit) of the integrator. When a lossy integrator is needed, it is relatively simple to include the finite gain of the OPAMP for calculating the component values, but in the general case of a filter model in which each element is lossless, it is still possible to modify the values of the passive prototype to include the loss [8]. This technique involves difficulties in determining the gain, a parameter that for some OPAMP topologies can be almost unpredictable, so that the resulting situation may well be one in which "the remedy is worse than the disease". Furthermore, even given well conditioned DC-gain characteristics, there is still a theoretical limit that happens in highly selective filters (High Q). This limitation becomes dominant in the case of high-frequency filters in which amplifiers must have a very fast response, which in turn dictates simple topologies yielding low voltage gain [9].

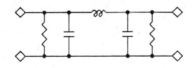

Figure 3a. Reactive ladder prototype

175

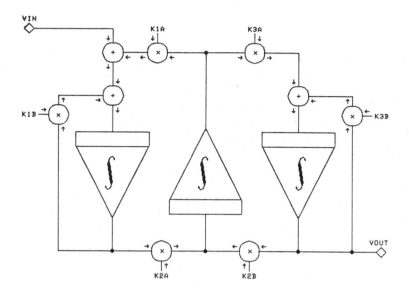

Figure 3b Analog simulation

Because of these considerations, it is customary to specify the loss caused by the amplifier ($\frac{1}{1+A_0}$) as being several times smaller than the *dispersion* of the response of the filter, that is, the passband ripple. For example a typical low-pass filter used for PCM telephony has a maximum allowed passband ripple in the order of 0.2dB. Hence an OPAMP with a gain of 1000, which represents a loss contribution of 0.01dB, is considered acceptable.

The amplifier must also settle to a prescribed tolerance within a semi-period of the system clock (assuming a 50% duty cycle). This tolerance, as in the case of the loss due to the finite DC gain, is also a function of the passband characteristics of the filter. To continue with the same example, the usual clock rates found in a PCM filter are between 8kHz and 1MHz, the most common case being 128kHz. In this case (128kHz), the OPAMP would have to settle to 0.1% in less than $4\mu sec$. It can be shown that the effect of having a residual error caused by a settling-time comparable to or greater than the clock semiperiod, appears as a Q enhancement or negative loss [10].

For an estimation of the required noise performance of the OPAMP, we can also use the previous example of the PCM filter which has a typical dynamic-range requirement of approximately 90dB, which is the ratio between the maximum and minimum signals within a frequency band. So given an output voltage equal to $2V_{rms}$, the maximum noise floor has to be less than $20\mu V_{rms}$. This figure is the sum

176

of the OPAMP noise and the corresponding thermal component $\frac{KT}{C}$ of the sampling capacitor [11]. As for the size of the capacitors generally used to implement SC integrators, experience shows that the first component dominates, hence it can be said that the noise performance of a SC filter is determined by the OPAMP performance.

Other *inner* specifications such as output impedance, output range will be discussed in detail later along with a design example.

From the second group of specifications, CMRR is rarely required for an amplifier used in SC filters because of the inverting characteristic of the integrators. On the other hand the implications of the PSRR performance in a SC filter are very important, as will be illustrated using the same example of the PCM filter. In a typical filter model, for example, an elliptic configuration, the Qs of the natural modes (poles) can be 10 or higher, implying that the internal gain at some frequencies is in that order. Any noise coming from the power supply that leaks into the output of an amplifier is amplified and also aliased down into the passband of the filter. The spectrum of this noise can be quite broad because of the switching transients caused by the digital logic present in the chip or in the same line card. What makes this problem worse is that the PSRR in an OPAMP generally decreases with frequency (lower loop gain), reaching very low values in the vicinity of the system clock frequency (e.g. an OPAMP with 1MHz unity-gain bandwidth used in a system clocked at 128kHz only gives 20dB of rejection).

Achieving a good PSRR requires a circuit design in which all the sources of coupling between the power supplies and the outputs are reduced, and the parasitics eliminated in as much as possible (it seems that in a single ended amplifier these cannot be totally removed). A fully differential architecture is a system solution that optimizes the PSRR performance, but it requires not only differential OPAMPs, but also duplications of every element in the network such as the capacitor arrays and switches. By simple symmetry considerations, it can be seen that the coupling is eliminated if the outputs are always sensed differentially, and that any residual coupling is due to mismatches between image elements. Unfortunately, as previously stated, this solution implies doubling the number of elements and connections.

3.2. Gain-setting Networks

The inverting and non-inverting configurations depicted in Figs. 2a (switched) and 2b (continuous) show typical cases in which the OPAMPs are subject to design specifications similar to those found in the case of SC filters, but with the additional requirement of reducing the loop-gain to correspond to the ratio of the elements $\frac{R_1}{R_2}$ or $\frac{C_1}{C_2}$. Using the same example of the PCM filter, most commercial devices have an input uncommited OPAMP on the transmit side which is generally used for gain setting. In the possible event that the gain is set to a value of 10 (20dB), some of the effects are the following: the output offset and noise increase by 10 times with respect to the input referred values, while the PSRR degrades and the unity gain bandwidth reduces also by 10 times. Additionally, input common mode signals can also be present, as in the case described next.

3.3. Buffers

Fig. 4 shows the configuration of a voltage follower where the capacitor C_1 can be any of the following depending on the application: a capacitor array of a digital-to-analog converter (D/A), part of a sample and hold circuit, etc. A voltage follower uses an OPAMP that must have the capability of handling input common mode signals, and depending on the type of load, might also require an output stage.

All the design considerations used for SC filters are also valid in this case, with the additional specification of CMRR. Depending on the particular design, sometimes a circuit that has a large input common-mode range has less flexibility in terms of an optimum PSRR configuration, as it will be seen later in the design example.

4. MOS MODELING

There are excellent treatments [12] for modeling the behavior of MOS transistors using pure physical considerations. By using theoretical and numerical techniques they can estimate the key parameters of the inversion layer (the channel) and from there the I-V characteristics of the transistor. Unfortunately though, in most cases these treatments are so mathematically involved that their use becomes impractical even for a computer simulator. Therefore we are interested in a simpler model capable of describing the electrical characteristics of the device from a *macroscopic* point of view. However, this simpler model should still reflect in as much as possible the physical nature of the device rather than being an arbitrary set of empirical equations.

For the purpose of describing the use of the transistor model, we will consider the design of an analog circuit to be composed of the following two phases:

(1) A stage in which the components are *calculated by hand* to meet the specifications based on basic principles.

(2) A simulation stage in which a computer program is used to *verify* the hand calculations and to observe if there are any unexpected discrepancies due to unforseen effects.

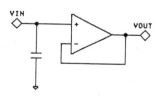

Figure 4. Voltage follower

These two phases require the availability of models capable of describing with *reasonable accuracy* the behavior of the device within a given operating range. It is clear that for our first design stage a simple model is a must, while for the second a more accurate and therefore more complex model can be used. What is meant by *reasonable accuracy* here is that we are able to fit the key model parameters with an accuracy comparable to the dispersion observed between real components.

One way of reducing the complexity of the models is to weigh the accuracy of the fit according to the importance of a parameter in a given application. For example, the output conductance of a transistor ($g_o = \dfrac{\partial I_{DS}}{\partial V_{DS}}$) plays an important role in determining the voltage gain, while the value of the current I_{DS} has little effect on the small signal behavior of the amplifier (the transconductance is a function of the square root of the current). Therefore in this case one would concentrate on optimizing the accuracy of the fit of the derivative rather than the absolute value of the current. The penalty for using this strategy is that the model loses generality for other applications. For example, in trying to characterize the propagation delay in a digital circuit, it is desirable to describe with accuracy the current as a function of the gate voltage (this is generally as large as the power supply), whereas the slope of the output characteristics has very little relevance. A possible solution for a situation in which the model parameters are part of a data base that is shared among users working in different areas is to create two or more different models.

For the analysis of the different amplifier configurations, we will use throughout this chapter the Shichman-Hodges [13] model, because of its mathematical simplicity and also adequate accuracy when used in circuit simulations. The model equations for the drain current are:

in the saturation region ($V_{DS} \geq V_{GS} - V_T$) by:

$$I_{DS} = \frac{\beta}{2} \frac{W}{L} (V_{GS} - V_T)^2 (1 + \lambda V_{DS}) \tag{1}$$

and in the triode region ($V_{DS} < V_{GS} - V_T$) as:

$$I_{DS} = \frac{\beta}{2} \frac{W}{L} [2(V_{GS} - V_T) V_{DS} - V_{DS}^2] \tag{2}$$

where V_{GS} is the gate-to-source voltage, V_T the threshold, V_{DS} the drain-to-source voltage, $\dfrac{W}{L}$ the ratio between the width and length of the transistor, λ an empirical parameter from which the output conductance can be evaluated (it is a function of the channel length, the substrate concentration and the oxide thickness of the transistor), and β the gain factor given by:

$$\beta = \mu C_{OX} \tag{3}$$

where μ is the mobility of the electron in silicon and C_{OX} is the oxide capacitance per unit area. Most of the circuits used in OPAMPs operate with the transistors in the saturation region, therefore all references to the equation of the current will be related to as (1) unless otherwise stated.

Another important parameter in our analog model is the variation of the threshold voltage with the body bias V_{BS}. For this we use the general expression of the threshold voltage [13]:

$$V_T = V_{T0} + \gamma(\sqrt{-V_{BS} + 2\varphi} - \sqrt{2\varphi}) \tag{4}$$

where V_{T0} is the threshold at zero body bias, φ is the Fermi potential, and γ is the body effect parameter given by:

$$\gamma = \frac{\sqrt{2\varepsilon q N_A}}{C_{OX}} \tag{5}$$

where ε is the permittivity of Silicon, q the charge of an electron, and N_A the substrate doping.

Because of the particular physical characteristics of a MOS transistor (i.e., high input and output impedances), the short-circuit admittance parameters are the most useful set for establishing a linear model. However if one considers the body bias effect, this model should be seen not as two, but rather, three port network where the gate is one input and the back-gate (substrate) another.

Since the transit time of electrons in the channel of a MOS transistor is negligible compared to the time constants generally found in this type of circuits, the dynamics of a transistor can be adequately modeled by lumped capacitors, and consequently merged with other capacitances that are always present in the circuit. Therefore the Y parameters can include only the real part (conductance). Furthermore, the device can be considered as being unilateral ($y_{12}=0$), and also as having an infinite input impedance ($y_{11}=0$), in which case we would not be too far from the reality of these devices. With these considerations in mind, the Y (G) parameters can be calculated from equations (1) and (4) as follows:

$$
\begin{aligned}
y_{21} &= g_m = \left.\frac{\partial I_{DS}}{\partial V_{GS}}\right|_{V_T=\text{const.}} = \sqrt{2\beta I_{DS}} \\
y_{21B} &= g_{mBS} = \left.\frac{\partial I_{DS}}{\partial V_T}\right|_{V_{GS}=\text{const.}} \\
&= g_m \frac{\gamma}{2\sqrt{V_{BS}+2\varphi}} = g_m \eta \\
y_{22} &= g_o = \left.\frac{\partial I_{DS}}{\partial V_{DS}}\right|_{V_{GS}=\text{const.}} = I_{DSS}\lambda
\end{aligned}
\tag{6}
$$

Fig. 5 shows the equivalent circuit of the linearized four-terminal device. The back-gate transconductance g_{mBS} is expressed as a linear function of the main transconductance g_m through a factor η defined here as the *gain loss* of the transistor. This factor, generally much smaller than unity, can be used as an indicator of the quality of an NMOS technology for analog applications.

5. NMOS TECHNOLOGY REVIEW

An NMOS technology used for analog LSI generally provides the following elements:

Enhancement transistor:
This device has a threshold ranging from 0.6 to 1V, a β factor ranging from 20 to 40 $\frac{\mu\Omega^{-1}}{V}$, and a γ factor ranging from 0.3 to 0.5$\sqrt{V}$. It is generally used as an active device, current mirror, etc.

Natural (unimplanted):
This transistor is characterized for having the lowest body effect in a given process. For this reason it is generally used as an analog switch, especially in cases in which the body bias is high. Since it does not receive any implant, the resulting number of surface states is also minimized yielding a low-noise transistor that can be used as the input pair of an OPAMP. Its threshold voltage can vary from -0.2 to 0.4V, while the β from 30 to 40 $\frac{\mu\Omega^{-1}}{V}$, and the body factor γ from 0.1 to 0.3$\sqrt{V}$.

Depletion:
With a threshold range between -3 and -1V, this transistor is used as a current source and also as a driven pull-up for output stages in which a large output drive is required. It generally has a lower body effect than the enhancement transistor, but a larger than the natural transistor (0.1 to 0.4$\sqrt{V}$). Its gain factor β is similar to the enhancement transistor. The evaluation of η (gain loss) becomes important for this transistor type.

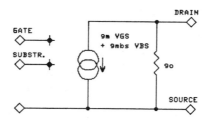

Figure 5. Linearized model

Assuming a Fermi potential of 0.3V, and considering that the load will generally be working with a body-bias equal to the quiescent output voltage of the stage (for example V_{BS} = 5V for a 10V supply), (6) gives values for η ranging from 0.02 to 0.1.

MOS capacitor:
In the case of a double polysilicon technology, the dielectric thickness ranges between 700 and 1200Å, while in the case of a single layer of polysilicon the range is between 400 and 1000Å. In the latter an additional diffusion for the bottom plate is required.

Polysilicon resistor:
This component has excellent voltage linearity although its resistivity is low (30 to 100 $\frac{\Omega}{\blacksquare}$). It finds its application in the implementation of timing constants, resistive dividers, etc. It is very seldom used inside the OPAMP circuits.

Diffusion resistor:
This transistor's resistivity is similar to that of the polysilicon (10 to 50 $\frac{\Omega}{\blacksquare}$), although its voltage linearity is not as good as the polysilicon resistor's. It is used primarily as a substitute for polysilicon when that item is not available (metal-gate NMOS).

6. BASIC NMOS AMPLIFICATION

The three basic amplifier configurations commonly used in NMOS OPAMP design are: common source, common drain, and common gate. The characteristics of each of these can be analyzed as an extension of analogous configurations found in other technologies, for example we can find a correspondence between the common source and common emitter in bipolar technology and so on. Besides quantitative differences, the analysis should also take into account the four-terminal behavior of the MOS transistor. As we will see, this additional effect has significance only for some configurations.

6.1. Common-source Amplifier

Fig. 6 shows a stage composed of transistor M_1 and a load composed of r_L.

In this configuration the back-gate effect on the active transistor can be disregarded because the incremental voltage between the source and the substrate is zero. The small-signal voltage gain is given by $-g_{m1}r_L$, where g_{m1} is the transconductance of M_1 at the operating point and r_L is the incremental impedance of the load. Assuming that the output conductance g_o is made arbitrarily small by a suitable choice of the transistor geometry, the load impedance will be determined by the load transistor. This load can be connected with the gate tied to the drain (its highest potential) or to the source (its lowest potential). Since the first choice is commonly used when only enhancement mode transistors are available, it is called enhancement load, while the second choice, because it requires a depletion transistor, is known as depletion load. Let us analyze each of these in more detail.

182

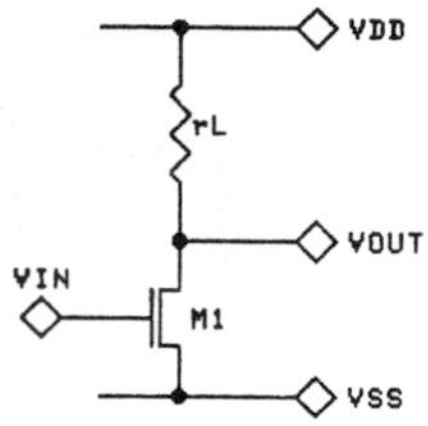

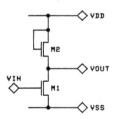

Figure 6. Common-source amplifier

6.1.1. Enhancement Load. The four-terminal linearized model can be used to calculate the impedance seen at the source when both the gate and back-gate (substrate) are grounded (Fig. 7). This calculation can be easily accomplished by using flowgraph techniques [15]. Although the simplicity of this circuit allows for the writing of equations by analytical inspection, flowgraph techniques prove to be a valuable tool for more complex circuits where analytical solutions become more prone to errors. For this purpose, the linear model is now shown in Fig. 8a with the nodes corresponding to the two gates and drain grounded, and the voltage source v_x connected between the output (the node corresponding to source) and ground. By including the relevant variables of the circuit in the flowgraph (Fig. 8b), the current across v_x can be computed and from this the impedance seen at the node, that is:

Figure 7. Enhancement-load amplifier

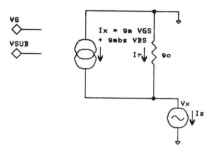

Figure 8a Linear model for E-L

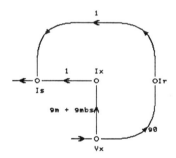

Figure 8b Flowgraph for E-L

$$\frac{v_x}{i_s} = r_L = \frac{1}{g_m + g_{mBS} + g_o} \tag{7}$$

Using (6) and ignoring the effects of the output conductance g_o, we obtain the expression of the voltage gain of the stage:

$$A_0 = \frac{1}{\eta+1} \left[\frac{(\frac{W}{L})_1}{(\frac{W}{L})_2} \right]^{\frac{1}{2}} \tag{8}$$

Since $\eta \ll 1$, the voltage gain approximates to the square root of the $\frac{W}{L}$ ratio. Practical considerations limit this ratio to less than 100, setting the maximum attainable gain to 10. This low figure shows that in order to achieve a gain of 1000, which is typical for SC OPAMPs, at least three of these stages are required [16].

6.1.2. Depletion Load. Flowgraph techniques can also be used to analyze the incremental conductance of this type of load (Fig. 9), giving:

$$g_L = g_{mBS} + g_0$$

ignoring the output conductance g_0, the voltage gain is obtained as:

$$A_0 = \frac{1}{\eta} \left[\frac{(\frac{W}{L})_1}{(\frac{W}{L})_2} \right]^{\frac{1}{2}} \tag{9}$$

The voltage gain here is $\frac{1}{\eta}$ times larger than that found in the case of enhancement load. Typical values of DC gain for a $\frac{W}{L}$ ratio of 20 range between 100 and 200. Another advantage of depletion load is that the charge across the gate

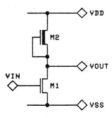

Figure 9. Depletion-load amplifier

185

capacitance does not change ($V_{GS} = 0$), whereas in the case of enhancement load the output voltage is developed across the gate capacitance of the load, increasing the phase shift of the stage.

The difference between the two types of loads can be summarized by plotting their I - V load lines together with the output characteristics of their active transistors and hypothetical resistive loads (Fig. 10). It should be emphasized that the departure of the loading characteristics of the depletion transistor from an ideal current source is not due to the output conductance g_o, which was already disregarded, but rather, to the back-gate effect that modulates the drain current. Another way of visualizing this effect is by considering the load as composed of two transistors (each modelled as a three terminal device) with the drains and sources tied together: one with a transconductance equal to g_m and its gate tied to the source, the other with a transconductance of g_{mBS} and its gate grounded.

6.2. Common-drain Amplifier

This configuration (Fig. 11) shows with respect to the common-source configuration, an inverse behavior regarding the back-gate effect, in other words, in this case the body bias influences only the active device leaving the load unaffected. The voltage gain of the stage is:

$$A_o = \frac{g_m r_L}{1 + (g_m + g_{mBS}) r_L} \tag{10}$$

As was expected, the gain is always less than one, but it is interesting to see that even when r_L is equal to infinity, the gain never reaches unity but becomes:

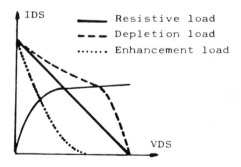

Figure 10. Different load lines

186

$$A_o = \frac{1}{1+\eta}$$

where the output conductance effects have been ignored. The output conductance is:

$$g_{22} = g_m + g_{mBS}$$

6.3. Common-gate Amplifier

This is a configuration that gives high input and low output conductances. Fig. 12 shows this arrangement and the voltage gain, the input and output conductances are respectively given by:

$$A_o = \frac{(g_m + g_{mBS})r_L}{1 + (g_m + g_{mBS})r_S}$$

$$g_{11} = g_m + g_{mBS}$$

$$g_{22} = \frac{g_o}{1 + (g_m + g_{mBS})r_S}$$

Unlike other configurations, the back-gate effect not only does not degrade the performance of the active transistor, but also to some extent even improves it by enlarging the effective transconductance. However, in this case the back-gate effect still influences the load transistor in the same way as it does in the common-source amplifier.

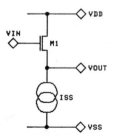

Figure 11. Common-drain amplifier

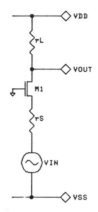

Figure 12. Common-gate amplifier

7. GENERAL-PURPOSE OPAMP DESIGN

At this point we have introduced all the arsenal available to the NMOS OPAMP designer. As a first step in the development of an NMOS depletion-load OPAMP, we have to define a circuit topology capable of meeting the requirements generally found in analog LSIs. As was previously stated, there are a wide variety of applications and hence different types of amplifiers possible for each of them. The criterion that we will follow is to develop an amplifier that meets the most stringent specifications, that is, a general purpose OPAMP, since others will most likely be particular cases of this one.

Here we define a general-purpose OPAMP as being a differential input, single-ended output circuit that is stable under any practical feedback configuration and meets the following specifications:

Supply voltage: V_{DD}: +4 to +7V, V_{SS}: −4 to −7V.

Gain: Minimum 1000.

Bandwidth: 3MHz (unity gain).

Phase margin: Minimum 60.

Slew rate: $3\dfrac{V}{\mu sec}$.

Load: C_L=50pF maximum, R_L=1kΩ minimum.

Input C-M range: Within 2V from each supply rail.

Output range: Within 2V from each supply rail.

PSRR: Minimum 40dB at 1kHz (unity gain).

CMRR: Minimum 40dB at 1kHz.

Let us analyze some of the basic properties of a classic two-stage OPAMP topology such as the one whose linear model is depicted in Fig. 13:

(1) The total DC amplification is the product of the gain of two stages. This situation makes it possible to reach the specified gain (1000) with NMOS common-source stages (the differential amplifier is considered as an extension of a common-source amplifier).

(2) The frequency compensation implemented by a single capacitor gives, by means of a pole-splitting reaction [14], the performance of a low-loss integrator yielding a phase margin close to 90° (neglecting other frequency singularities of the circuit). In this type of topology the intermediate nodes (not the output) have very little voltage swing, thus all the parasitic capacitors see no voltage change across them, or in other words the signal is propagated internally as a *current*.

(3) The high-bandwidth, unity-gain buffer isolates the load from the gain-stage, making the DC gain and frequency response of the amplifier independent of the load.

In any case, the logical steps in the development of an OPAMP for an LSI are better appreciated if we impose a configuration that has no output stage. Of course this precludes the possibility of loading the amplifier with a resistor of

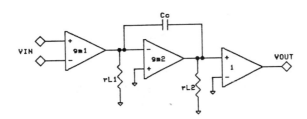

Figure 13. Two-stage OPAMP

189

magnitude comparable to the output resistance of the second stage. But unless the amplifier has to drive off-chip loads, this results in the most common situation. In any case, the inclusion of the unity-gain buffer can be regarded as a particular design case in which both the capacitive and resistive loading become negligible. Obviously in this case the design of this unity-gain buffer must be treated separately. Thus, for our design we will change the loading condition to:

C_L=5pF (max).

R_L=∞.

We must also specify the key parameters characterizing the devices provided by the technology available that can be considered typical of a modern NMOS process.

Threshold (V_T):   Enhancement: 0.7V, depletion: −2.5V, unimplanted: 0.0V.

Gain Factor (β):   Enhancement, depletion and natural $30\frac{\mu\Omega^{-1}}{V}$.

Body effect (γ):   Enhancement: 0.4√V, depletion and natural: 0.25√V.

Gain loss (η):   Depletion: 0.06 at V_{BS}=−5V, 0.13 at V_{BS}=0V.

Oxide thickness: 700Å.

We begin our "descent" along the hierarchical *top-down* design by expanding the block diagram of Fig. 13 (without output stage) into a simplified circuit schematics at a device level such as the one shown in Fig. 14.

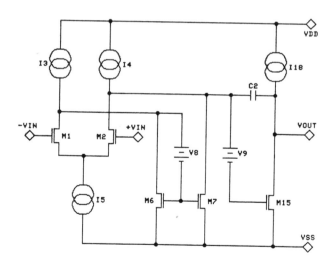

Figure 14. Simplified circuit schematics

190

The first transconductance stage is implemented by combining an input differential amplifier M_1-M_2, a pair of active loads I_3-I_4 (here indicated as ideal current sources), a differential-to-single-ended converter M_6-M_7, and a level shifter (indicated as ideal voltage sources V_8-V_9). The second stage (also called gain stage) is formed by M_{15} and the active load I_{18}. The phase compensation is implemented by capacitor C_2. Obviously this is not the only possible implementation of the block diagram of Fig. 13, for example Fig. 15 shows a different arrangement of the level shifters [17].

Our example will be based on the scheme shown in Fig. 14. For reasons that will become obvious once the design is completed, we will analyze the stages in the following order:

(1) Second stage (gain stage).

(2) Input stage and differential-to-single-ended stage.

(3) Bias stage.

7.1. Gain-stage Design

Fig. 16 shows this stage in more detail, where the ideal current source I_{18} is replaced by M_{18}. Based on the data available, we must establish for this stage the value of C_2 and the geometries of M_{15} and M_{18}.

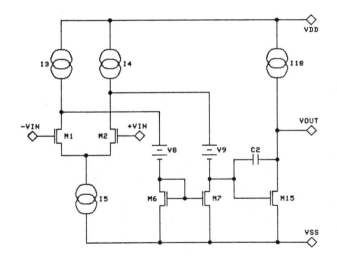

Figure 15. Alternative circuit schematics

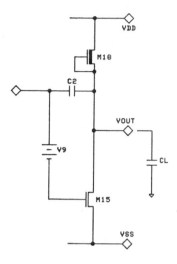

Figure 16. Circuit schematics of the second stage

For the geometry of M_{18}, we must consider that its DC current has to be large enough to supply both the maximum current delivered by the first stage under the positive slewing condition and the current required to charge the loading capacitor. If this criterion is not met, the transistor M_{15} will turn off under the positive slewing condition, giving an undesirable transient response. The ratios commonly used between the current of the first stage and the current of M_{18} range from 2 to 10.

The next objective is to choose the value of C_2. The criterion that we will follow is based on the fact that only an upper boundary, and not the actual value for the capacitive loading C_L is known. Thus a situation in which the performance is not significantly affected by this uncertainty is desirable. This, as a intuitive consideration would suggest that C_2, the only other capacitor in the circuit (parasitics ignored) must be several times larger than C_L. Of course this very conservative (and impractical) strategy should not be taken literally but only as a point of departure. Furthermore, it might happen that in a subsequent optimization phase, C_2 becomes considerably smaller, but this situation would evolve from a more complex evaluation of the circuit where all the second order effects are considered.

Choosing for C_2 a value of 10pF and a safety margin of 2, the current I_{18} becomes:

$$I_{18} = 2(C_2 + C_L)SR = 90\mu\text{A}.$$

where SR is the slew-rate.

With the value given for β and threshold of the depletion device V_{TD}, equation (1) gives the $\dfrac{W}{L}$ ratio of M_{18} as being equal to 1.

The geometry of M_{15} is a parameter that can be considered *adjustable*. It is usually used to control the total amount of phase shift and to obtain the desired overall gain of the OPAMP once the dimensions of the first stage are known. It will be shown later that there is less flexibility in controlling the voltage gain changing the geometry of the first stage. However the adjustability of the $\dfrac{W}{L}$ of M_{15} is limited in the upper and lower bounds according to the following considerations:

Upper bounds:
> The voltage gain of the stage is augmented by increasing the g_{m15}, which for operating at a constant current can be calculated using (6). At the same time the parasitic capacitance of the transistor is $\propto W$. Therefore it can be easily seen that the bandwidth of the transistor is $\propto \dfrac{1}{\sqrt{W}}$ (once the minimum channel length is established by the technology). It generally happens that this bandwidth is several times larger than the overall bandwidth of the amplifier, therefore its maximum size is commonly limited by die area allocation.

Lower bounds:
> This is mainly determined by considering the phase shift contribution of the second stage. Therefore in the next topic we will analyze the dynamics of this stage.

7.1.1. Frequency Response of a Common-source Stage. Let us recall the linear model of the transistor and merge its components with the other elements of the circuit into the form of lumped impedances. Fig. 17 shows this linear network in which the elements are distinguished as follows:

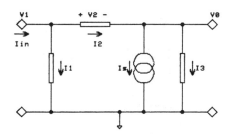

Figure 17. Linear model of second stage

193

(1) An input signal I_i that represents the output current of the first stage,

(2) A voltage-controlled current source I_s representing M_{15},

(3) An impedance z_1 that includes both the resistive and capacitive components of the input stage,

(4) An impedance z_2 that represent the feedback components of the stage,

(5) An impedance z_3 that includes the output capacitive loading ($C_3=C_L$) and the resistance at the node (mainly due to the body-effect).

Fig. 18 shows a flowgraph of the linear model, that after some manipulation [15] yields the transfer function expressed as a transimpedance:

$$z_{21} = \frac{z_1 z_2 z_3 (g_m z_2 - 1)}{(z_1 + z_2)(z_2 + z_3) + z_1 z_2 (g_m z_2 - 1)}. \tag{11}$$

Equation (11) contains all the information about the dynamics of the circuit. If the values of the impedances z_1, z_2 and z_3 are replaced by their capacitive and resistive component, the magnitude and the phase response of the circuit can be obtained. However, the information that we need can be obtained without the need of complex algebraic manipulations if we consider that in the vicinity of the unity gain frequency the magnitude of the impedance $z_1 \gg z_2$, then (11) approximates to:

$$z_{21} = \frac{\dfrac{g_m}{sC_2} - 1}{g_m + sC_3} \tag{12}$$

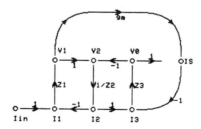

Figure 18. Flowgraph of the linear model

194

Equation (12) shows that the second pole is at $s_2 = -\dfrac{g_m}{C_3}$. This second pole gives an additional 45° of phase shift at s_2, therefore g_m has to be calculated to place this pole at a frequency considerably larger than the unity gain frequency. That is:

$$g_m > 2\pi f_0 C_3 = 2\pi\ 3\text{MHz}\ 5\text{pF} = 100\mu\Omega^{-1}.$$

Using the value of the DC current I_{18} and the β of the enhancement transistor, equation (6) shows that a transistor with a $\dfrac{W}{L}$ of 2 gives the desired g_m. However, equation (9) gives for a $\dfrac{W}{L}$ ratio of 2 a gain equal to 9 (at $V_{BS}=-2\text{V}$) that is obviously too low.

There is still another source of phase shift in the circuit that is caused by a right-half-plane (RHP) zero that is also included in equation (11). This RHP zero is at a frequency in which $g_m z_2 = \dfrac{g_m}{C_2} = 1$. This contribution to the phase-shift is in this case more significant than the one corresponding to the second pole $(C_2 > C_3)$. However, this can be eliminated as will be shown next.

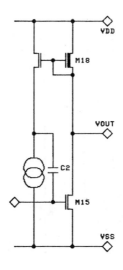

Figure 19. Source follower RHP zero cancellation

7.1.2. Right-half Plane Zero Cancellation. The three most common methods to overcome the effect of the RHP zero are:

(1) Using a buffer (source follower) between the output of the stage and the capacitor to eliminate the forward path of the current (Fig. 19). This method has a drawback in that it consumes additional power [16].

(2) Adding a resistor of value $\dfrac{1}{g_{m15}}$ in series with the capacitor C_2. The cancellation of the RHP zero can be readily observed in equation (11) by modifying z_2. From the same equation it can also be seen that this zero can be easily shifted towards the LHP so as to introduce what is known as *lead* compensation for cancelling any excess phase shift. This resistor is generally implemented by a self-biased depletion-mode transistor. The geometry of this transistor (M_{19}) is easily derived by combining the following equations: (1) to express the current I_{18} as a function of the $\dfrac{W}{L}$ of M_{18}, (2) to find the ON conductance of M_{19} using the derivative as a function of V_{DS} (and $V_{GS} = V_{DS}$), and (6) to find the transconductance of M_{15}. From (11) the RHP zero is cancelled when the ON conductance of M_{19} is equated to the transconductance of M_{15}, which gives (assuming that all the betas are identical):

$$(\frac{W}{L})_{19} = \sqrt{(\frac{W}{L})_{18}(\frac{W}{L})_{15}} \qquad (13)$$

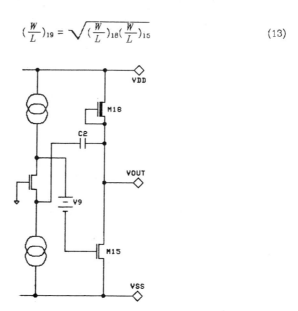

Figure 20. Common-gate RHP zero cancellation

196

(3) A common gate stage (Fig. 20) also serves as a buffer in a similar way does a source follower. It will be shown later that this circuit can be merged with other portions of the OPAMP to avoid extra power consumption. Furthermore, this circuit in the absence of signal has a voltage across the compensation capacitor that is independent of the power supply voltage, a very desirable feature for having a good PSRR at frequencies greater than zero.

In the simple circuit shown in Fig. 16, there are no options to increase the DC gain. This can be seen in equation (9), where the small signal amplification is a function of the $\frac{W}{L}$ ratio and the DC potential of the output. Furthermore the output conductance of M_{15}, that has a relatively large value of $\frac{W}{L}$ and as a consequence a small L, can significantly reduce this gain. The effect of the output conductance of M_{18} can be ignored because its $\frac{W}{L}$ is generally ≤ 1, giving more flexibility in the absolute dimensions of W and L. A classical approach to solving this is by changing the configuration to a *cascode* stage that gives a much smaller output conductance. Another advantage of the cascode configuration is that it allows for increasing the transconductance of M_{15} without increasing the back-gate transconductance (g_{mBS}) of M_{18} by means of a *feeder* which augments the bias current of M_{15}, keeping I_{18} constant. Fig. 21 shows the modified stage. It remains as an exercise to show that the voltage gain of the modified gain-stage is:

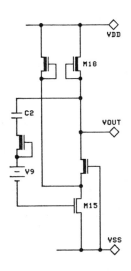

Figure 21. Modified gain-stage

$$A_o = \frac{1}{\eta} \left[\frac{I_{15}(\frac{W}{L})_{15}}{I_{18}(\frac{W}{L})_{18}} \right]^{\frac{1}{2}}$$ (14)

7.2. Transconductance-stage Design

The design of this stage will be separated into two parts:
(1) the input source-coupled pair M_1-M_2, and
(2) the level-shifters V8-V9 in conjunction with the differential-to-single-ended converter.

7.2.1. The Differential Amplifier. As was previously stated, a MOS differential amplifier can be considered an extension of the simple common-source configuration. In fact by applying equation (1) on both M_1 and M_2 with the condition $I_5 = I_1 + I_2$, we obtain the differential output current $(I_1 - I_2)$:

for $X_{in} \geq 0$

$$I_x = + I_5 \sqrt{1 - (1 - X_{in})^2},$$

for $X_{in} < 0$

$$I_x = - I_5 \sqrt{1 - (1 - X_{in})^2},$$

(15)

where X_{in} indicates a normalized input signal equal to:

$$X_{in} = \left[\frac{\beta \frac{W}{L}}{2I_5} \right]^{\frac{1}{2}} V_{in}$$ (16)

where V_{in} is the differential input signal $(V_1 - V_2)$. Fig. 22 shows a plot of equation (15).

Some interesting observations can be made about equations (15) and (16):
(1) The differential output current is an odd function, that is $I_x(-X_{in}) = -I_x(X_{in})$, which means that all the even order terms in a series expansion are cancelled. This represents an advantage with regard to the harmonic distortion of the amplifier, and also gives symmetric slewing characteristics under large input signals of opposite polarities.
(2) By taking the derivative of (15) as a function of V_{in}, we obtain the transconductance of the stage, which has the identical form of (6).
(3) There is a saturation voltage of the stage V_{sat}, (which should not be confused with the saturation voltage of the transistor) that results from making $X=1$ in (15), yielding:

198

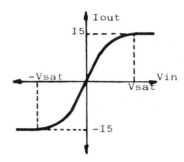

Figure 22. Transfer curve of differential amplifier

$$V_{sat} = \left[\frac{2I_5}{\beta \frac{W}{L}} \right]^{\frac{1}{2}}$$

This voltage is an indicator of the linearity range of the amplifier and can also be used as an additional design parameter.

The amplifier goes into the slewing condition at the saturation voltage of the input stage, in which the maximum output current becomes $+I_5$ or $-I_5$. So the slew-rate is simply given by:

$$SR = I_5 C_2$$

which gives:

$$I_5 = 30\mu A$$

The $\frac{W}{L}$ of the pair M_1-M_2 is calculated based on the specified value of unity-gain bandwidth, which can be calculated by equating the product of the transconductance of the first stage and the transimpedance of the second stage (equation 11) to 1, that is $\frac{g_{m1}}{2\pi BWC_2} = 1$. For the sake of simplicity, the magnitudes of z_1 and z_3 are assumed to be much larger than z_2 at the unity gain frequency. This yields:

199

$$g_{m1} = 190\mu\Omega^{-1}$$

which gives a $\dfrac{W}{L}$ for M_1-M_2 of 20. The calculation of the gain of this stage can be done once the geometries of M_3-M_4 are known. The dimensioning of M_3-M_4 will be done together with that of the differential-to-single-ended converter, which is the subject of our next topic.

7.2.2. Differential-to-single-ended (DSE) Converter.

In the simplified schematics of Fig. 14 the pair M_6-M_7 is functionally equivalent to a conventional DSE converter. The only difference is that the level shifter V_8 is interposed in the feedback path of M_6. Since the input-common-mode voltage of the first stage can vary over a wide range, this level shifter must have enough voltage so that the transistors M_1-M_2 will operate in the saturation mode. For example, if the power supplies are V_{DD}=5V, V_{SS}=−5V, and the maximum input common-mode voltage is 0V, the value of V_8 must be at least $-V_{SS}-V_{GS(6,7)}$. The maximum common-mode compliance of the amplifier results when V_8 is set to:

$$V_8 = V_{DD} - V_{SS} - V_{GS(6,7)} - V_{min(3,4)} \tag{17}$$

where $V_{min(3,4)}$ is the minimum operating voltage for the current sources I_3, I_4 (generally equal to V_{TD}). Since the potential of the gates of M_{15} and M_6, M_7 are the same (the current density of M_{15} and M_6, M_7 should also be the same), V_9 is also equal to V_8. Fig. 23 shows how these level shifters are implemented with the source followers M_8, M_9.

From this schematics it can be seen that the voltage drop across the gate and source of M_8-M_9 is not only a function of their dimensions but also of the current of transistors M_{12}-M_{13}. The generation of this current is done with a *replica* type of bias-circuit that will be shown later.

For the calculation of the dimensions of M_8-M_9 we can assume an approximate value for $V_{GS(6,7)}$ of 1.2V, and the saturation voltage of M_3-M_4 to be equal to V_{TD}=−2.5V (in reality V_{TD} is smaller in magnitude because of the body effect), thus according to (16) this source follower needs a V_{GS}=6.3V for V_{DD}=5V and V_{SS}=−5V. If we budget a current of 25μA for each transistor, (1) gives a $\dfrac{W}{L}$ of 0.05 for each. This small $\dfrac{W}{L}$ makes the output resistance of the source follower large enough to create a significant additional phase shift. However, it is possible to compensate for this effect by placing a feed-forward capacitor in parallel with the gate and source of M_8-M_9.

The quiescent current of M_6-M_7 should be made equal to that of the pair M_1-M_2. Otherwise, if it were made smaller, the DSE conversion would not be symmetrical under large input signals, while if it were made larger, waste of power would occur. The transconductance of M_6-M_7 also should not be made neither too small relative to that of M_1-M_2, because of PSRR and CMRR considerations, nor too large, because of a resulting degradation in noise performance [14]. A good compromise therefore is to make the value of the $\dfrac{W}{L}$ of M_6-M_7 equal to one half of the $\dfrac{W}{L}$ of M_1-M_2.

200

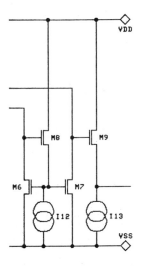

Figure 23 Schematics of the DSE converter

Now the total quiescent current flowing in M_3-M_4 is known ($30\mu A$ each), therefore we can proceed to calculate the $\dfrac{W}{L}$, which results in being equal to 0.3. With the values of the geometries of M_1-M_2 and M_3-M_4, we can further calculate the DC gain of the stage. Equation (14) can be used, but it should be noted that in this configuration the current through the load is larger than that of the active device, resulting in a situation opposite from that found in the case of the second stage. The calculation gives a gain of 130, but this figure becomes considerably lower if the effect of the output conductances of both pairs M_1-M_2 and M_6-M_7 are included (In this case there is no cascode effect to boot-strap the output impedance).

A common-gate stage interposed in the DSE (Fig. 24) serves the function of isolating the noise at the power supplies from the compensation capacitor for PSRR enhancement (this was mentioned when we studied the cancellation of the RHP zero). This common-gate stage combined with the DSE converter performs as a cascode stage, reducing the loading over the first stage.

Fig. 25 shows the detailed circuit schematics of the OPAMP.

7.3. Bias Stage

The bias stage must provide the gate voltages for transistors M_{12}-M_{13} and M_5 (the transistor that implements the current source I_5). The circuit shown in Fig.

201

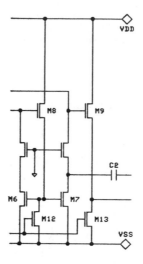

Figure 24. DSE with cascode buffer

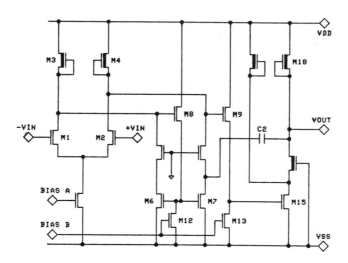

Figure 25. Circuit schematics of the OPAMP

26 is a *replica* type of a bias circuit that has the two required outputs A and B.

202

The operation of this circuit can be described as follows: Transistor M_{24} generates a current which is taken as a reference. The current mirror composed by M_{25}-M_{26} (and level-shifter M_{29}) reflects this current into M_{23} which is operating in the triode mode. Using (1) and (2) it can be shown that the voltage drop across M_{23} becomes equal to $-V_{TD}$ when its $\dfrac{W}{L}$ is one third of the $\dfrac{W}{L}$ of M_{24}. This voltage at the source of M_{23} contains the information about the threshold of the depletion transistor when its body bias is equal to one depletion threshold below V_{CC}. It is converted into a current by a long transistor M_{28} (which has the same geometry as M_8-M_9 and M_{29}) and further reflected by the current mirror M_{30}-M_{31}. Node A is used to bias the gate of M_5 while node B is used to bias the gates of M_{12}-M_{13}.

7.4. Unity Gain Buffer

A simple source follower can perform quite well when the impedance of the load is not too low and the amount of dissipated power is not an issue. In CMOS, like in bipolar technology, a complementary pair is a good solution when good driving and low power consumption are required. In NMOS however, a stage that performs as an equivalent P-channel source follower must be created. By looking at the typical operating conditions of this pseudo P-channel transistor, we can see that it is equivalent to a whole OPAMP connected as a voltage follower but without the capability of draining current (the N-channel source follower does that). Of course circuits much simpler than an OPAMP can be used for this function [17], but they suffer from dynamic range problems. In any case, an LSI requires fewer

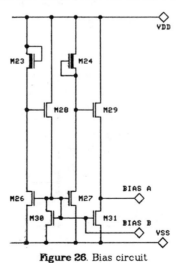

Figure 26. Bias circuit

OPAMPs to drive resistive loads (hopefully only those connected to off-chip loads), so that it pays off to have a better quality circuit for the function.

What kind of OPAMP should we use for this voltage follower?

(1) Its phase shift at the unity gain frequency of the core amplifier should be small,

(2) It must have a slew-rate faster than that of the core amplifier,

(3) The quiescent current of the output should be kept at a reasonable low level (class AB operation).

The first condition implies that the OPAMP must have a unity gain bandwidth larger than that of the core section. However, with proper designing it is possible to make this amplifier behave like a second order system with complex natural modes. In this way, sharper phase transition regions are obtained, in other words, the phase shift can be kept constant within the desired range of frequencies. If the second condition is not met, oscillations can occur.

The circuit topology can be similar to that of the core amplifier. It is also possible to use other configurations like the one shown in Fig. 27.

This circuit operates in the following way: the level shifter M_{8B} feeds the signal at the drain of M_{2B} back to the gate of the current source M_{5B}, which in turn forces the current at the output of M_{2B} to remain constant even in the presence of signal. This current also passes through M_{1B} doubling the effective signal current, and thus yielding the desired DSE conversion. Although this circuit is

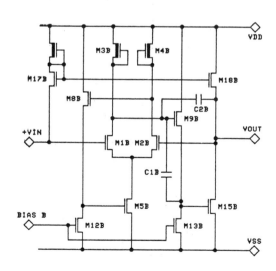

Figure 27. Unity gain buffer

considerably simpler than the one used for the core amplifier (it does not have the current mirror), it has the drawback of having square-law current characteristics, which means that the signal current at the drain of M_{1B} is an even function of the input voltage. This makes this configuration not very suitable for the core stage because it has higher harmonic distortion and asymmetrical slew-rate characteristics, which under large input signals can absorb all the current of the second stage through the compensation capacitor, thereby turning off the gain transistor (M_{15B}). However, when this configuration is used for an output buffer, the gain is always unity, which minimizes the absolute distortion which is further reduced by the negative feedback around the amplifier. The asymmetrical slew-rate performance is not a problem either, because the charging current for the compensation capacitor (on positive going signals) is supplied by a source follower and not by a constant current source.

This design methodology is similar to the one used for the core stage, although the design efforts must concentrate on having large transconductance for M_{15B}, large bandwidth, etc, while the open loop gain can be quite low (a DC gain of 50 to 100 suffices) [4].

8. LIMITED-PURPOSE OPAMP DESIGN

The development of a general-purpose OPAMP takes into account all the possible operating conditions in which the circuit can be used. However as was previously mentioned, most amplifiers used in an LSI device should be designed with the application in mind in order to simplify the design. Therefore the question to ask is: how should the system be conceived in order to simplify and improve the OPAMP circuits? The main answers are:

(1) If the input common-mode voltage is fixed and known, the level shifters and bias circuits should be simplified.

(2) If the loading is only capacitive, there is no need for the unity gain buffer, in other words, resistors should be avoided.

(3) If possible, all the amplifiers should operate in a mode in which they do not have to be reset (This will be explained later).

We will analyze an NMOS OPAMP that does not require level shifters and has an extremely simple bias circuit. This differential-input, differential-output circuit can be used as the gain block of a fully differential system. As previously stated, the advantages of fully differential architectures are: the dynamic range is doubled, and the rejection to spurious common-mode signals is improved (e.g. PSRR). Additionally there is an enhanced flexibility for the design of bilinear SC filters [4], chopper-stabilized amplifiers [5], etc. However the cost of these added features is that they require not only duplicating the number of interconnections and peripherical components such as switches and capacitors, but also creating a common-mode feedback circuit which further requires more switches and capacitors.

8.1. Differential OPAMPs

A single ended OPAMP is a system that can be considered as having one *degree of freedom* because its output voltage changes proportionally to its input voltage. In a differential amplifier, the equivalent variable is the differential output voltage, while the common mode voltage of the outputs (or average) is unknown. Therefore it is necessary to create another input to control this common mode output. In Fig. 28 is depicted the differential amplification path (A_d) and the common-mode amplification path (A_c).

For most applications, the common-mode amplifier operates in a voltage follower configuration where the input signal is the average of the two differential outputs. The most difficult task in designing differential OPAMPs is to control the way in which this average voltage is taken. Conceptually, the simplest way to perform this function is by means of a resistive divider as shown in Fig. 29. The problems associated with this approach are mainly two:

(1) Resistive loading,

(2) The input level of the common-mode path is at ground level, imposing the need for a level-shifter.

The resistors can be replaced by source followers (Fig. 30), but this approach results in non-linear behavior in the presence of large output swings due to possible source follower turn-off.

Our solution is to attack the problem from a system point of view. The reasoning behind this is: Why not use for the common-mode feedback the same methods used for processing the differential signals? Fig. 31 shows a symbolic

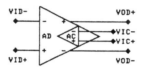

Figure 28. Differential amplifier

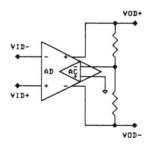

Figure 29. Resistive common-mode feedback

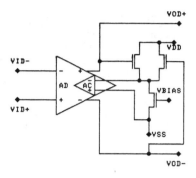

Figure 30. Source-follower common-mode feedback

representation of this differential SC integrator.

In this circuit, the common-mode feedback is implemented by a pair of integrating capacitors C_{C_1} and C_{C_2}, and a pair of switched-capacitors C_{S_1} and C_{S_2}. These two pairs of capacitors take the average of the two voltages V_o and $-V_o$. This common-mode configuration can be recognized by SC filter designers as a lossy integrator. Its integrating constant should be made larger (faster) than the one in the differential signal path.

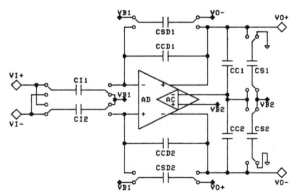

Figure 31. Differential SC integrator

Let us analyze how a proper choice of the bias voltages V_{B1} and V_{B2}, which define the quiescent voltage at the inputs of the differential and common-mode amplifiers, simplifies the internal circuit design. In other words, let us observe how all the level shifting operations can also be done by the switched-capacitors. Fig. 32 shows the circuit of the differential OPAMP.

M_1-M_2 is the input pair for the differential path, with loads M_3-M_4. The second stage consists of another differential pair M_6-M_7. There is a local common-mode feedback between the second and first stages implemented by transistor M_5. The input to the common-mode amplifier is a single transistor M_{10}. It should be noted that the open-loop gain of the differential path is larger than that of the common-mode path (a cascade of two stages versus one). However since in the system the signal is always sensed differentially, common-mode errors are unimportant.

By simple inspection of the circuit, it can be seen that the voltage at the drains of M_1-M_2 is equal to $V_{GS(6,7)} + V_{GS5}$. If V_{B1} is made equal to this value then the pair M_1-M_2 will always work in the saturation region (there is also a margin of two gate voltages). V_{B2} should be made equal to V_{GS10}.

The different quiescent potentials of the inputs and outputs make a reset operation (short-circuiting inputs to outputs) undesirable for this particular type of OPAMP.

The component values can be calculated in the same order as in the case of the single-ended amplifier:

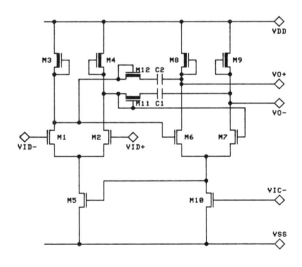

Figure 32. Differential OPAMP

208

(1) The output bias current,

(2) The minimum g_m of the second stage,

(3) The compensation capacitor,

(4) The input differential amplifier and DSE converter.

9. CONCLUSION

Complementary devices play an important role in the development of both bipolar and MOS OPAMPs. In the case of bipolar OPAMPs, which were first built exclusively with NPN transistors, the introduction of PNP transistors helped a great deal in achieving higher performance circuits. Similarly, the additional use of P-channel transistors in MOS OPAMPs yielded such improvements that it seems that in the future all analog LSIs will be constructed using CMOS technology. One might then ask: Why bother learning how to design in NMOS if CMOS is available? The answer to this question lies in the following two arguments:

(1) NMOS is still widely used for purely digital systems because of its higher density. When it is necesary to include a few OPAMPs in such LSIs, they obviously have to be NMOS.

(2) If one learns to optimize all the options in NMOS which is limited to a single transistor, then one will be more likely to be able to take full advantage of CMOS to achieve even better circuits.

The OPAMPs circuits presented in this chapter do not intend to be comprehensive but rather serve as examples of some of the many possible approaches to circuit design. As a partial list of other NMOS configurations we have:

(a) All enhancement amplifiers: [16], [18], [19],

(b) Enhancement/depletion amplifiers: [22], [3], [17],

(c) Amplifiers with impedance bootstrapping: [20], [21],

(d) Chopper stabilized amplifiers: [5],

(e) Differential amplifiers: [5], [4], [23],

(f) Dynamic amplifiers: [19].

REFERENCES

[1]   J. L. McCreary, and P. R. Gray, "All-MOS charge redistribution analog-to-digital conversion techniques" -part I," *IEEE Journal of Solid-State Circuits*, Vol. SC-10, No. 6, pp. 371-379, December 1975.

[2]   R. E. Suarez, P. R. Gray, and D. A. Hodges, "All-MOS charge redistribution analog-to-digital conversion techniques" -part II," *IEEE Journal of Solid-State Circuits*, Vol. SC-10, No. 6, pp. 379-385, December 1975.

[3]   B. J. Hosticka, R. W. Brodersen, and P. R. Gray, "MOS sampled-data recursive filters using switched capacitor integrators," *IEEE Journal of Solid-State Circuits*, Vol. SC-12, No. 6, pp. 600-608, December 1977.

[4]   D. Senderowicz, S. F. Dreyer, J. M. Huggins, C. F. Rahim, and C. A. Laber, "A family of differential NMOS analog circuits for a PCM codec-filter" chip," *IEEE Journal of Solid-State Circuits*, Vol. SC-17, No. 6, pp. 1014-1023, December 1982.

[5]   K. C. Hsieh, P. R. Gray, D. Senderowicz, and D. G. Messerschmidt, "A low-noise chopper-stabilized differential switched-capacitor" filtering technique," *IEEE Journal of Solid-State Circuits*, Vol. SC-16, No. 6, pp. 708-715, December 1981.

[6]   B. J. Hosticka, "Dynamic CMOS amplifiers," *IEEE Journal of Solid-State Circuits*, Vol. SC-15, No. 6, pp. 887-894, December 1980.

[7]   J. A. Guinea, and D. Senderowicz, "A differential narrow-band switched-capacitor filtering technique," *IEEE Journal of Solid-State Circuits*, Vol. SC-17, No. 6, pp. 1029-1038, December 1982.

[8]   A. I. Zverev, **Handbook of filter synthesis**. Wiley: New York, 1967. pp. 157.

[9]   T. C. Choi, R. T. Kaneshiro, R. W. Brodersen and P. R. Gray, "High frequency CMOS switched-capacitor filters for communications," *ISSCC Digest of Technical Papers*, pp. 246-247, February 1983.

[10]   K. Martin, and A. Sedra, "Effect of operational amplifier finite gain and bandwidth on" switched-capacitor filter performance," *IEEE Transactions on Circuits and Systems*, August 1981.

[11]   R. W. Brodersen, P. R. Gray and D. A. Hodges, "MOS switched-capacitor filters," *IEEE Proceedings*, Vol. 17, pp. 61-75, January 1979.

[12]   A. S. Grove, **Physics and Technology of Semiconductor Devices**. Wiley: New York, 1967.

[13]   H. Shichman, and D. A. Hodges, "Modeling and simulation of insulated-gate field-effect transistors," *IEEE Journal of Solid-State Circuits*, Vol. SC-3, pp. 285-289, September 1968.

[14]   P. R. Gray, "Basic MOS operational amplifier design - An overview," **Analog MOS Integrated Circuits**. IEEE Press, pp. 28-49, 1980. Vol. SC-16, No. 4, pp. 253-260, August 1981.

[15]   W. Heinlen, and H. Holmes, **Active Filters for Integrated Circuits** Englewood Cliffs, NJ, Prentice-Hall, 1974.

[16]   Y. P. Tsividis, and P. R. Gray, "An integrated NMOS operational amplifier with internal compensation," *IEEE Journal of Solid-State Circuits*, Vol. SC-11, pp. 748-753, December 1976.

[17]   D. Senderowicz, D. A. Hodges, and P. R. Gray, "A high-performance NMOS operational amplifier," *IEEE Journal of Solid-State Circuits*, Vol. SC-13, No. 6, pp. 760-766, December 1978.

[18]   I. A. Young, "A high-performance all-enhancement NMOS operational amplifier," *IEEE Journal of Solid-State Circuits*, Vol. SC-14, No. 6, pp. 1070-1076, December 1979.

[19]   J. T. Caves, C. H. Chan, S. D. Rosenbaum, L. P. Sellars, and J. B. Terry, "A PCM voice codec with on-chip filters," *IEEE Journal of Solid-State Circuits*, Vol. SC-14, No. 1, pp. 65-73, February 1979.

[20]   D. Senderowicz, and J. M. Huggins, "A low-noise NMOS operational amplifier," *IEEE Journal of Solid-State Circuits*, Vol. SC-17, No. 6, pp. 999-1007, December 1982.

[21]   E. Toy, "An NMOS operational amplifier," *ISSCC Digest of Technical Papers*, pp. 134-135, February 1979.

[22]   Y. P. Tsividis, and D. L. Frazer, "A process insensitive NMOS operational amplifier," *ISSCC Digest of Technical Papers*, pp. 188-189, February 1979.

[23]   W. L. Eversole, D. J. Mayer, P. W. Bosshart, M. De Wit, C. R. Hewes, and D. D. Buss, "A completely integrated thirty-two-point chirp Z transform," *IEEE Journal of Solid-State Circuits*, Vol. SC-13, No. 6, pp. 822-831, December 1978.

7

ANALOG-DIGITAL CONVERSION TECHNIQUES FOR

TELECOMMUNICATIONS APPLICATIONS

PAUL R. GRAY and DAVID A. HODGES
Department of Electrical Engineering
and Computer Sciences
University of California
Berkeley, California

1. INTRODUCTION

A key element of the continuing progress in improving the performance capabilities of voice and data communications systems has been the application of digital VLSI logic and memory technology to the switching, transmission, storage, and processing of voice, data, and video information. The representation of data in digital form is fundamental to the operation of these systems. However, such systems must often interface with analog signal sources, the most important example being voice information. Other examples are voiceband data signals which have complex phase, frequency, and/or amplitude modulation applied as well as phase and amplitude distortion introduced by the transmission medium, data signals transmitted on wire pairs which have experienced severe phase and frequency distortion and thus must be extensively processed prior to re-interpretation as digital information, and finally video information, including both image information and data which has been phase, frequency, and/or amplitude modulated on a video channel in a broadband local area network system.

This chapter covers techniques for the circuit implementation of several different classes of MOS analog-digital converters which find application in telecommunications systems. In section 2, the various requirements placed on such converters in different types of applications is discussed. In Section 3, circuit techniques for implementing linear and companding successive approximation converters are discussed. In section 4, the design of algorithmic converters is discussed. In section 5, the important topic of oversampled coders using noise shaping is covered. In section 6, parallel converters for video applications, and in chapter 7, several important limits to performance in A/D converters are discussed.

2. THE ROLE OF A/D CONVERTERS IN TELECOMMUNICATIONS SYSTEMS

Because of the large number of analog-digital interfaces in modern telecommunications systems, the development of low-cost monolithic MOS analog-digital conversion techniques has been an important element in reducing the cost of digital transmission and switching systems, particularly in telephony where a very large number of interfaces are required. The role of these interfaces in VLSI digital systems in general is depicted in Fig. 1. Important types of interfaces include the interface to transducers and actuators as used in industrial control, instrumentation and robotics, the interface to storage media such as tape, disk, and bubble, and the interfaces to transmission media. In this chapter we will focus on those interfaces of particular importance in telecommunications.

Perhaps the most important analog-digital interface from a commercial viewpoint is that to the analog subscriber loop in digital switching systems. Presently, most new telephone digital switching systems are implemented with per-line codec/filter chips on the line card, the chip containing the A/D, D/A, anti-alias filtering, line driver, and miscellaneous digital control, sequencing, and data storage functions. Presently most such codec chips utilize companding charge-redistribution coders with switched capacitor filters, but as MOS technology feature sizes continue to fall oversampling coders with digital filters will become more competitive

In voiceband data modems, half duplex modems at data rates above 1200bps typically utilize adaptive equalization to compensate for the phase and amplitude distortion of the line, and as a result are best implemented with at least the adaptive portion of the required signal processing realized digitally. An analog-digital converter with resolution in the 8-10 bit range is thus required. For full duplex modems, the echo problem requires either that extensive analog bandsplit filtering be included or, if the filtering is to be done digitally, that a 12 to 14 bit linear A/D converter be incorporated in front of the digital filter.

In digital subscriber loops and digital PBX's, the digital transmission of baseband data over long twisted pairs introduces phase distortion, which in turn can introduce large intersymbol interference lasting over many symbol intervals. For half duplex transmission, the equalization required may be most easily implemented in the analog domain, but for full duplex transmission the combination of the echo cancellation function together with the equalization required dictates that extensive digital processing be used. The optimum architecture for such a full-duplex transceiver is still debatable but will likely involve analog-digital and/or digital-analog converters operating at the symbol rate, which is in the range of 100 to 200khz.

Digital processing, storage, and transmission of video information requires analog-digital converters operating at sample rates of 20Mhz and above and resolution levels of 8 bits.

Each of these applications requires a conversion function with a unique combination of sampling rate, quantization noise as a function of signal level (ie companding or non-companding), and linearity error. In the next section circuit techniques for implementing several classes of such converters will be described.

3. SUCCESSIVE APPROXIMATION DATA CONVERTERS

A typical analog-digital converter system is illustrated in fig. 2. Usually some level of anti-aliasing filtering is required at the input, and a gain function, often programmable, is often required in the signal path in order to match the signal distribution to the converter input range. A multiplexer is incorporated in multi-channel systems, and a reference potential is required to establish the full-scale potential. A sample-hold function is required to determine the sampling instant, and for the case of a successive approximation converter as shown here the combination of voltage comparator, digital-analog converter(DAC), and successive approximation logic are used to carry out a quantization of the sampled value. All of these functions are important in determining the performance of the overall interface, but due to space limitations the discussion will be limited to the digital-analog converter, sample/hold, and voltage comparator.

213

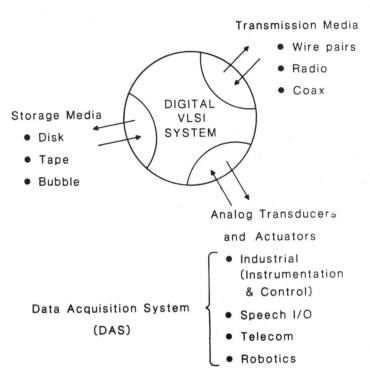

Transmission Media

- Wire pairs
- Radio
- Coax

DIGITAL
VLSI
SYSTEM

Storage Media

- Disk
- Tape
- Bubble

Analog Transducers
and Actuators

Data Acquisition System
(DAS)

- Industrial
  (Instrumentation
  & Control)
- Speech I/O
- Telecom
- Robotics

1. Role of Analog Circuits in LSI Digital Systems

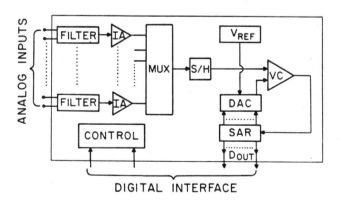

ANALOG INPUTS

FILTER → IA

FILTER → IA

MUX → S/H

V_REF

VC

DAC

SAR

D_OUT

CONTROL

DIGITAL INTERFACE

2. Typical Successive-approximation Converter System

214

3.1 High-speed DACs in MOS Technology

For digital-analog converters realized in bipolar technology, the most prevalent implementation is an array of binary weighted current sources which are selectively switched to the output under control of the logic input. The weighted currents are usually generated using an R-2R ladder of resistors, which in the case of high-resolution DACs are thin film resistors laser trimmed at wafer sort. This type of DAC will be referred to a a current-switched DAC. It is well-suited to bipolar technology because bipolar transistor emitter-coupled pairs can be used to implement very effective high-speed current switches. It is not particularly well suited to MOS technology however, because the resistance values available in practical MOS technologies are low enough that very large transistor switch geometries are required to carry the necessary currents. The most common implementations of MOS DACS utilize resistor strings, capacitor arrays, or some combination of the two.

3.1.1 Resistor-string DACs: MOS transistors are particularly useful as zero-offset analog switches, and the resistor-string DAC makes good use of this property. A typical resistor string DAC is shown in fig. 3. The tree of analog switches is used to connect the output to one of the taps on the string. For an n-bit DAC, 2^n resistor elements and 2^{n+1} transistor switches are required. However, the resistor string itself can be laid out in an x-y addressed array which is very economical in area(1). For example, a typical 8-bit resistor string DAC occupies about 1000 square mils in a 5 micron technology.

The principal advantages of the resistor string are small size for moderate resolution levels, high speed, and inherent monotonicity (ie each output level is inherently higher than the previous one). An important disadvantage is that the linearity of the transfer characteristic can be strongly affected by the differential temperature coefficient and the voltage coefficient of the resistors in the string, which are usually implemented with a diffused region. This type of DAC is particularly useful in technologies which contain no capacitor, and is used extensively in NMOS microprocessors with on-board analog-digital converters(2,3)

3.1.2 Charge-redistribution DACs: A second useful property of MOS transistors is the fact that the input resistance of the device is virtually infinite. As a result, the transistor can be used to sense the voltage on a small capacitor continuously for a period of milliseconds without discharging the capacitor and destroying the stored information. This property allows the implementation of dynamic MOS memory, switched capacitor filters, and charge-redistribution analog-digital converters.

A typical charge-redistribution ADC is shown in fig. 4. It consists of a binary weighted set of capacitors and associated switches, and a comparator. Initially, the top plate of the array is grounded and the bottom plate is connected to the input signal. When the top plate switch is opened, the signal is sampled in the form of a charge on the array.(fig. 4a) Next the bottom plates are all connected to ground, causing the top plate to go to a potential equal to $-v_{in}$.(fig. 4b) Next the largest capacitor bottom plate is attached to the reference voltage. Since this capacitor together with the rest of the array forms a 2:1 voltage divider, the top plate goes to a voltage $-v_{in}+V_{ref}/2$. By comparing this bit to ground the most significant bit can be determined. This process continues through the remaining bits.(fig. 4c) When the process is completed. the charge on the array has been redistributed and now resides only on those capacitors whose corresponding bit value is a one(fig. 4d).

The principal advantage of the charge-redistribution adc is the fact that it incorporates the sample-hold function, which would otherwise have to be implemented separately. Another important advantage, discussed later, is the fact that the memory of the array can be used to carry out an auto-calibration cycle to remove the effects of component ratio errors. Also, the tempera-

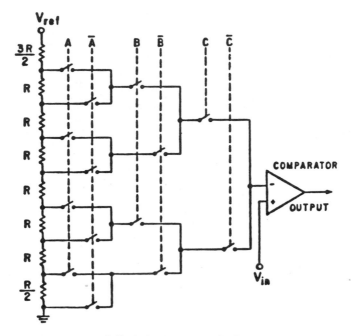

3. Typical resistor-string DAC

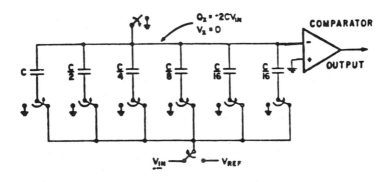

4. Charge-redistribution ADC

a. Sample mode

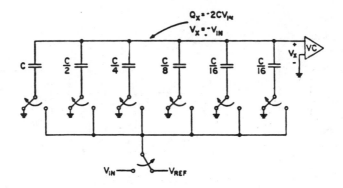

b. Hold mode

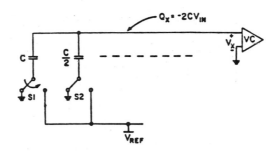

c. Redistribution/comparison

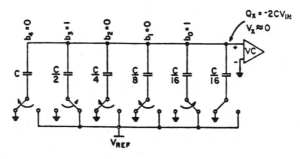

d. Final state

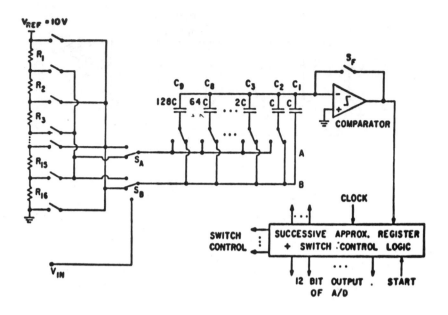

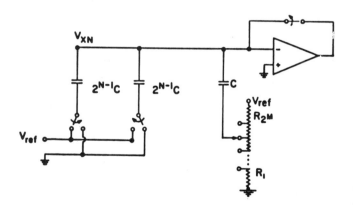

5. Extended resolution DACs

    a. Resistor string with interpolation capacitor array

    b. Capacitor array with resistor-string sub-dac

ture coefficient long-term stability of the linearity of the adc is primarily dependent on the stability of the SiO_2 capacitors. Since this dielectric is a form of quartz, the long-term stability and temperature stability is very good.

An important disadvantage of charge-redistribution converters is the requirement that the technology produce a SiO_2 capacitor with two conducting electrodes. Such capacitors occur naturally in metal gate MOS technologies and in silicon gate technologies with two layers of polysilicon. However, in technologies with one layer of polysilicon, an additional process step must usually be added to produce a capacitor of suitable quality.

One disadvantage of charge-redistribution ADCs is the fact that in its simplest form a capacitor ratio of 2^n to one is required. For a 12-bit converter, for example, this would imply a ratio of smallest to largest capacitor of 4096 to one. This disadvantage can be overcome in a number of ways, two of which are illustrated in figs. 5 a and b. In Fig. 5a, a resistor string is used as the primary DAC, and a capacitor array is used to interpolate between the taps on the resistor string(4). This configuration has the advantage that the ADC is monotonic as long as the low-resolution capacitor array is monotonic, and has been used for example to implement a 16-bit data converter system for signal processing applications. In Fig. 5b, a capacitor array is used as the main DAC and a resistor string is used to drive the bottom plate of the smallest capacitor. This configuration is not inherently monotonic.

3.1.3 Companding Converters: A basic property of human speech perception is that a relatively wide dynamic range of signal amplitudes must be accomadated. If a linear coder were used to encode speech signals, approximately 13 bits of resolution would be required to obtain acceptable quantization noise at the lowest signal level in telephone systems. Fortunately, at higher signal levels the quantization noise can be allowed to increase without significantly reducing perceived speech reproduction quality. Because of this fact the quantization step size is allowed to increase with increasing sample amplitude in voiceband coders used in telephone PCM switching and transmission systems. The so-called u=255 law is standard in North America and Japan, while Europe uses a slightly different standard, the A-law. Both provide a wide dynamic range over which the signal-to-quantizing ratio is approximately constant, and a small step size near the origin so that the idle channel is quiet and crosstalk signals are not enhanced. The μ 255 chord structure, shown in fig. 6, consists of segments of 16 steps each, with each successive segment having twice the step size of the previous one as one moves away from the origin. There are a total of 256 steps and 8 bits are required to specify one unique level from this set.

An important development in integrated circuits for digital telephone systems was the development of a simple DAC which implements a companding characteristic. This was first done in bipolar technology using a binary weighted current source array and a 16-step current divider as shown in fig. 7 (6). The same type of characteristic can be realized in MOS by combining an 8-bit binary weighted capacitor array and a 4-bit resistor string in a configuration like the one shown in fig. 5b(7). Here the resistor string is first used to step the voltage applied to the bottom plate of the smallest capacitor from zero to Vref in 16 steps. Then the bottom plate of the smallest capacitor is tied to Vref and the resistor string is used to step the bottom plate of the next smallest capacitor from zero to Vref. The step size observed on the capacitor top plate has doubled. This process is repeated for the rest of the capacitors, constructing the u-law characteristic. This circuit is used in the majority of commercially available PCM codecs at the present time.

4. ALGORITHMIC ANALOG-DIGITAL AND DIGITAL-ANALOG CONVERTERS

The basic function of the capacitor array in the converters described above was to develop voltage levels related to each other by a precise ratio of two. A successive approximation type

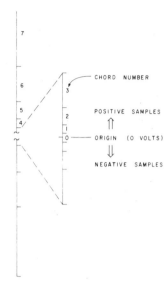

CHORD NUMBER

POSITIVE SAMPLES
⇑
ORIGIN (0 VOLTS)
⇓
NEGATIVE SAMPLES

6. u255 Chord Structure

$\downarrow I_{OUT}$

$(1 - \frac{n}{16}) I_x \downarrow$ $\downarrow \frac{n}{16} I_x$

16-STEP CURRENT DIVIDER } 4-BIT DIGITAL STEP #

$\downarrow I_x$

128 I | 64 I | 32 I | 16 I | 8 I | 4 I | 2 I | I

CURRENT SOURCE ARRAY V^-

ANALOG OUT

AT THIS POINT I GOES DIRECTLY TO OUTPUT

I

DIGITAL IN

7. Current-switched Companding DAC

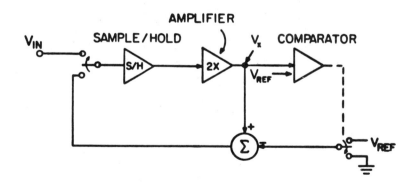

V_{IN}

SAMPLE/HOLD AMPLIFIER V_x COMPARATOR

S/H 2X V_{REF}

Σ + − V_{REF}

8. Block Diagram of Algorithmic Converter

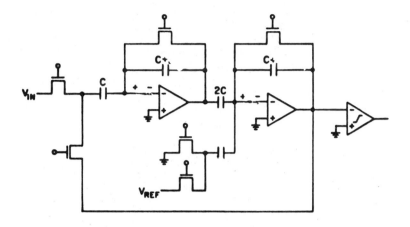

9. Typical Circuit Implementation of an Algorithmic Converter

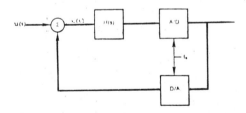

10. Block diagram of differential feedback coder.

221

algorithm can also be carried out serially in time by sucsessively multiplying the signal by two, comparing the result to the reference voltage, and if it is greater subtracting the reference potential off and setting that corresponding bit to a 1. The process is then repeated on the remainder until all of the bits are decided. A block diagram of a typical algorithmic converter is shown in fig. 7 This approach to analog to digital conversion was first proposed in the 1950s, and was first implemented in a monolithic prototype by McCharles.[8] Due to its simplicity, it offers the potential of significant area reductions relative to other approaches to linear coding, and has the additional advantages that it is inherently a floating point converter, that it inherently incorporates the precision amplifier function since the signal can be recirculated an arbitrary number of times before A/D conversion begins, and that various self-calibration techniques can be easily applied to it. A typical implementation of an algorithmic converter is shown in fig. 9.

Algorithmic conversion has not received wide application to date for several reasons. The first is that it is heavily dependent on the settling time and gain of MOS operational amplifiers, and it has only been recently that the performance of amplifiers has reached a reasonable level. Second, the linearity of the converter can be severely degraded by offsets from the operational amplifiers and from charge injection in the loop. Thirdly, the linearity of the converter depends on capacitor ratios just as in the case of capacitor arrays. These problems can be addressed using improved offset cancellation and automatic calibration approaches, to be discussed later.[9]

5. OVERSAMPLED CODERS

The quantization noise introduced in the a/d conversion process is approximately white and is uniformly spread from dc to half the sampling rate, assuming the signal amplitude is large compared to the step size. Since the total energy contained in the quantization noise is a constant determined by the step size, that portion of the noise lying in a given spectrum can be reduced by increasing the sample rate, at the rate of 3dB per octave of sample rate increase. Viewed another way, if the number of samples taken per unit time is doubled and each pair of samples produced is averaged to produce new samples at the original sample rate, the signal component of the samples will add linearly while the quantization noise component in the samples will add as uncorrelated random variables. Thus the signal-to-noise ratio will improve by 3dB.

In principle, one could utilize the approach of simply increasing the sampling rate to directly improve the snr in, for example, speech encoders. There are two drawbacks to this, however. The first is that the rate of improvement is relatively slow; In order to get the equivalent of 1 additional bit of resolution the sample rate must be increased by a factor of 4. A second reason is that the underlying assumption that the signal amplitude is much larger than the step size will be violated for small signals if this process is carried very far. Both of these drawbacks can be solved by incorporating the quantizer in a feedback loop and having it operate on the difference between the signal and a recent estimate of the signal (differential coding). A general block diagram of this type of coder is shown in fig. 10. By incorporating an integrator or more complex transfer function in the forward portion of this feedback loop it is possible to make the noise transfer function to the output different from that of the signal and effectively introduce a zero at the origin in the noise transfer function.

Commercial application of this type of coder to date has focussed principally on two variations, interpolative coders[11] and delta-sigma coders.[12] The principal difference between them is that the interpolative coder incorporates a companding DAC in the feedback loop which allows a lower sampling rate. Delta-sigma coders utilize a 1-bit DAC as the feedback element and thus must operate at higher sampling rates.

5.1 Interpolative Coders

A block diagram of a typical interpolative coder is shown in fig. 11. On each clock cycle the comparator decides if the integrator output is positive or negative, and increments the shift register up or down. During the following clock period, the integrator integrates the difference between the instantaneous input signal and the current DAC output value. At the start of the next clock cycle the shift register is again clocked up or down. The result is a waveform at the DAC output which constantly switches, or interpolates, between the output levels of the DAC, as shown in fig. 12. The average value of this waveform is a representation of the signal. The sample rate for such coders is in the 256khz range for voiceband signals.

The principal advantage of interpolative coders is that the effective sampling rate is much higher than in the successive approximation case. This introduces the possibility of using digital filtering to perform the anti-aliasing function as the sampling rate is decremented to 8khzfor switching or transmission. The analog filtering required prior to signal sampling then becomes very simple. The hardware complexity of an interpolative coder is similar to that of a conventional charge-redistribution codec. A key disadvantage is the fact that the DAC used in the feedback loop is companded. This means that the coder displays a quantization noise which increases with signal level, just as in the case of a successive approximation companding ADC. In some important applications for oversampling coders, such as all-digital full duplex modem front ends, this is an important disadvantage. The interpolative coder has been incorporated in certain commercially available single-chip half-duplex modems and codec/filters.(11,12)

5.2 Delta-Sigma Coders

The principle difference between delta-sigma coders and the interpolative coder is the fact that the DAC in the feedback loop in the delta-sigma case is a one-bit DAC producing only two levels. The resulting waveform when a sinusoid is encoded is shown in fig. 14. Note that this waveform is a square wave of amplitude equal to the full scale reference voltage, and as a result the total energy in the waveform is quite large. When the signal is small, the quantization noise energy is larger than the signal by a factor which is much larger than in the interpolative case.

The delta-sigma coder has advantages in hardware simplicity and high sampling rate, which allows more use of digital anti-alias filtering. Most of the quantization noise can be moved out-of-band by the use of an appropriate filter in the forward path. However the large amplitude of the out-of-band quantization noise requires a more complex anti-alias filter prior to decimation than in the case of interpolative coding. Delta-sigma coders must sample significantly faster than interpolative coders to achieve adequate in-band signal-to-noise ratios for telephony application. Another important aspect of delta-sigma coder performance is the fact that for low signal amplitudes the DAC output waveform can contain " noise" components which are not random, but which are concentrated at one frequency and result in spurious tones. It is necessary to randomize these signals and this is often accomplished by the introduction of a jitter signal.

The principal practical advantage of oversampled coders is the fact that their sampling rate is high and more digital anti-alias and reconstruction filtering can be used. Since digital filters can directly take advantage of scaled technologies, it is likely that oversampled coders with digital filters will play a more important role in future telecommunications analog-digital interfaces. The rate at which this occurs, however, is dependent on to what extent current circuit approaches involving some analog filtering can take advantage of technology scaling and preserve their current area advantage.

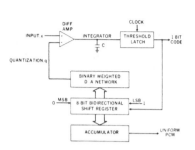

11. Block diagram of interpolative coder

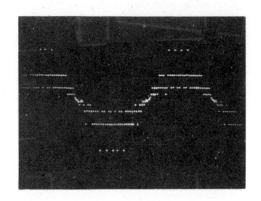

12. Typical DAC output waveform resulting

from quantization of a sine wave.

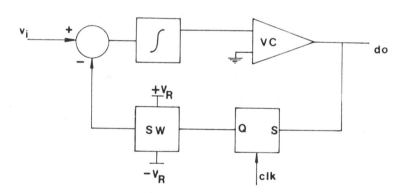

13. Block diagram of a typical delta-sigma coder

6. LIMITS TO CONVERSION RATE AND ACCURACY IN ANALOG-DIGITAL CONVERTERS

Independent of the particular approach used, properties of the technology used to implement the analog-digital converter place upper limits on the performance achievable in the analog-digital converter implemented in it. While one can identify a few fundamental limits, the performance of currently available or projected converters is far from these limits. Certain practical limits are likely to be more important in the short term. In this section we review some of the practical and fundamental limits to achievable performance.

6.1 Some Fundamental Limits to Converter Performance

In the context of implementation in MOS technology, it is possible to identify ultimate limits to dynamic range and speed of analog-digital converters. We first consider noise.

6.1.1 Limits to dynamic range: The process of sampling a signal requires that it be represented as a packet of energy, and in a monolithic sample/hold the packet of energy is normally stored on a capacitor as shown in fig. 16. The thermal noise in the switch resistance is sampled onto the capacitor and appears as a random component of the sampled value with a variance equal to kT/C. The only factor affecting the value of this noise is the value of the sampling capacitor. The value of the signal energy stored on the capacitor is $CV^2/2$ while the value of the noise energy stored on the capacitor is $kT/2$. This thermal noise phenomena represents a fundamental limit on the per-sample dynamic range achievable in monolithic sample/hold amplifiers as the technology is scaled down and capacitor size is decreased. The value of kT/C for various capacitor sizes is shown in fig. 17. At the present time, a/d converters used for telecommunications applications utilize step sizes which are far larger than these values.

6.1.2 Limits on Comparator Delay Time: An essential element of all analog-digital converters is a comparator. The time of conversion is limited by many factors but can be no shorter than some multiple of the comparator delay. This multiple is unity for parallel converters, n for successive approximation converters where n is the number of bits, and 2^n for serial converters.

The design of a practical comparator is a complex process process involving many tradeoffs. The number of gain stages must be selected, and in the MOS case an offset cancellation strategy must be selected. The response of the comparator when initially precharged to a condition far from the threshold and subsequently driven by a small overdrive must be optimized. All these factors will have an effect on the comparator delay actually achieved.

A lower limit on comparator delay can be obtained by considering a simplified case as shown in fig. 18. The following set of simplifying assumptions are made:

The comparator is considered to be a series of identical MOS source-coupled pairs.

Parasitic capacitances are neglected as are transistor drain-gate capacitances so that the only capacitances present are the transistor gate- source capacitances.

The internal nodes are initialized to the balanced condition at t=0, and an input voltage v_{in} is then applied.

During the portion of the transient of interest the differential stages remain in their linear region of operation with approximate constant g_m.

The load impedances are high enough that the time constants at all drain nodes is much longer than the delay time of the comparator.

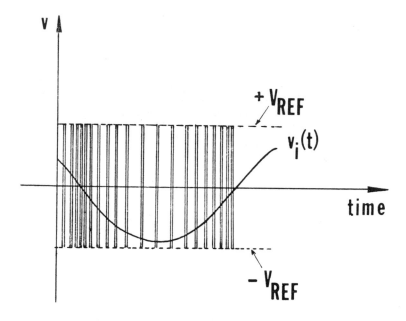

14. Typical DAC output waveform resulting from quantization of a sine wave.

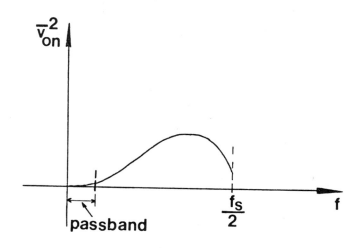

15. Typical output noise spectrum of delta-signa coder.

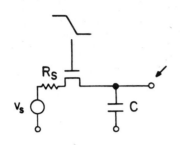 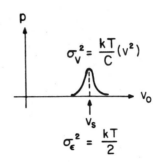

16. Typical MOS sample/hold and probability distribution of resulting sample.

C	$\sqrt{kT/C}$
100 pF	$6.4\mu V$
10 pF	$20.2\mu V$
1 pF	$64\mu V$
0.1 pF	$202\mu V$

17. Residual thermal noise voltage vs capacitor size.

Under this last assumption the individual stages can be considered to behave as an integrator for the time period of interest, as illustrated in fig. 18. It is then possible to obtain a closed-form expression for the output voltage as a function of time, the number of stages n, and the input drive voltage, and from this determine the optimum number of stages for minimum delay with a given drive level at the input and desired logic level at the output.(14) This optimum number of stages is illustrated in Fig. 19. The minimum comparator delay achievable under these idealized conditions is illustrated in fig. 20. It is approximately equal to the transit time of the devices used multipled by the natural log of the desired gain. It is interesting to note that this result is the same as would be obtained if the single-stage latching comparator were used.

The results of this simple analysis indicates that a comparator delay of a small multiple of the device transist time should be achievable. For a 5 micron channel length nmos transistor operated at a V_{dsat} of 1 volt, the transit time is approximately 0.3ns, giving a comparator delay on the order of 2ns for a gain of 1000. Practical comparators display delay times much longer than this. Factors such as parasitic capacitance and overdrive recovery play a predominant role in limiting achievable delay.

6.2 Some Practical Limits to Analog-Digital Converter Performance

As can be seen from the above discussion, monolithic analog-digital converters are at the present time far from the identifiable fundamental limits to speed and dynamic range imposed by comparator delay and kT/C noise. Presently achievable performance levels tend to be dictated more by practical circuit, technology, and packaging limitations. Examples include undesired coupling between the digital and analog portions of the circuitry within the converter, problems of charge injection from MOS analog switches, offsets in operational amplifiers and comparators, and so forth. Space does not permit discussion of these, but the following section addresses one particular practical limitation, that of component matching.

6.2.1 Component Mismatch Effects in Monolithic DACs: Perhaps the single most important factor limiting the linearity of successive approximation analog-digital converters is the component mismatch problem. Virtually all successive approximation converters utilize a digital-analog converter which displays a linearity which is dependent on the precise matching of passive components. A binary weighted capacitor array is shown as an example in fig. 21. Here, a mismatch between the largest capacitor and the rest of the capacitor array results in a nonlinearity in the transfer characteristic at the major carry point. In order to achieve 1/2 lsb differential linearity at the major carry, approximately 0.4% matching is required for 8-bit resolution, 0.1% for 10 bits, and 0.025% for 12 bits. In the as-fabricated state with no trimming of ratios, a typical capacitor array would display an average mismatch of largest capacitor to the rest of the array of perhaps 0.1 to 0.2% due to processing gradients, random processing variations, and limited lithographic resolution. Thus the fabrication of DACs at resolutions of 10 bits or above usually requires some kind of trimming or calibration.

For most telecommunications applications, integral linearity on the order of 8 bits is adequate. Thus the fabrication of PCM companding codecs, for example, does not require trimming of DAC linearity. However, applications in full duplex data communications over wire pairs, both in voiceband modems and in subscriber loops, may require substantially higher resolution and linearity, particularly if the echo cancellation or band separation filtering is to be done digitally.

Two recent developments appear to offer promise for elimination of trimming with its associated cost and technology constraints in high-resolution DACs. The first, called self-calibration, utilizes the memory capability of a binary weighted capacitor array to measure the ratio errors of the individual capacitor elements so that a correction can be made to the DAC output using a separate correction DAC. The basic concept of the measurement is illustrated in fig. 22, where two

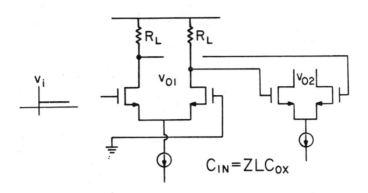

$$C_{IN} = ZLC_{OX}$$

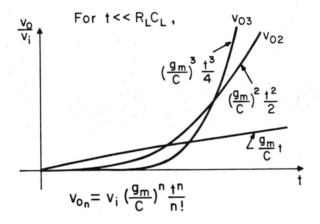

For $t \ll R_L C_L$,

$$\left(\frac{g_m}{C}\right)^3 \frac{t^3}{4}$$

$$\left(\frac{g_m}{C}\right)^2 \frac{t^2}{2}$$

$$\frac{g_m}{C} t$$

$$V_{O_n} = V_i \left(\frac{g_m}{C}\right)^n \frac{t^n}{n!}$$

18. Simplified approximate model for multi-stage comparator.

229

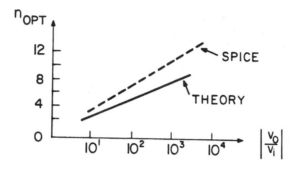

19. Optimum number of stages as a function of desired gain from a comparator delay standpoint under idealized conditions.

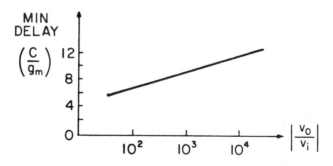

20. Minimum comparator delay time as a function of desired gain.

supposedly equal capacitors are connected such that their top plates are at ground, one bottom plate is connected to a reference voltage, and the other bottom plate is connected to ground. If the top plate switch is opened and the bottom plate voltages are interchanged, the top plate voltage should be unchanged if the two capacitors are equal. If there is a ratio error, the residual top plate voltage will be proportional to this error, and this voltage can be measured and used to quantize the ratio error. (16) In a binary weighted capacitor array DAC, the ratio error in the largest capacitor is measured by doing the above experiment using the largest capacitor as one element and the rest or the array as the other. The residual error remaining on the top plate of the array is encoded by carrying out a normal A/D conversion following the switching sequence described above. The result is stored in a RAM on chip. The same process is repeated for the remaining bits as required. Subsequently, for a given bit pattern supplied to the DAC, the digital correction values for the non-zero bits are summed and applied to an auxiliary correction DAC. to correct the linearity of the output. Experimental versions of such converters have achieved integral linearity of 1/2 lsb at 15 bits (16). The block diagram of such an experimental converter is shown in fig. 23.

A second approach to realizing high integral linearity without trimming is to utilize the inherent capability of a pair of capacitors to multiply a signal by a factor of two without dependence on component ratios. This can be accomplished, for example, by charging a pair of capacitors in parallel to a voltage, and then connecting the capacitors in series to give twice the voltage. Unfortunately this process is sensitive to capacitive parasitics present at the nodes of the capacitors. A more practical approach, illustrated in fig.24, utilizes a switched capacitor integrator. The input signal is integrated twice, and then the sampling and integrating capacitors are interchanged. This gives an output voltage which is twice the input independent of the capacitor ratio. (17,9)

The ratio-independent multiply-by-two is most applicable in algorithmic converters. When combined with careful cancellation of offsets due to charge injection and operational amplifier offset, it is capable of achieving linearity in the 13 bit range with sampling rates in the 30ksamples/second. (9)

It appears that these techniques or others which may evolve will allow the realization of monolithic A/D and D/A converters of high linearity without the need for thin film resistors or costly, time consuming laser wafer sort trimming.

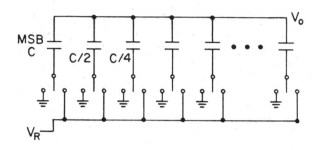

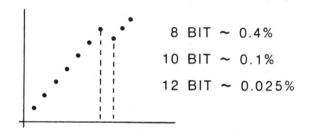

8 BIT ~ 0.4%

10 BIT ~ 0.1%

12 BIT ~ 0.025%

DIFF. NONLIN. ~ $\dfrac{\Delta C}{2C}$

AT MAJOR CARRY

21. Mismatch effect on linearity of a binary weighted capacitor array DAC.

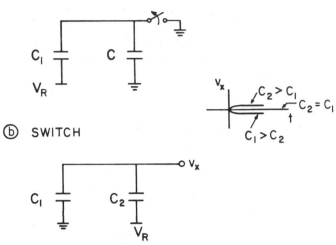

22. Conceptual example of measurement of the ratio error between two nominally identical capacitors.

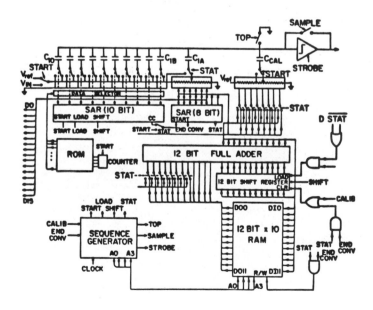

23. Block diagram of experimental self-correcting analog-digital converter.

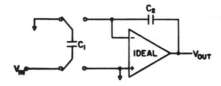

- **INTEGRATE V_{IN} TWICE**

$$V_{OUT} = 2\frac{C_1}{C_2}V_{IN}$$

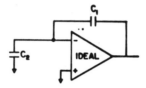

- **EXCHANGE CAPACITORS**

$$V_{OUT} = \left(2\cdot\frac{C_1}{C_2}\right)\cdot\left(\frac{C_2}{C_1}\right)V_{IN}$$

$$= 2\cdot V_{IN}$$

24. Idealized ratio-independent multiply-by-two

REFERENCES

1. M. E. HOFF, J. HUGGINS,and B. M. WARREN, " An NMOS Telephone Codec for Transmission and Switching Applications" , IEEE Journal of Solid-State Circuits, vol SC-24, February, 1979

2. M. E. HOFF and M. TOWNSEND, " An Analog Input-Output Microprocessor" , Digest of Technical Papers, 1979 International Solid State Circuits Conference, February, 1979

3. Intel 8022 Application Note, Intel Corporation, 3065 Bowers Avenue, Santa Clara, California, 95051

4. B. FOTOUCHI and D. A. HODGES, "An MOS 12b Monotonic 25us A/D Converter." Digest of Technical Papers, 1979 International Solid- State Circuits Conference, February, 1979

5. D. HESTER, K. S. TAN, and C. R. HEWES, " A Monolithic Data Acquisition Channel" , Digest of Technical Papers, 1983 International Solid State Circuits Conference, February, 1983

6. J. A. SHOEFF, " A Monolithic Companding D/A Converter" , Digest of Technical Papers, 1977 International Solid State Circuits Conference, February, 1977

7. Y. P. TSIVIDIS, P. R. GRAY, D. A. HODGES, and J. CHACKO, " A Segmented μ 255 Law PCM Encoder Using NMOS Technology" , IEEE Journal of Solid State Circuits, vol SC-10, December 1975

8.R. H. MC CHARLES and D. A. HODGES, " Charge Circuits for Analog LSI," IEEE Transactions on Circuits and Systems, Vol SC-25, No. 7, July, 1978

9. P. W. LI, M. CHIN, P. R. GRAY, and R. CASTELLO, " A Ratio-Independent Algorithmic Analog-Digital Conversion Technique, Digest of Technical Papers, 1984 International Solid-State Circuits Conference, February, 1984

10. B. A. WOOLEY and J. C. CANDY, " An Integrated Per-channel Encoder Based on Interpolation" , IEEE Journal of Solid-State Circuits, vol SC-14, February, 1979

11. J. D. EVERHARD, " A Single-Channel PCM Codec," IEEE Transactions on Communications, COM-27, February, 1979

12. Application Note, AMD 7901/7902 Subscriber Line Audio Processing Circuit, Advanced Micro Devices, 901 Thompson Place, Sunnyvale, California, 94086, November, 1891

13. Application Note, AMD 7910 Modem, Advanced Micro Devices, 901 Thompson Place, Sunnyvale, California, May, 1983

14. B. L. TIEN, " Optimum Design of MOS Voltage Comparators" , MS Thesis, Department of Electrical Engineerng and Computer Sciences, University of California, Berkeley, California, January, 1982

15. P. KWOK, " Optimum Architectures for MOS Voltage Comparators" , to be published

16. H-S LEE, D. A. HODGES, and P. R. GRAY, " A Self-Calibrating 12b, 12us CMOS ADC" Digest of Technical Papers, 1984 International Solid-State Circuits Conference, February, 1984

17. C. C. LEE," A New Switched-Capacitor Realization for a Cyclic A/D Converter" , Proceeding of the 1983 International Symposium on Circuits and Systems, Newport Beach, California, April, 1983.

8

MOS DIGITAL FILTER DESIGN

WALTER ULBRICH
SIEMENS AG
CENTRAL RESEARCH AND DEVELOPMENT
OTTO-HAHN-RING 6
D-8000 MÜNCHEN 83

1. INTRODUCTION

Since the early seventies a lot of research has been done in the area of sampled-data filters. The volume of publications, both in the theoretical field and in circuit design, gives proof of the efforts made. It has been shown that MOS technology is well-suited for an efficient implementation of analog filters using CCD, switched capacitor, or continuous time techniques.

At the present time, progress in MOS technologies and design time for custom digital VLSI tend to favor the realization of digital filters. Besides the known technical advantages (greater insensitivity to technology changes, better long time and temperature stability, definite signal to noise ratio), economic reasons increasingly support the extensive use of digital signal processing techniques. Although we restrict ourselves to the design of digital filters in this chapter, we have to recognize that decisions about analog/digital, multiplexing, architecture, technology, design style and so on will be influenced by global considerations when implementing a whole system on a chip.

Nevertheless, digital filters are good examples for demonstrating different approaches for an efficient implementation of digital signal processing algorithms into silicon. There are a lot of real-time applications at sample rates ranging over about four orders of magnitude, from 8 kHz for PCM or voice systems and speech recognition, approximately 50 kHz for audio, a few MHz for telecommunication applications, 20 MHz for video and imaging and up to 70 MHz for fast digital modems. It is obvious that different architectures and multiplexing schemes must be developed to match such different system requirements with a given MOS technology. The goal is to use an available technology that will meet the speed requirements and to optimize silicon area, power consumption, and design time.

A rough classification of fundamentally different implementations of digital filters together with general remarks about the necessary interactions between filter synthesis and MOS integrated circuit design is given in Section 2. Hardwired FIR and IIR digital filters for the highest sample rates utilizing parallelism and extensive pipelining are described with detailed chip-level information in Section 3. Architectures for customized signal processors tailored to digital filter applications with low and medium sample rates are discussed in section 4. These are based on a macrocell design approach.

2. BASIC CONSIDERATIONS

We start with the simple example of a finite-impulse response (FIR) filter (also called transversal filter) to demonstrate common aspects of digital filter implementations in VLSI MOS technology. The filter is characterized by the input/output relation

$$y_n = \sum_{k=1}^{m} c_k \, x_{n-k} \tag{1}$$

where $x_{n-k} = x(t_n - kT_s)$ are the sampled input signals and $y_n = y(t_n)$ is the corresponding output signal. T_s is the sampling period, $t_n = nT_s$ are the sample instants, and $f_s = 1/T_s$ is the sample rate.

Applying the transformation $z^{-1} = \exp(-sT_s)$ of the complex frequency s, we obtain the transfer function

$$G(z) = \frac{Y(z)}{X(z)} = \sum_{k=1}^{m} c_k z^{-k} \tag{2}$$

as a function of z. Usual representations of (1) are signal flowgraphs as shown in Fig. 1 (actually m = 5). They are composed of adders, multipliers, and delay elements. The two flowgraphs of Fig. 1a) and 1b) are equivalent with respect to the transfer function but they will be different with respect to their implementation in silicon.

Eq. (1) describes a sampled-data filter, where only time is discrete while the other quantities are continuous. When applied to digital filters, all the terms of (1) must be expressed by digital words of finite length, i.e. they must be quantized. For example, the data wordlength at the input in Fig. 1 is 10 bit. The number of bits for the data words as well as for the coefficients c_k has a strong effect on the amount of hardware required to build the

237

digital filter and must be minimized very carefully by filter
synthesis and simulation programs.

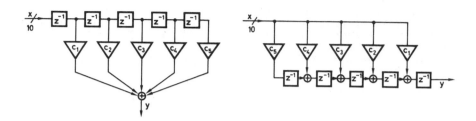

a) serial-in/parallel-out   b) parallel-in/serial-out
   structure   structure

Fig. 1: Equivalent signal flowgraphs of a FIR filter

2.1 Elementary Operations

This section is devoted to a brief review of relevant
problems with emphasis on hardware considerations. Some supple-
ments to the classical papers /1,2,3/ are necessary in order to
take into account actual ways of realizing dedicated digital
filters in silicon.

2.1.1 Data Format – For most filter applications fixed-point
arithmetic with a data wordlength of less than 24 bits is sufficient. Coefficient wordlength may differ from this. Numbers
can be represented as fractions or as integers. Floating-point
arithmetic is simpler in the case of multipliers but more complicated for add or subtract operations and therefore confined
to special filters with high Q values or an extraordinary dynamic range. Two's complement notation is usually used to handle
positive and negative fixed-point numbers. This is because
addition and subtraction are much simpler with two's complement
than with sign and magnitude notation.

In parallel-data systems, the different bits of the data
word appear on separate wires at the same time. In serial-data
systems, the bits appear sequentially, usually least significant
bit first, on one signal wire. The clock rate of serial-data
systems is d times the word rate, if d is the data wordlength.

2.1.2 Arithmetic Elements – Real-time digital processing algorithms should be implemented on powerful arithmetic units to

obtain a data throughput as high as possible. On the other hand, the arithmetic resources are very expensive in terms of silicon area and power dissipation and should be exploited as much as possible. Therefore, architecture and circuit design have to concentrate on optimal arithmetic structures tailored to the given algorithm. However, a change of the algorithm or an optimization of the flowgraph and the coefficient values can also save hardware. Only a very close interaction between filter specialists and MOS chip designers leads to an efficient realization of the arithmetic units. This will be shown later in more detail.

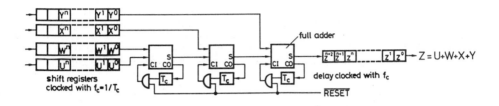

a) serial-data adders

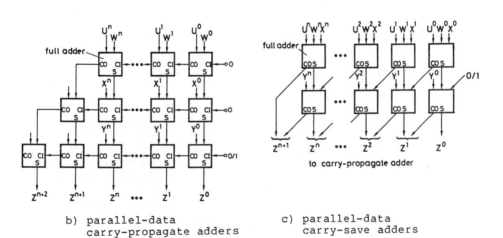

b) parallel-data
   carry-propagate adders

c) parallel-data
   carry-save adders

Fig. 2: Different schemes of adder arrays

The basic arithmetic element of digital filters is the adder. Adders are commonly classified according to whether they

handle data in serial or parallel form. Fig. 2 shows three
usual schemes of single-bit full adders. Each of them perform
the operation Z = (U+W+X+Y). The serial-data adders of Fig. 2a)
contain an additional register which takes the carry from one
bit position and applies it to the next higher bit position
as required. Parallel-data adders require as many full adders
as there are bits in the input words, including sign. Figs.
2b) and 2c) show two fundamentally different ways to connect
the carry. The carry-propagate adder computes the final result
at the sum outputs within a propagation delay that depends
on the data wordlength. The simplest realization is the ripple
adder as described in Section 3. A variety of clever additional
circuits can be used to speed up the carry propagation. Several
techniques for carry-look-ahead have been developed, as well as
for carry-select adders /4/. The arithmetic unit (AU) of a
microprogrammed processor usually contains one such fast carry-
propagate adder as shown in Fig. 3a) to perform one addition
together with a register transfer in one clock period. Three
processor cycles are necessary to accumulate Z as mentioned
above.

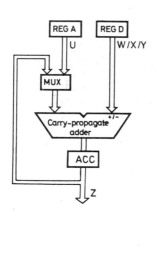

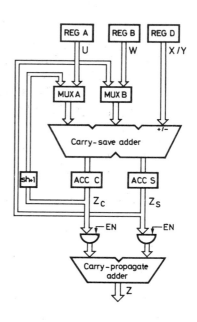

a) AU based on a   b) AU based on a
   carry-propagate adder   carry-save adder

Fig. 3: Different schemes of arithmetic units

240

Digital signal processing tasks often include a fixed sequence of additions, where only the final sum is of interest i.e. all the partial or intermediate sums need not be fully calculated. For example, Z may be the sum of products at the output of the transversal filter of Fig. 1a). In such cases the carry-save adder is recommendable. Fig. 2c) demonstrates the wiring of two rows of carry-save adders /4/. The corresponding arithmetic unit of a signal processor is shown in Fig. 3b). As the accumulation $((U+W+X)+Y)$ is performed, the partial sum is given by two parallel data words Z_S and Z_C representing sum and carry. These two processor cycles may be very short because the delay of a carry-save adder is that of a simple single-bit full adder. The final result Z in two's complement form is summed up only once with a carry-propagate adder which is stimulated by the control signal ENable, for example.

A subtraction $(U-Y)$ is realized according to the rules of two's complement arithmetic. For hardwired filters all bits Y_i of the second word Y have to be inverted, for programmable filters the EX-OR operation $(Y_i \oplus SUB)$ is previously executed at the second input with SUB = 0 for addition and SUB = 1 for subtraction. Moreover, for subtraction the carry input belonging to the least significant bit is set to logical 1.

The most important decision about the effective use of silicon area is the realization of the multiplications. A parallel-data multiplier is such an expensive arithmetic element that we can usually implement only one of it on a chip by applying available MOS technologies. Hence, we have to look for alternative, more economical solutions.

For custom fixed-coefficient designs, it is unnecessary to multiply the signal with a coefficient bit which is zero. Coefficients can further be reduced to a minimum of nonzero bits using the canonical signed digits (CSD) code /5/. Positive and negative bits are allowed because the realization of an adder or an subtractor is equivalent. Transforming coefficients given as two's complement numbers of M bit wordlength into the CSD code, the number of nonzero bits reduces to an average number of M/3. Beyond that, the filter designer must be urged to optimize the filter transfer function within the given specification in order to get coefficients with a global minimum of nonzero bits. In connection with this objective the question of the sensitivity of the filter function to coefficient variations arises in a new manner. Filter structures with low sensitivities (which are derived from classical resistive loaded reactance filters) will allow a larger change of coefficients from their nominal value to discrete values with a minimum number of nonzero bits. Special CAD programs have been developed to solve this discrete optimization problem /6,7, 8/ and help us to save adders or subtractors for hardwired filters (see Section 3) or a number of cycles for shift-and-add-based signal processors (see Section 4).

Stripping of multipliers to the minimal number of shift-and-add operations obviously cannot be utilized for adaptive filters where the coefficients are not fixed. A recoding of time-variant coefficients to CSD code would not only require the implementation of a recursive algorithm /5/ but also generates a fluctuating number of shift-and-add operations. In this case, the modified Booth's algorithm /9/ is much better suited. In its original so-called second order form the number of shift-and-add operations is reduced to the constant number of M/2. The coefficient can be recoded in parallel always taking three bits at the same time. In comparison to CSD recoding, an addition with zero is not fully avoidable. Instead of going into details, a simple example can demonstrate the differences. The multiplication x times c with the coefficient c = 15/64 (c = .001111 in binary form and c = .01000-1 in CSD code) is performed as

$$\frac{x}{64} + \frac{x}{32} + \frac{x}{16} + \frac{x}{8} + 0\frac{x}{4} + 0\frac{x}{2} \quad \text{with a general-purpose multiplier,}$$

$$\frac{x}{64} + \frac{x}{32} + \frac{x}{16} + \frac{x}{8} \qquad \text{using stripped multiplication,}$$

$$-\frac{x}{64} + \frac{x}{4} \qquad \text{utilizing CSD code representation,}$$

$$-\frac{x}{64} + 0\frac{x}{16} + \frac{x}{4} \qquad \text{applying the Booth algorithm.}$$

A last aspect concerns the wordlength and the accuracy of the product. The product has a wordlength equal to the sum of data and coefficient wordlengths, whereas it is desirable for input and output data words to be of the same length. Therefore the output must be rounded or truncated as described in Section 2.4. To save hardware it is desirable to restrict the wordlength during the execution of the multiplication to the wordlength of the data words without any loss of accuracy in comparison to a truncation of the output after finishing the complete multiplication. Horner's scheme /10/ or nested multiplication is recommended in this context. The essential point is that the partial sums are shifted without any change of the sign instead of the input data word as is done in the former example. The multiplication with the same CSD recoded coefficient c is carried out as

$$\frac{1}{4}(x + \frac{1}{16}(-x)).$$

Many papers have been published in the last 20 years about implementations of multipliers in silicon. All combinations of data formats are of interest to obtain the required through-put with the minimal amount of hardware: serial /11/, serial-parallel /3/ and parallel multipliers /12,13/ together with different pipelining schemes. The full adder (FA) together with an AND-gate to prepare the partial product is the main

242

logic cell. Architectures are preferred which allow a very regular and modular layout as per /14/ and avoid carry propagation within a clock cycle. Considerations of this last aspect are very similar to the comparison of carry-ripple and carry-save techniques associated with the parallel adder discussed previously and are discussed in /15/, for example. Attractive and frequently used architectures are the classical serial-parallel multiplier and the carry-save array multiplier combined either with Booth's algorithm to shorten the time for a complete multiplication (16x16 bit in 90 ns, 20 mm^2 typical for a state-of-the-art 2 μm-CMOS technology /16/) or highly pipelined to obtain maximum throughput rates (simulations and verifications of basic circuits show that a sample rate of about 70 MHz is attainable with a standard 2 μm NMOS technology).

Special attention should be directed to the handling of two's complement data words and/or coefficients. Usually sign extension and/or complementer circuits after the final multiplier stage are recommended. Some logic and silicon area can be saved with customized on-chip parallel multipliers by the use of slightly modified cells at the periphery of the array /11/.

Some attempts have been made to make efficient use of read-only memories (ROMs) /17/ to store pre-calculated results. Because of the exponential growth of the memory sizes these solutions are restricted to short wordlengths.

2.1.3 Sample Delay Elements- Digital filters necessarily need not only arithmetic elements but also delay elements. Past values of input samples or intermediate results must be stored during a sampling period T_s as shown in the signal flowgraphs of Fig. 1a) and Fig. 1b), respectively. It is assumed that a flowgraph may not contain any closed signal path that does not contain a sample delay operation. But this general realizability condition must be examined in a more quantitative manner with respect to hardware speed constraints /18/. Each arithmetic operation and each access to a memory is time consuming. The overall time delay along a hardwired signal path may not be larger than the sum of sample delays specified in the corresponding signal path of the flowgraph where arithmetic elements are assumed to be delay free. This statement must be extended and generalized if a hardware unit is shared or multiplexed for a given filter algorithm. All signals or state variables of the digital filters must be updated and stored within one sampling period by solving the given system of difference equations.

There are different ways to implement delay and storage elements in MOS. From the architectural point of view sample delay storage can be organized as shift registers or as random-access memories (RAMs). The decision about memory organization is strongly related to the question of the type of filter implementation as discussed in Section 2.3. Shift registers and arithmetic cells are joined together for highly pipelined digital

filters with a hardwired regular data flow. Memories and arithmetic units are better kept separate for programmable signal
processors or for highly multiplexed systems if stored samples
are seldom fetched within a sequence of program steps. RAMs
can be combined with sophisticated address generating units
as discussed in Section 4 where pointer movements instead of
data movements are preferred. This is one way to reduce the
dynamic power consumption. It applies mainly to CMOS circuits.

From circuit design considerations we have the choice between static or dynamic registers and memories. The former ones
need more silicon area, while the dynamic ones are restricted to

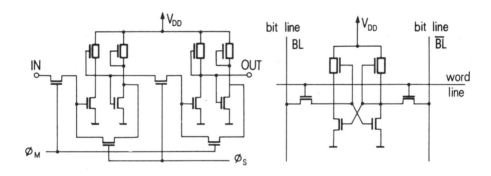

a) static master/slave register cell   b) static RAM cell

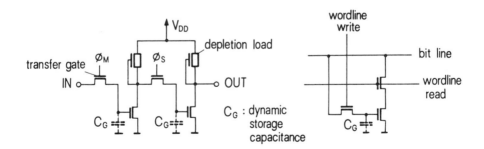

c) dynamic shift register cell   d) three transistor
   with depletion loads   dynamic RAM cell

Fig. 4: Typical memory cells as NMOS circuits

sample frequencies larger than 10 kHz if additional refresh
cycles are to be avoided. Fig. 4 shows four typical memory cells
designed for NMOS. The dynamic RAM cell of Fig. 4d) with 3 transistors is more compatible with MOS technologies for logic circuits than the 1-transistor cell of commercial DRAMs. Changing
to CMOS, the transfer gate is mostly replaced by a transmission
gate with n- and p-channel devices and the depletion load is
replaced by a p-channel device in order to minimize the static
power consumption. Shift registers as shown in Figs. 4a) and 4c)
can directly be connected to arithmetic-logic circuits and
perform reads and writes with two non-overlapping clocks $\emptyset_M$ and
$\emptyset_S$ during one clock cycle T_c. RAM cells are arranged as an
array and require additional circuits such as address decoders,
precharge units, sense amplifiers etc., and additional ports and
lines for simultaneous read and write operation. Last but not
least the access time to a separate memory can be crucial. Conveniently, the arithmetic unit of a signal processor is usually
supplemented by a scratchpad memory for fast temporary storages.

2.1.4 Miscellaneous Operations - Digital filters and digital
implementations of stable differential equations may manifest
limit cycle oscillations or steady-state errors. A lot of papers
deal with the effects of the finite wordlength available for the
representation of signals (see, e.g. /19,20/ for a first general overview). In this contribution hardware aspects are mainly
discussed, first with respect to quantization, and secondly
with respect to overflow.

Common quantization characteristics together with their
two's complement arithmetic evaluation are depicted in Fig. 5.
Roundoff quantization is performed by substituting the nearest
possible word that can be represented by the limited number of
B bits below the binary point. Therefore, a minimum of quantization error or noise is achieved. A simple truncation of signals represented by sign and magnitude leads to magnitude truncation quantization which is recommended in /20/ to guarantee
the implementation of wave digital filters without any limit
cycles. Magnitude truncation of signals with two's complement
format requires the subsequent addition of the sign bit as
shown in Fig. 5b), which in turn requires additional logic
circuits or program steps. This can be saved if and only if
the filter designer tolerates value truncation according to
Fig. 5c).

Yet another consequence of using a fixed-length word to represent data samples is that the result of some arithmetic operations will occasionally call for a value which lies outside the
allowed range of values. This effect, called overflow, results
in an incorrect in-range number being produced whose value depends on the details of how the arithmetic operation in question
is performed. In the case of commonly used two's complement adders, the result is an overflow characteristic like that depicted in Fig. 6a). It has been found that when this type of overflow occurs in the feedback loop of second-order sections or

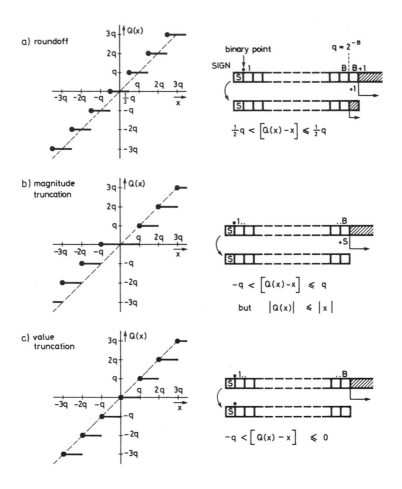

Fig. 5: Three possible fixed-point quantization characteristics

other recursive filters with certain coefficient values, full-scale oscillations ensue which are stable and which persist regardless of what input sequence is subsequently applied to the filter /3/. Another aspect is the forced response stability of digital filters /20/. Although overflow can be prevented by providing several extra bits at the most significant end of the data word, this is a costly and unnecessary waste of dynamic range

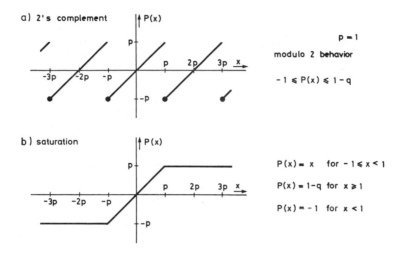

Fig. 6: Possible overflow characteristics

and is in any case unsatisfactory, since momentary overflow can still occur due to a power turn-on transient or a soft error within the memory. A more effective way of dealing with overflow is the use of the saturation characteristic depicted in Fig. 6b). The word that causes the overflow is replaced by a word having the same sign, but a magnitude corresponding to the overflow level. Other possibilities are discussed in /19/. There are different ways to realize such an overflow detection and correction circuit, whether the adder is extended by one additional bit or the overflow criteria is derived from the signs of the input signals and the sign of the sum. In any case a subsequent multiplexer must be realized which consumes delay processing time and chip area. If no special hardware for saturation is provided in an arithmetic unit of a signal processor, several processor cycles including conditional branch instructions must be spent. This is another reason to emphasize the need for a good interaction between the designers of filters and circuits. They have to decide about the necessity and the most efficient way to implement saturation or another kind of overflow management.

The optimization of the dynamic range of the filter asks for a careful scaling of data words. Scaling can be accomplished in serial-data systems through the use of shift registers and in parallel-data systems by an intentional misalignment of bit lines or by the use of a programmable shifter. Further miscel-

laneous operations may be concerned with the resetting of registers, with switching for different schemes of multiplexing,
or with changing the sampling rate as discussed in Section 2.3.
Questions about clocking, programmability and control must be
omitted in this overview but they are also relevant to a decision about architecture and design style.

2.2 Pipelining, Parallelism, and Multiplexing

The real-time implementation of one and the same filter
function may be quite different for different sample rates. An
essential task for a chip designer is to match hardware processing rate to the required signal sample rate. Some papers, e.g.
/21/, discuss the different concepts with regard to digital
signal processing. It may be useful to take our first example
of the FIR filter of Fig. 1 again and demonstrate how the flowgraph can be modified to meet the speed constraints with a
given MOS technology characterized by the delay of a single-bit
full adder, for example. It should be mentioned in this context
that the circuit designer could reduce such a critical delay
time by altering the transistor sizes but this is only possible
within a small domain and at the expense of silicon area.

An essential way to improve the throughput of a digital
circuit is the use of concurrency. This may be explained by
means of a small FIR filter with a parallel-in/serial-out structure as shown in Fig. 7a). Parallel arithmetic is assumed. A
one-to- one implementation of multipliers, adders and shift
registers would be very area-consuming. Furthermore, the sample
rate is limited by the sum of the delays for multiplication,
addition and storage. An important saving of hardware elements
is possible if the coefficients are fixed and optimized with
respect to a minimal number of non-zero bits. In our example
the critical delay decreases to two adds and one storage operation within the sample period (see Fig. 7b)). A further reduction is obtained by introducing two additional delay elements
as shown in Fig. 7c). An improvement of the throughput rate
is paid for by a greater latency time and the expense of two
registers. This manipulation, called pipelining, can be used
efficiently when the algorithm is applied repeatedly to a stream
of input data which is mostly true for high speed filter applications. However, the introduction of additional sample delays
obviously cannot be allowed for recursive filters with feedback
loops. For such cases methods described in /22/ and carry-save
techniques /23/ are recommended. Pipelining can be extensively
used but the profit decreases more and more. The critical delay
of one add- and-storage operation in Fig. 7c) cannot further
be reduced with two additonal registers as shown in Fig. 7d).
The throughput would be improved if the parallel data words are
split into bytes for the most significant and the least significant bits. The delay of a carry-propagate adder with half of
the wordlength is nevertheless greater than half of the delay
of the original one.

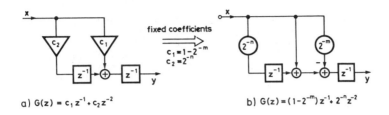

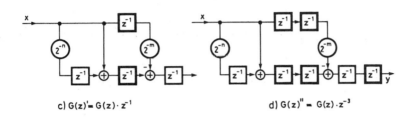

a) $G(z) = c_1 z^{-1} + c_2 z^{-2}$

b) $G(z) = (1-2^{-m}) z^{-1} + 2^{-n} z^{-2}$

c) $G(z)' = G(z) \cdot z^{-1}$

d) $G(z)'' = G(z) \cdot z^{-3}$

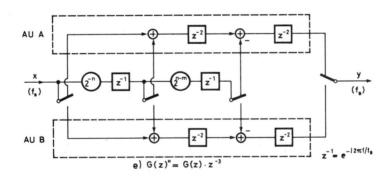

e) $G(z)'' = G(z) \cdot z^{-3}$

$z^{-1} = e^{-j 2\pi f/f_s}$

Fig. 7: Example to illustrate pipelining and parallelism
a) given FIR filter b) multipliers stripped to hardwired scalers and adders c) pipelining with one
additional delay stage d) pipelining with two delay
stages e) AUs A and B operate in parallel with $f_s/2$

Moreover, functional parallelism can be used to perform
operations on very fast signals, by using a larger number of

hardware units running in parallel, with appropriate techniques
for combining their partial computations. The arithmetic units
(AUs) A and B of Fig. 7e) operate at half of the sample frequency f_s. A and B are said to be ping-ponged. The switches
always connect alternate input samples to the same AU. The
throughput is increased by a factor of two at the expense of
twice the hardware.

On the other hand, multiplexing is significant for accomo-
dating many separate operations on relatively slow signals
by sharing a small amount of fast hardware. This is the typical
application area of microprogrammable signal processors. Fig. 8
illustrates the hardware implementation of the FIR filter of
Fig. 1a) on a multiplier-based AU of a signal processor. The
switch is realized by controlled data transfers along a bus
from a RAM to the AU. The coefficient memory is usually a ROM.
The accumulation of all the products is accomplished with a
carry-propagate adder as shown in Fig. 3a). The register holds
all the results to provide the filter output.

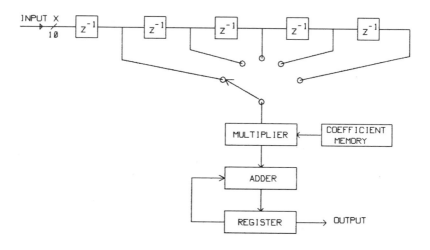

Fig. 8:   Time-shared implementation on a multiplexed AU

Not only arithmetic operations can be multiplexed, but
also complete filters for multiplexed signals on different
channels. Whereas the filter arithmetic is time-shared, the
state variables of the different filters must be stored in
different memory locations.

2.3 Multirate Digital Filters

Since the complexity of digital filters depends on the number of arithmetic operations to be carried out during one sampling period, it is desirable to use a sampling rate as low as possible obeying the sampling theorem. An economical use of the arithmetic units should provide for different sampling rates for different parts of the system in order to cope with the spectral contents of the signals to be processed. Therefore sampling rate decrease (decimation) and sampling rate increase (interpolation) are operations frequently required in digital signal processing systems /24,25/. From the hardware point of view it is essential that the corresponding digital filter operates at the lower sampling rate that is the frequency at the output of the decimator and at the input of the interpolator. If the FIR filter of Fig. 1 is used to alter the sampling rate by factors of 2, corresponding signal flowgraphs are given in Fig. 9 together with an illustration of the interpolation and decimation process in time. A possible realization of the switches together with the necessary clocks is shown in Fig. 10.

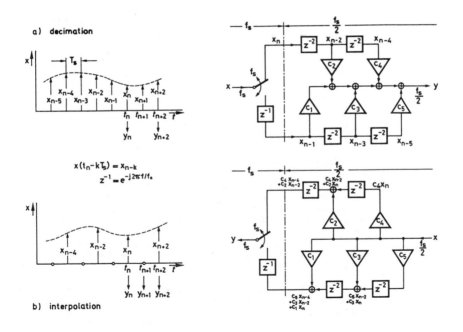

Fig. 9: Signal flowgraphs and typical waveforms of multirate transversal filters

251

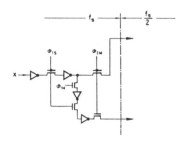 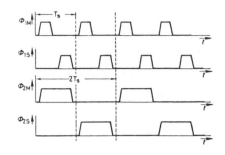

a) logic diagram   b) timing diagram of the clocks

Fig. 10:   Realization of the decimation stage

2.4 Hardwired Versus Microprogrammed Implementation

Besides the technical aspects of matching the speed of
arithmetic units and the required sample rate as discussed in
Section 2.2 the question arises how flexible or how dedicated
MOS circuits for digital filters should be designed. This is
a fundamental problem of economics. The design and successful
fabrication of a chip are difficult and expensive enough so
that it is clearly desirable to develop a flexible general-purpose device that can be mass produced. This is practicable
without a major loss of performance for applications at lower
sample rates if programmable arithmetic units and RAMs may be
time-shared with a suitable instruction set. In this case the
architecture and the design style are not determined only by
digital filter requirements since several other tasks are executed with the same hardware.

High volume speech processing and telecommunication applications will stimulate the use of several customized digital
signal processors with bus-oriented architectures and tailored
memories, arithmetic units and interfaces. To save design time
and costs such single-purpose chips should be assembled from
a library of predefined modules (silicon macro-cells). Some
examples are mentioned in Section 4.

The higher the sample rate of the signals, the higher the
degree of pipelining and parallelism, the rarer the possibility
of sharing programmable hardware and the more lucrative architectures based on dedicated hardwired signal paths will be. If we
restrict ourselves to a well-known class of algorithms such as
FIR filters or special cases of IIR filters, we can imagine
efficient top-down design tools like "silicon compilers" to
obtain sufficiently optimal designs within a short time.

The following two sections describe real implementations
of digital filters in MOS technology with special emphasis on
the chip level.

3. DIGITAL FILTERS WITH DEDICATED SIGNAL PATHS

High volume and/or high-throughput applications require
optimized algorithm-specific implementations. The regular data-
flow typical for digital filters permits a time-sequential
execution with a high degree of pipelining and therefore an
effective use of silicon. Arithmetic elements and sample delay
storages interlace more and more especially for filters with
sample rates within the megahertz range. In order to keep the
signal propagation time as short as possible the signal data
lines should be hardwired instead of using transfers over
switched busses. The minimal propagation time is achieved by
an abutment of layout cells without any additional wiring. Such
a design will be conform with the concept of "systolic arrays"
/14,26/.

In this context it is significant to distinguish between
bit-serial and bit-parallel architectures. Both methods have
their advantages and disadvantages which are discussed e.g. in
/21,27/. In general the two solutions will require data memories
of the same size, if each implements the same algorithm. Bit-
serial architectures have the attractive fundamental property
that the arithmetic elements are heavily exercised due to their
bit-level pipelining. But they need a clock that is d times the
word rate if d is the data wordlength. Although a bit-serial
full adder is essentially faster than a bit-parallel carry
propagation adder, the frequency range of bit-serial architectures is more restricted because of the higher system clock.
One should not ignore the problems of clock generation and
clock distribution. The mapping of an algorithm onto the layout
level necessitates more consideration even down to the bit-
level. But some useful work is already done, and this is briefly
summarized in the following.

3.1 Bit-Serial Architectures

On the basis of /1,11/ a bit-serial IC design methodology
is proposed in /28/ which leads to the FIRST silicon compiler
/21,29/ and to more optimized and compact IC realizations of
dedicated digital filters /30/. Some other papers presented in
/31/ or recently published /32,33,34/ are indications that a
number of research and development groups are investigating
this technique which can certainly be used for efficient algorithm-specific implementations of digital signal processing
tasks within the voice and audio frequency range.

The hardware implementation of a FIRST circuit consists of a network of interconnected bit-serial operators, laid out according to a relatively fixed floorplan. Each bit-serial operator is implemented as a separate function block, which is, in turn, assembled from a library of hand-designed leaf cells. Some silicon area is wasted by this approach, since function blocks may differ in height /29/. But this work points out the direction to an automatic and error-free silicon compilation of dedicated digital filters.

A method of systematic hardware minimization is given in /30/ on the following three levels: algorithm, architecture, and layout. Filters with fixed coefficients take advantage of the CSD coefficient minimization as described in Section 2.1.2. Weakly programmable filters require additional basic circuits and a more sophisticated control. Special CAD tools have been developed to accomplish the delay management /35/. The two

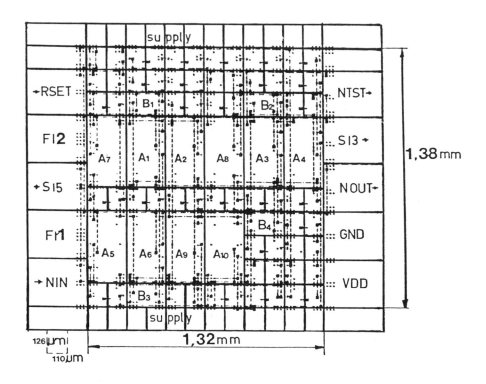

Fig.11: "LEGO" style layout of a wave digital filter

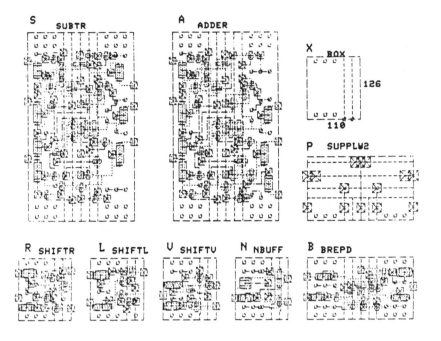

Fig. 12: Symbolic layouts of bit-serial primitive cells

Figures 11 and 12 from /30/ illustrate the "LEGO" layout style.
Geometrically related cells can be stacked unconstrained in
both X and Y directions since the primitive cells are compacted
in the frame of a standard pitch with abutting supply lines
(power distribution and clocks) and dummy tracks. Horizontal
and vertical tracks and dummy contact areas are reserved for
signal interconnection lines using horizontal polysilicon and
vertical metal lines respectively. Fig. 11 shows the floorplan
of a third-order wave digital filter which is designed for
a sampling rate of 312 kHz and a data wordlength of 16 bits
using a 6 um NMOS technology with two-phase depletion load
circuitry and dynamic shift registers as shown in Fig. 4c).

The layouts of the cells are generated by the symbolic layout program CABBAGE-Leuven, which translates interactively
edited, layout-rule-independent stick diagrams into compacted
layouts. The symbolic editor also allows for imposing constraints on the distance between lines in order to maintain
the strategy of abutting cells in both directions. Fig. 12

shows symbolic layouts of cells of the library. The lengths and widths are integer multiples of a basic pitch in both directions, such as "LEGO" blocks. The layout density is about 500 tr/mm^2 for 6 μm NMOS which is very compact for layout based on a cell library. Further developments favor CMOS technologies. The racefree dynamic CMOS technique NORA /36/ is very attractive for pipelined logic circuits and suggests bit-serial implementations of digital filters with less power consumption. Fig. 13 /36/ shows a NORA serial full adder containing only 32 transistors.

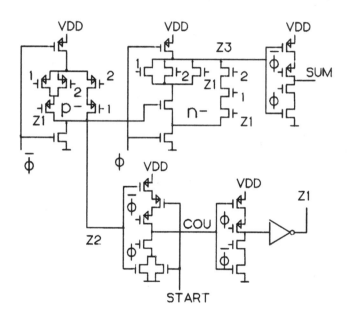

Fig. 13: Circuit diagram of a NORA-CMOS serial full-adder

Serialized datapaths are very promising because of the simple wiring of one line between cascaded bit-serial operators. CAD tools for two-layer-wiring of standard cells should be practical too. The situation will be quite different for bit-parallel structures as described in the following subsection. Before we leave the bit-serial technique it should be mentioned that conversion to a bit-parallel form can be carried out step by step by the introduction of parallelism or by partitioning of the data word into serially processed bytes.

3.2 Bit-Parallel Architectures

The highest throughput rates of about a few hundred Mbit/s for sample rates of typically 20 MHz can only be obtained using highly pipelined bit-parallel architectures. Even more attention must be paid to the wiring problem when favoring short regular local interconnections within the cell array mainly in metal and not in polysilicon. Modified bit-slice architectures /14/ like linear semi-systolic arrays /26/ will be quite useful.

A first suggestion for a bit-parallel implementation of a transversal filter with a parallel-in/serial-out structure according to Fig. 1b) is given in /37/. A realization of such a large array of pipelined carry-save adders was still of academic interest in 1981. In the meantime a modular block of adaptive transversal filters has been designed /38/ for easy expansion of filter length, the size of the coefficients, and the size of the signal data word. High-volume applications for digital video signal processing only require digital filters with fixed coefficients. Therefore some work has been done to find economical solutions with NMOS technologies available at that time. A low-pass filter for separation of the luminance and chroma spectra is realized using a two-phase race-compensated MOS logic in a 4 μm E/D NMOS technology /39/. More general considerations result in two different design styles. The first one is a "gate-array"-like filter technique for a fast first shot /40/, the second one is a bit-sliced datapath concept which is optimized with respect to a minimal consumption of area /23,41/. This will be described briefly in the following.

The starting-point is the signal flowgraph of a pipelined transversal filter as shown in Fig. 7c). Investigations on different parallel-data carry-propagate adders show that the simple ripple adder fulfils the requirements with regard to speed, modularity and minimal chip area very well as long as the data wordlengths do not exceed a limit of typically 12 bit. To save inverter delays in the carry path, we allow inverted carry outputs at each second full adder cell within the ripple adder. A cascade of m ripple adders which concludes with a wordlength of n bits is characterized by a critical path with the delay

$$\tau_{ADD} = 2m\tau_{HA} + \frac{n-1}{2}\tau_{c} \qquad (3)$$

τ_{HA} is the delay of an EX-OR half adder and τ_{c} is the carry propagation time through two full adders. Typical delays for a 2 μm NMOS technology are $\tau_{HA} = 5.5$ ns and $\tau_{c} = 4.5$ ns. The required sample rate and the wordlength n determine how many ripple adders can be cascaded between synchronously clocked dynamic shift registers. One inverter of such a register can be saved if the register is combined with the EX-OR half adder

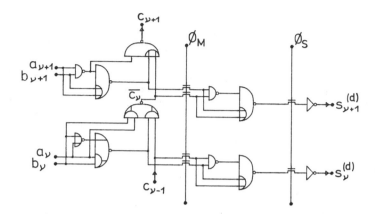

Fig. 14: Logic diagram of full-adder cells with latches

at the sum output as shown in the logic diagram of Fig. 14. The
additional delay of a shift register is about 2 ns. Subtraction
is performed by inversion of the sum outputs of the preceding
adder and a carry input c_{-1} = 1 at the least significant bit.

Fig. 15 shows the layouts of two full adder circuits including shift registers which correspond to the logic diagram
of Fig. 14. The interconnection of the clock lines and the
carry signals are performed by an alternate abutting of these
cells to obtain a ripple adder. The layout design style is
illustrated in Fig. 15. Power and clock lines are routed parallel to the x-axis in metal and parallel to the y-axis in polysilicon, respectively. Ripple adders and additional shift registers are arranged in columns. Predefined dummy tracks over
cells enable the transfer of one signal bus parallel to the
x-axis in metal without altering cells. Relative shifts (scaling) of data words are performed in polysilicon within channels
between ripple adders and registers columns. The width of each
channel is minimized by applying river routing.

The signal flowgraph and the layout of a transversal filter
module with the transfer function

$$G(z) = (1 - 4z^{-1} + z^{-2})z^{-2}$$

are shown in Fig. 16 and Fig. 17, respectively. The characteristic data are: 20 full adders including shift registers and
16 additional shift registers, 480 transistors, 2 µm NMOS E/D
NMOS technology, 0.18 mm^2 silicon area, 44 mW static power

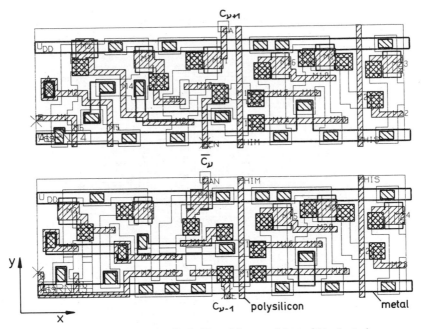

Fig. 15: Layouts of full-adder cells with latches

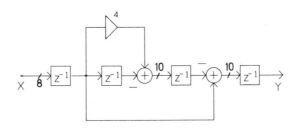

Fig. 16: Signal flowgraph of a transversal filter of degree 2

consumption, maximal sample rate 17.6 MHz specified, 27 MHz measured /23/. The design style described above is further used to implement a more complex digital filter for component encoding of color TV signals /42/. Novel test strategies are also described in that paper.

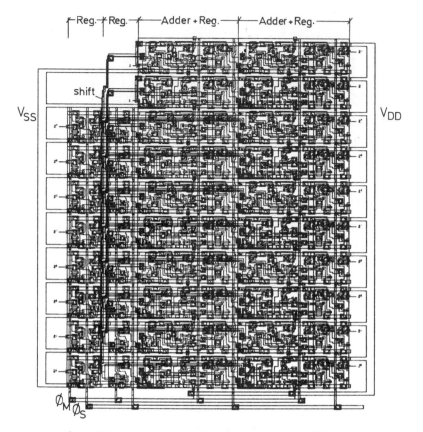

Fig. 17: Layout of the transversal filter

The same design style is applied to implement recursive filters. The floorplan of a simple wave digital allpass section is published in /40/. The filter structure determines how many dummy tracks must be provided in the full adder and shift register cells. To obtain more flexibility in this point and with respect to prospective CMOS designs it could be appropriate to route signal paths outside of the cells between the power supply lines. Another possibility is to route the two clock lines in metal parallel to the x-axis as shown in /28/ or in a second layer if it is available.

In contrast to transversal filters it is not possible to introduce further delays into the recursive loops to moderate

the time condition for the pipelined system. Carry-look-ahead
architectures can be used to speed up the adder. But their
regularity is much less than that of carry-select-adders /4,43/
which seem well-suited for MOS-bit-slice configurations. An
alternative is the carry-save approach as discussed in Section
2.1.2. In this case the carry does not ripple through the word
within one sampling period but is saved in an additional register. The final processing of carry and sum bits can be done
behind the feedback loop with modified pipelined carry-save
adders /41/. A comparison of the carry-ripple and the carry-save
approach is given in Fig. 18 on the basis of floorplans for a
simple digital integrator with the transfer function

$$G(z) = z^{-2}/(1 - z^{-1}).$$

The time-critical paths are marked by dash-dotted lines. It is
obvious that the carry-save delay time is equal to the delay of
one full adder including the shift registers and is independent
of the data wordlength. If a recursive filter contains more than
one feedback loop, the use of simple full adders with three
inputs and two outputs is no longer sufficient. The carry-save
approach is still applicable but we need a more general type
of adder cells with up to six inputs and three outputs.

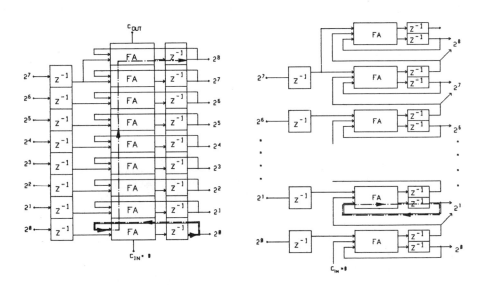

a) ripple-adder approach   c) carry-save approach

Fig. 18: Floorplans of the digital integrator

261

The implementation of fast adaptive filters where the coefficients c_k of Eq. (1) vary in time participates in advanced parallel multiplier implementations. Characteristic data and some interesting details are published in some papers of the IEEE Journal of Solid-State Circuits and the Proceedings of the ISSCC Conferences.

For some alternative designs of dedicated digital filters and arithmetic functions we refer to /44,45,46/.

4. CUSTOMIZED SIGNAL PROCESSORS FOR DIGITAL FILTER APPLICATIONS

Of course, microprogrammable general-purpose digital signal processors, available on the market /47,48/, can be used to implement digital filter algorithms. Although they are very powerful, they have certain disadvantages /49/. In order to build a complete signal processing system, many other circuits are necessary to interface the signal processor to its environment. The performance of some chips for speech processing or tele-communication applications has been significantly improved with a tailored on-chip A/D converter utilizing oversampling principles /51/. A general-purpose digital signal processor offers only a fixed set of features, such as an array multiplier and a fixed amount of RAM and ROM. Many applications, mainly digital filters with fixed coefficients, do not require a fast parallel multiplier, and need varying amounts of RAM and ROM. For such applications customized microprogrammable signal processors, assembled from parametrisable building blocks, can offer an attractive solution. With this approach, low design costs can be achieved if efficient techniques are developed for the optimal mapping of DSP algorithms into a chip architecture constructed from a restricted but efficient library of modular macro-cells. In this direction promising design strategies for signal processing ICs are offered in /27,49,52/.

It would go beyond the limits of this section to discuss all aspects of digital signal processor architecture. Fundamentals are given in /53,54/. Harvard-type architectures are generally prefered with separated but simultaneously operating arithmetic units for signal processing and address computation as well as with separated memories for sample delay storage, microprogram code and coefficients. The essential elements of a specific digital signal processor architecture must be deduced from a careful consideration of the algorithms to be executed. Since a fast arithmetic unit is shared according to the microprogrammed sequence of instructions it is less important what filter structure (e.g. transversal or recursive, second-order sections, LDI or wave-digital filters) should be implemented. It is more critical whether the coefficients are constant or variable and, what additional linear or non-linear operations, e.g. normalization of signal levels, should be carried out with the same hardware etc.

Whereas the hardware level is dominant for the design of digital filters with dedicated signal paths as discussed in Section 3, software aspects must also be taken into account when the architecture of a digital signal processor is to be outlined. The given signal flowgraph is a graphical representation of the filter algorithm. In the first instance general simulation programs, e.g. /55/, are useful to analyse the digital filter on the basis of such a signal flowgraph, to calculate the transfer function, to determine finite wordlength effects etc. Next the algorithm must be transformed into a sequence of program steps suitable for evaluation by shared hardware. A special microinstruction code and a suitable assembler language must be defined to support the process of program development and the process of translating algorithm representations into programs. Some simulation programs based on the microinstruction code of the specific processor will be useful at the higher design level to verify the algorithm and, others on the register transfer level to verify the processor hardware.

After these general remarks some hardware units are mentioned which usually exist in each digital signal processor.

4.1 Execution Unit

The execution unit usually consists of the arithmetic/logic unit and a scratch pad memory of some registers for temporary storage. In /54/ an essential distinction is introduced between multiplier-based and adder-based (or shifter-based) arithmetic units. General-purpose signal processors favor the former one. On the other hand the machine cycle time of an adder-based digital signal processor can be shorter in comparison to a multiplier-based one /56/. Another argument against the use of parallel array multipliers is the desired physical compactness of one arithmetic unit for multiprocessor designs /27/. Besides this, execution of a division operation, necessary for normalization, cannot be provided with a multiplier.

An adder-based arithmetic unit performs a parallel-serial multiplication. The sequence of partial products is generated in a bit-parallel, word-serial form, and accumulated by a single accumulator. The adder can propagate or save the carry as depicted in Fig. 3. A suitable way to minimize the number of shift-and-add cycles is to use the CSD code for multiplication with fixed coefficients and the modified Booth algorithm for multiplication with variable coefficients or for data multiplication. Both recoding schemes were described in subsection 2.1.2.

Fig. 19 shows the block diagram of the execution unit of the digital signal processor PROFI (PROcessor for FIltering) /54/ as an example. It contains a 4-port register file RA, RB,

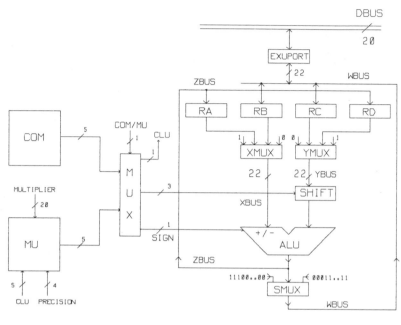

DBUS

EXUPORT

ZBUS   WBUS

RA   RB   RC   RD

COM/MU   CLU

COM

MULTIPLIER

MU

XMUX   YMUX

XBUS   YBUS

SHIFT

SIGN

+/-

ALU

ZBUS

SMUX

WBUS

CLU PRECISION

COM: COEFFICIENT MEMORY   CLU: CONTROL LOGIC UNIT
MU: MULTIPLICATION UNIT

Fig. 19:   Execution unit of the signal processor PROFI

RC, RD, a barrel shifter SHIFT that shifts the data word of the
YBUS with a programmable amount of 0 to 7 bit to the right, and
a multiplexer SMUX that performs saturation according to the
characteristic of Fig. 6b). A multiplication with a fixed coefficient is coded by an instruction "MULCM" stored in the program
memory PRM and a sequence of CSD codes stored in the coefficient
memory COM (1 bit sign, 3 bits shift amount, 1 control bit
to terminate the sequence). It should be pointed out that the
accumulated partial sum (stored in RC or RD) is shifted right
and the least-significant partial product is accumulated first
to realize Horner's scheme /10,27/. Another instruction "MUL"
starts the multiplication with a variable coefficient which
is controlled by the multiplication unit MU. The coefficient
(or data word) stored in a multiplier register is decoded step
by step to carry out shift-and-add codes as described above.
Some other features of this execution unit are given in /56/.

4.2 Data Storage and Address Computation

The number of data samples to be stored during the execution of a program usually motivates a separate data RAM. For smaller RAMs it may be useful to stretch the width of the memory cell and fit it to the width of the bitsliced data path /57/. The throughput of the processor can be improved by introducing a pipeline segmentation so that memory access and arithmetic operations can be carried out concurrently. Another aspect is whether writing or reading of the RAM is possible within the same processor cycle or not.

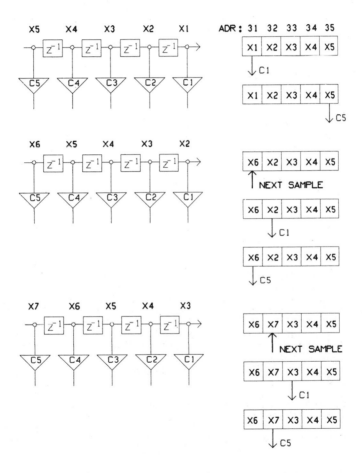

Fig. 20: Memory management via cyclic pointers

A main reason for the increased throughput rate or a larger
number of MOPs (mega operations per second) of signal processors
in comparison with standard microprocessors is that address
computation is not performed with the arithmetic unit but concurrently with suitably designed address generating units.
Direct addressing is the simplest case but not very attractive.
Signal processing algorithms often involve the unconditional
execution of the same sequence of operations. Thus the address
sequence is also fixed and can be provided with a simple incre-
mentation for each read or write operation to simulate a shift
register. This is reasonably effective when only single-sample
delays are needed in the signal flow. For longer delays, e.g.
for a transversal filter with a larger number of taps, it is
more efficient to implement a circular shift register. This
method (also called "precessing" in /27/) is realized in the
signal processor PROFI /56/ with cyclic pointers and should
be explained with the aid of Fig. 20. Previously a cyclic range
of addresses is defined including the addresses 31 to 35 for
an example. The state variables x_1 to x_5 of the transversal
filter in serial-in/parallel-out structure should be stored
at places as shown in the first row on the right. During the
multiply-accumulate operations the pointer is always incremented
and it stops at address 35. The next incremented address is
now the base address 31 according to the definition of a cyclic
address sequence. The input sample x_6 of the next sampling
period overwrites the old sample x_1. The execution of the filter
algorithm starts with address 32 and ends with address 31. The
n sample delays are performed with one simple pointer shift
instead of shifting all (n-1) data words within the RAM.

4.3 Program Control

The control sequencer of a signal processor usually loops
through a section of code several times. Since the control
flow is data-independent, the number of iterations is predetermined and does not vary from sample to sample. But some miscellaneous operations could require conditional branch instructions. The goal for customized signal processors tailored for
an efficient implementation of digital filters should be to
preserve an unconditional control flow with the aid of flags
for Boolean state variables at the expense of executing more
instructions. Flags are well-suited to program e.g. multirate
filters or filter banks together with multiplier-free modulation
schemes /58/.

Pipelining between the operations: instruction fetch,
decoding, execution, e.g. can improve the throughput once more.
But the complexity of the control unit increases more and more.
Subroutines may be very useful to implement digital filters
with regular subsections and save code lines.

Control structures have the nature of centralized or distributed finite state machines. Software tools for an automatic

generation of Programmable Logic Arrays (PLAs) including partitioning, minimization, and folding, based on a suited cell
library, are very useful. Counters, LIFOs, FIFOs, and some kind
of pure random logic will be employed to complete the control
part and to design the interfaces of such a signal processor.

5. CONCLUSIONS

Filters are usually only parts of signal processing
systems. A comparison between analog and digital realizations
must consider the whole system configuration including the A/D
and/or D/A converters. Since digital circuit techniques directly
benefit from the continued shrinking of device dimensions and
the addition of further interconnection layers, we can expect
that some digital solutions will require a smaller chip area
than analog ones. Further work must be done to reduce the power
consumption. Advanced CMOS, efficient use of transmission gates
and the support of highly sophisticated circuit optimization and
design centering programs should moderate this problem.

Design methodologies based on the use of sophisticated
CAD tools need to be developed for a fast compilation of digital
filter algorithms into silicon to close the gap between the
classical theories of digital signal processing and modern
MOS technologies.

ACKNOWLEDGEMENT

The author wishes to thank his collegues, especially
Mr. Noll, Mr. Rainer, and Dr. Pfleiderer, as well as Dr. Gazsi
from the Ruhr-Universität Bochum, and members of the technical
staff of ESAT at the Katholieke Universiteit Leuven, for valuable discussions and their basic work, referred to in this paper.

6. REFERENCES

/1/   L.B.JACKSON, J.F.KAISER, H.S.McDONALD, "An approach to the
       implementation of digital filters", IEEE Trans. Audio
       Electroacoust., AU-16, Sept. 1968.

/2/   A. PELED, B.LIU, "A new hardware realization of digital
       filters", IEEE Trans. Acoust., Speech, Signal Proc.,
       ASSP-22. Dec. 1974.

/3/   St.L.FREENY, "Special-purpose hardware for digital filter-
       ing", Proc. IEEE, April 1975.

/4/   K.HWANG, "Computer arithmetic. Principles, architecture,
       and design", John Wiley, 1979.

/5/ G.W.REITWIESNER, "Binary arithmetic. Advances in computers, Vol. I", Academic Ed., 1960.

/6/ L.GAZSI, S.N.GÜLLÜOGLU, "Discrete optimization of coefficients in CSD code", IEEE Mediterranian Electr. Conf., Athen, May 1983.

/7/ Y.Ch.LIM, S.A.PARKER, "FIR filter design over a discrete powers-of-two coefficient space", IEEE Trans. Acoust., Speech, Signal Proc., ASSP-31, June 1983.

/8/ R.JAIN, J.VANDEWALLE, H.DeMAN, "Efficient CAD tools for the coefficient optimization of arbitrary integrated digital filters", Proc. ICASSP, March 1984.

/9/ L.P.RUBENFIELD, "A proof of the modified Booth's algorithm for multiplication", IEEE Trans. Conputers, C-24, Oct. 1975.

/10/ L.GAZSI, "Hardware implementation of wave digital filters using programmable digital signal processors", Proc. ECCTD, The Hague, 1981.

/11/ R.F.LYON, "Two's complement pipeline multipliers", IEEE Trans. Commun., COM-24, April 1976.

/12/ J.DEVERELL, "Pipeline iterative arithmetic arrays", IEEE Trans. Computers, C-24, March 1975.

/13/ S.WASER, "High-speed monolithic multipliers for real-time digital signal processing", Computer, Oct. 1978.

/14/ C.MEAD, L.CONWAY, "Introduction to VLSI systems", Addison-Wesley, 1980.

/15/ J.R.JUMP, S.R.AHUJA, "Effective pipelining of digital systems", IEEE Trans. Computers, C-27, Sept. 1978.

/16/ J.M.ANDERSON, B.L.TROUTMAN, R.A.ALLEN, "A CMOS LSI 16x16 multiplier/multiplier-accumulator", ISSCC 1982.

/17/ H.LING, "High-speed computer multiplication using a multiple-bit decoding algorithm", IEEE Trans. Computers, C-19, August 1970.

/18/ M.RENFORS, Y.NEUVO, "The maximum sampling rate of digital filters under hardware speed constraints", IEEE Trans. Circuits and Systems, March 1981.

/19/ Th.A.C.M.CLAASEN, W.F.G.MECKLENBRÄUKER, "Effects of quantization and overflow in recursive digital filters", IEEE Trans. Acoust., Speech, Signal Proc., ASSP-24, Dec. 1976.

/20/ A.FETTWEIS, "Realisierungsprobleme von Digitalfiltern
- Durch Wortlängenbegrenzung bedingte Effekte sowie Fil-
terstrukturen", NTZ Archiv, Dec. 1980.

/21/ P.B.DENYER, "An introduction to bit-serial architectures
for VLSI signal processing", in "VLSI architecture",
Ed. by B.RANDELL, P.C.TRELEAVEN, Prentice Hall, 1983.

/22/ A.L.MOYER, "An efficient parallel algorithm for IIR filters", Proc. ICASSP 1976.

/23/ W.ULBRICH, T.NOLL, B.ZEHNER, "MOS-VLSI pipelined digital
filters for video applications, Proc. ICASSP, March 1984.

/24/ R.E.CROCHIERE, L.R.RABINER, "Interpolation and decimation
of digital signals - a tutorial review", Proc. IEEE,
March 1981.

/25/ A.FETTWEIS, J.A.NOSSEK, "Sampling rate increase and decrease in wave digital filters, IEEE Trans. Circuits and
Systems, CAS-29, Dec. 1982.

/26/ H.T.KUNG, "Why systolic architectures?", Computer,
Jan. 1982.

/27/ St.P.POPE, R.W.BRODERSON, "Macrocell design for concurrent
signal processing", Third CALTECH Conf. on Very Large
Scale Integr., Ed. by R.BRYANT, Springer, 1983.

/28/ R.F.LYON, "A bit-serial VLSI architectural methodology for
signal processing", in "VLSI 81", Ed. by. J.P.GRAY,
Academic Press, 1981.

/29/ N.BERGMANN, "A case study of the F.I.R.S.T. silicon compiler", Third CALTECH Conf. on Very Large Scale Integr.,
Ed. by R.BRYANT, Springer, 1983.

/30/ J.K.J.VanGINDERDEUREN, H.J.DeMAN, N.F.GONCALVES,
W.A.M.VanNOIJE, "Compact NMOS buidling blocks and a
methodology for dedicated digital filter applications",
IEEE J. Solid-State Circ., SC-18, June 1983.

/31/ H.T.KUNG, B.SPROULL, G.STEELE (Ed.), "VLSI systems and
computations", Computer Science Press, 1981.

/32/ N.R.STRADER II, "VLSI bit-sequential architectures for
digital signal processing", Proc. ICASSP, April 1983.

/33/ Ch.CARAISCOS, B.LIU, "Bit-serial VLSI implementations
of FIR and IIR digital filters", Proc. ISCAS, May 1983.

/34/ H.BARRAL, N.MOREAU, "Circuits for digital signal processing", Proc. ICASSP, March 1984.

/35/ L.CLAESEN, H.DeMAN, J.VANDERWALLE, "Delay management algorithms for digital filter implementations", Proc. ECCTD, Sept. 1983.

/36/ N.F. GONCALVES, H.J.DeMAN, "NORA: A racefree dynamic CMOS technique for pipelined logic structures", IEEE J. Solid-State Circ., SC-18, June 1983.

/37/ P.B.DENYER, D.J.MYERS, "Carry-save arrays for VLSI signal processing", in "VLSI 81", Ed. by J.P.GRAY, Academic Press, 1981.

/38/ F.WILLIAMS, "Video-speed filtering gets its own digital IC", Electronics, Oct. 20, 1983.

/39/ H.J.M.VEENDRICK, "An NMOS dual-mode digital low-pass filter for color TV", IEEE J. Solid-State Circ., SC-16, June 1981.

/40/ P.DRAHEIM, "Digital signal processing for video applications", IEEE J. Solid-State Circ., SC-18, June 1983.

/41/ T.NOLL, W.ULBRICH, "Digital filter structures with parallel arithmetic for custom designs", Proc. ECCTD, Sept. 1983

/42/ H.-J.GRALLERT, F.MATTHIESEN, B.ZEHNER, "An integrated filter for component encoding of color TV-signals", SIEMENS Res. Development Reports, Sept. 1984.

/43/ O.J.BEDRIJ, "Carry-select adders", IRE Trans., EC-11, June 1962.

/44/ N.OHWADA, T.KIMURA, M.DOKEN, "LSI's for digital signal processing", IEEE J. Solid-State Circ., SC-14, April 1979.

/45/ P.F.ADAMS, J.R.HARBRIDGE, R.H.MACMILLAN, "An MOS integrated circuit for digital filtering and level detection", IEEE J. Solid-State Circ., SC-16, June 1981.

/46/ L.WANHAMMER, "A comparison of wave digital lowpass filters from an implementation point of view", Proc. ISCAS, May 1982.

/47/ T.NISHITANI, R.MARUTA, Y.KAWAKAMI, H.GOTO, "A single-chip digital signal processor for telecommunication applications", IEEE J. Solid-State Circ., SC-16, Aug. 1981.

/48/ S.S.MAGAR, E.R.CLAUDEL, A.W.LEIGH, "A microcomputer with digital signal processing capability", ISSCC, Febr. 1982.

/49/ B.R.MEARS, "A modular method for designing custom signal processing integrated circuits", Proc. ICASSP, March 1984.

/50/ R.APFEL, H.IBRAHIM, R.RUEBUSH, "Signal-processing chips enrich telephone line-card architecture", Electronics, May 5, 1982.

/51/ D.VOGEL, P.PRIBYL, "CMOS digital signalprocessing codec-filter with high performance and flexibility", ESSCIRC, Sept. 1984.

/52/ K.KNAUER, M.NIEMEYER, A.RAINER, "Modular design of digital signal processors with macro cells", Proc. ICCD, Oct. 1984

/53/ J.ALLEN, "Computer architecture for signal processing", Proc. IEEE, April 1975.

/54/ J.S.THOMPSON, S.K.TEWKSBURY, "LSI signal processor architectures for telecommunication applications", IEEE Trans. Acoust., Speech, Signal Proc., ASSP-30, Aug. 1982.

/55/ L.J.M.CLAESEN, H.J.DeMAN, J.VANDEVALLE, "DIGEST: A digital filter evaluation and simulation tool for MOSVLSI filter implementations, IEEE J. Solid-State Circ., SC-19, June 1984.

/56/ A.RAINER, W.ULBRICH, L.GAZSI, "Adder-based digital signal processor architecture for 80 ns cycle time", Proc. ICASSP, March 1984.

/57/ P.A.REUTZ, St.P.POPE, B.SOLBERG, R.W.BRODERSON, "Computer generation of digital filter banks", ISSCC, Febr. 1984.

/58/ A.FETTWEIS, "Multiplier-free modulation schemes for PCM/FDM and audio/FDM conversion", Archiv Elektr. Übertr., AEÜ-32, Nov. 1978.

9

SWITCHED-CAPACITOR FILTER SYNTHESIS

ADEL S. SEDRA
DEPARTMENT OF ELECTRICAL ENGINEERING
UNIVERSITY OF TORONTO
TORONTO, ONTARIO, CANADA M5S 1A4

1. INTRODUCTION

There are currently two approaches for the implementation of precision analog filtering functions in monolithic integrated-circuit form. Although both approaches use MOS technology, they differ in a fundamental way: while in one the analog signal is processed directly in its continuous-time form, the other (switched-capacitor) approach is based on processing samples of the analog signal and thus the resulting circuits are discrete-time or sampled-data systems. The continuous-time approach is described by Yannis Tsividis in another chapter of this book. The object of the present chapter is to provide a concise exposition of some of the practical methods of designing switched-capacitor (SC) filters.

Switched-capacitor filters have grown from active-RC filters. The latter utilize op amps together with resistors and capacitors and, depending on production volume, are implemented either on printed circuit boards using IC op amps, metal-film resistors and polystyrene capacitors, or as thick or thin film hybrid circuits. Active-RC is now a mature filter implementation technology with a comprehensive, well-understood design theory. These filters are extensively employed in telecommunication and instrumentation systems and excel in low-frequency (< 100 kHz) applications.

Attempts to *directly* fabricate active-RC filters in monolithic form have not been successful for two reasons: (1) the need for large-valued resistors and capacitors (especially for low-frequency filters), and (2) the need for accurate RC time constants. It will be seen that the SC filter technique circumvents both problems.

Switched-capacitor filters are based on the principle that a capacitor C periodically-switched between two circuit nodes at a sufficiently high rate (f_c) is *approximately* equivalent to a resistor $R = 1/Cf_c$ connecting the two nodes. It is thus possible to realize filter functions using op amps, capacitors and periodically-

272

operated switches. Since MOS technology provides high-quality capacitors, offset-free switches, and moderate-quality op amps it is eminently suited for the realization of SC filters.

Switched-capacitor filters have two other features that make them particularly suited for MOS IC technology:

(1) Large resistors can be simulated using small-sized capacitors. This is a result of the inverse relationship between the value of the switched capacitor C and the realized equivalent resistance. As an example, a 1 pF capacitor, which occupies about 3 mil^2 of chip areas, when switched at the rate of 100 kHz simulates the operation of a 10 MΩ resistor.

(2) The precision of the realized frequency response is dependent on the precision of the clock frequency and the tolerance to which capacitor ratios are implemented. To see how this comes about consider a time constant $C_2 R_1$ and let R_1 be realized as a capacitor C_1 switched at the rate of f_c. Thus the time constant realized is $\dfrac{1}{f_c}\left(\dfrac{C_2}{C_1}\right)$. Since f_c can be accurately controlled using a crystal resonator in the clock oscillator and since capacitor ratios can be realized to a high accuracy (as good as 0.1%) in MOS technology, MOS switched-capacitor filters can achieve very precise frequency responses.

Although the principle of generating frequency-selective responses using periodically-switched capacitors has been known since at the least the mid 1960s, integrated circuit switched-capacitor filters became a reality only in the late 1970s [1,2].

The focus of this chapter is on synthesis methods for SC filters. It is important, however, that the designer acquire an appreciation of MOS technology limitations, especially the effect of such limitations on the characteristics of SC circuit components (op amps, switches and capacitors). For such a study we refer the reader to [3,4]. In addition, a number of chapters in this book deal with relevant and related topics, e.g. the design of CMOS and NMOS op amps. Of special importance is Paul Gray's chapter on the limitations and fundamental limits of SC filters.

2. FIRST-ORDER BUILDING BLOCKS

One of the most popular methods for designing active-RC filters utilizes op amp-RC integrators as basic building blocks. A second-order filter section, known as a *biquad*, is realized using a feedback loop containing one inverting and one noninverting integrator. A high-order (i.e.> 2) filter is realized either as a cascade of these two-integrator-loop biquads or, if a low-sensitivity circuit is desired, as a simulation of the operation of an LC ladder prototype. These ladder simulation circuits also utilize op amp-RC integrators for simulating the operation of the L and C elements of the ladder prototype.

As in the active-RC case, integrators are the basic building blocks of SC filters. Fig. 1a shows the inverting and Fig. 1b shows the noninverting SC integrator circuits that have become standard building blocks of SC networks [5,6]. It will be seen that these integrator circuits are insensitive to the stray capacitances present between the source and drain diffusions of the MOS transistors and the substrate. This feature turns out to be the most important in the selection of a building block. For this reason we shall chapter to SC filter design methods that employ the pair of strays-insensitive integrators of Fig. 1.

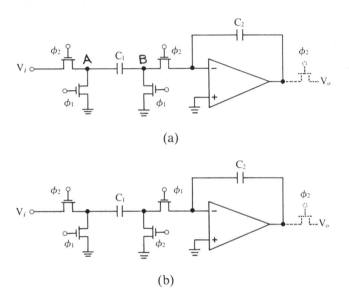

(a)

(b)

Fig. 1   Standard strays-insensitive SC filter building blocks: (a) inverting integrator, (b) noninverting integrator.

2.1 Basic Operation

Both integrator circuits utilize a two-phase nonoverlapping clock having the waveform sketched in Fig. 2. Operation of the inverting integrator circuit of Fig. 1a is illustrated in Fig. 3. During clock phase ϕ_1 both plates of C_1 are grounded (Fig. 3a) and thus C_1 is fully discharged. Meanwhile the op amp is isolated and thus C_2 maintains its charge. Then, during clock phase ϕ_2 capacitor C_1 is connected between the source v_I and the virtual ground input terminal of the op amp (Fig. 3b). Current flows in the direction indicated and C_1 charges to the instantaneous value of v_I. Since the same current flows through C_2, the charge on C_2 will change by an amount equal to the charge deposited on C_1. It follows that during one clock period, T, a

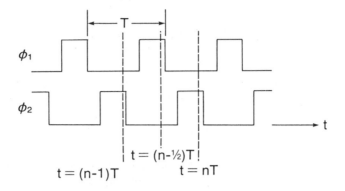

Fig. 2   Waveform of the two-phase nonoverlapping clock used in SC filters.

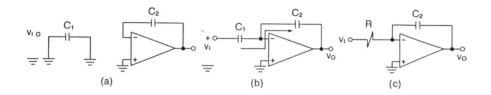

(a)   (b)   (c)

Fig. 3   Operation of the inverting integrator circuit of Fig. 1a: (a) during ϕ_1, (b) during ϕ_2, and (c) continuous-time equivalent circuit.

charge of $C_1 v_I$ is transferred to C_2 in the direction indicated by the arrow in Fig. 3b.

Now, if clocking is performed at a rate $(f_c = 1/T)$ much higher than the rate of change of the input signal the charge transfer process appears almost continuous and thus can be modeled by a continuous current I flowing between v_I and the op amp input with $I = C_1 v_I /T$. This current can be obtained by connecting a resistance R between v_I and the op amp input as indicated in Fig. 3c. The value of this *equivalent resistance* R is given by

$$R \equiv \frac{v_I}{I} = \frac{T}{C_1} \tag{1}$$

The continuous-time equivalent circuit in Fig. 3c is obviously that of an inverting (Miller) integrator. Its transfer function is given by

$$\frac{V_o(\omega)}{V_i(\omega)} = -\frac{1}{j\omega C_2 R} \tag{2}$$

Substituting for R from (1) gives the transfer function of the SC circuit of Fig. 1a as

$$\frac{V_o(\omega)}{V_i(\omega)} = -\frac{1}{j\omega T}\frac{C_1}{C_2} \tag{3}$$

Thus, the circuit of Fig. 1a when clocked at a frequency much higher than the signal frequencies implements an inverting integrator function with a time constant of $T(C_2/C_1)$ or, equivalently, $\dfrac{1}{f_c}\dfrac{C_2}{C_1}$.

The SC circuit of Fig. 1a will have the transfer function derived above irrespective of the value of parasitic capacitances between nodes A and B and ground. This follows because during ϕ_1 these stray capacitances are fully discharged, and during ϕ_2 one is connected across v_I and the other is connected between the op amp input terminals. The end result is that neither of the two parasitic capacitors participates in the charge transfer process and the circuit is fully strays-insensitive.

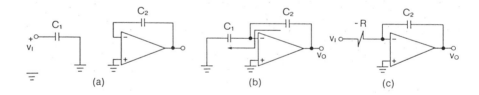

Fig. 4   Operation of the noninverting integrator circuit of Fig. 1b: (a) during ϕ_2, (b) during ϕ_1, and (c) continuous-time equivalent circuit.

The operation of the circuit in Fig. 1b is illustrated in Fig. 4. During ϕ_2, C_1 is charged to the value of v_I and C_2 maintains its previous charge. During ϕ_1, C_1 is fully discharged, thus transferring a charge of $C_1 v_I$ to C_2 in the direction indicated in Fig. 4b. Following a reasoning process similar to that used above for the inverting circuit, one can show that if clocking is done at a sufficiently high rate the circuit has the continuous-time equivalent shown in Fig. 4c. This latter circuit has a transfer function similar to that in (2) except with a positive sign. Thus the SC circuit of Fig. 1b when clocked at a high rate implements a noninverting integrator function with a time constant of $T(C_2/C_1)$. Note that unlike the active-RC case where an additional op amp is usually needed to implement a noninverting integrator, here a simple reversal of clock phases achieves the required inversion. Finally, the reader can easily verify that this circuit too is fully insensitive to stray capacitances.

2.2 Exact Transfer Functions

In deriving the transfer functions above it was assumed that the clocking frequency is much higher than the signal frequencies so that the discrete-time operation of the circuits appears almost continuous. Under this approximation, a switched capacitor C_1 is equivalent to a resistance (T/C_1). The equivalent resistance is positive if the switch phasing of Fig. 1a is used and is negative with the phasing of Fig. 1b. This switched-capacitor — resistor equivalence has been used as the basis for the design of SC filters, as will be illustrated in the next section. In many cases, however, the approximation inherent in this equivalence leads to filter circuits whose responses deviate from the desired. In order to understand the reason for such performance deviation we shall derive the exact transfer functions of the basic building blocks of Fig. 1.

Consider the inverting circuit of Fig. 1a and refer to the clock waveform in Fig. 2. The switch drawn in broken line indicates that the output is sampled at the end of clock phase ϕ_2, that is at the instant $t = nT$. The value of the output voltage at this instant, $v_O(n)$, can be expressed (using the circuit in Fig. 3b) as the sum of the output voltage just before ϕ_2 goes high, $v_O(n-\frac{1}{2})$, and the change in the output voltage that occurs as a result of the charge transferred to C_2 during ϕ_2. This charge is $C_1 v_I(n)$ and the transfer is in the direction indicated in Fig. 3b. Thus,

$$v_O(n) = v_O(n-\frac{1}{2}) - \frac{C_1 v_I(n)}{C_2} \tag{4}$$

The value of $v_O(n-\frac{1}{2})$ can be found from Fig. 3a which shows the circuit configuration during ϕ_1. We see that during ϕ_1, C_2 maintains its charge, thus

$$v_O(n-\frac{1}{2}) = v_O(n-1) \tag{5}$$

Substituting in (4) results in

$$v_O(n) = v_O(n-1) - \frac{C_1}{C_2} v_I(n) \tag{6}$$

Since the operation of the circuit in Fig. 1a is repeated identically every clock period, the first-order difference Eqn. (6) provides a complete description of the operation. To obtain the frequency response of the circuit, the z transform is applied to (6) yielding

$$V_o(z) = z^{-1}V_o(z) - \frac{C_1}{C_2}V_i(z)$$

which after rearranging results is

$$\frac{V_o(z)}{V_i(z)} = -\frac{C_1}{C_2}\frac{1}{1-z^{-1}} \tag{7}$$

This transfer function can be written in the alternate form

$$\frac{V_o(z)}{V_i(z)} = -\frac{C_1}{C_2}\frac{z^{\frac{1}{2}}}{z^{\frac{1}{2}} - z^{-\frac{1}{2}}} \tag{8}$$

To obtain the response for physical frequencies ω we substitute $z = e^{j\omega T}$ which corresponds to evaluating the transfer function in (8) along the unit circle in the complex z plane (see Fig. 5). This results in

$$\frac{V_o}{V_i}(\omega) = -\frac{C_1}{C_2}\frac{e^{j\omega T/2}}{j2\,\sin(\omega T/2)} \tag{9}$$

This transfer function is obviously not that of an ideal integrator; apart from the excess phase lead represented by the factor $e^{j\omega T/2}$ the magnitude is inversely proportional to $\sin(\omega T/2)$ and not to ω as is the case for an ideal integrator. It will be seen shortly that the excess phase lead is not important. The dependence on $\sin(\omega T/2)$, however, unless taken into account in the synthesis process, causes deviations in the filter response. These deviations can be kept small by clocking the filter as a sufficiently high rate so that $(\omega T/2) \ll 1$ and thus $\sin(\omega T/2) \simeq \omega T/2$ which leads to

$$\frac{V_o}{V_i}(\omega) \simeq -\frac{C_1}{C_2}\frac{e^{j\omega T/2}}{j\omega T} \tag{10}$$

Note that apart from the excess phase lead, this transfer function is identical to that derived above (Eqn. (3)) using the SC-resistor equivalence.

A similar analysis can be performed on the noninverting circuit of Fig. 1b and results in the transfer function

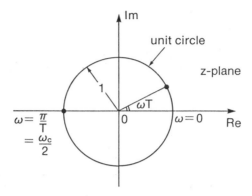

Fig. 5   The frequency response of discrete-time circuits is found by evaluating the transfer function along the unit circle in the z plane.

$$\frac{V_o(z)}{V_i(z)} = \frac{C_1}{C_2} \frac{z^{-\frac{1}{2}}}{z^{\frac{1}{2}} - z^{-\frac{1}{2}}} \tag{11}$$

For physical frequencies, $z = e^{j\omega T}$,

$$\frac{V_o}{V_i}(\omega) = \frac{C_1}{C_2} \frac{e^{-j\omega T/2}}{j2 \sin(\omega T/2)} \tag{12}$$

Except for the expected polarity reversal and the fact that the excess phase here is a lag, this transfer function is identical to that of the inverting circuit. Thus, the comments made above regarding the nonideality of the inverting circuit as an integrator apply equally well to the noninverting circuit.

As will be seen in later sections, SC filters are designed using feedback loops comprising one inverting and one noninverting integrators. We note that the circuits in Fig. 1 can be directly connected to form these two-integrator loops because the input of each circuit is sampled during ϕ_2 and it has been assumed that the output of each integrator is also sampled during ϕ_2. We shall, therefore, consistently use the switch phasing scheme of Fig. 1 in all our designs. Furthermore, under this phasing scheme, the two complementary circuits have excess phase shifts of equal magnitude and opposite polarity. It follows that the excess phase around every two-integrator loop will be zero and we need not concern ourselves with the excess phase in the transfer functions of the integrators.

The transfer functions of the SC circuits of Fig. 1 can be expressed in yet another useful form: Defining the complex variable γ as

$$\gamma \equiv \tfrac{1}{2}(z^{\frac{1}{2}} - z^{-\frac{1}{2}})$$

Eqns. (8) and (11) can be written as

$$\frac{V_o}{V_i} = -\frac{C_1}{2C_2} \frac{z^{\frac{1}{2}}}{\gamma} \tag{13}$$

and

$$\frac{V_o}{V_i} = \frac{C_1}{2C_2} \frac{z^{-\frac{1}{2}}}{\gamma} \tag{14}$$

respectively. Thus, apart from the unimportant numerator terms, the SC circuits behave as perfect integrators in terms of the γ variable. In the digital filter literature, γ is known as the Lossless Digital Integrator (LDI) variable [7]. Physical frequencies ω map to locations on the imaginary axis of the complex γ plane according to

$$\mathrm{Im}(\gamma) = \sin(\omega T/2) \tag{15}$$

We note that $\omega = 0$ maps to $\gamma = 0$ and $\omega = \pi/T$ maps to $\mathrm{Im}(\gamma) = 1$. We shall make use of the γ variable in SC filter synthesis.

2.3 Damped Integrators

Damping an SC integrator can be accomplished by connecting a switched-capacitor across the integrating capacitor C_2, as illustrated in Fig. 6 for the noninverting integrator. Here it is important to note that the clock phasing of the switches in the damping branch results in an equivalent positive resistor. The transfer function of the damped integrator of Fig. 6 can be derived directly in the z domain as follows. We consider the circuit as an integrator with two inputs: V_i through the switched capacitor C_1 and V_o through the switched capacitor C_3. Thus, we can express the output voltage $V_o(z)$ as the sum,

$$V_o(z) = \frac{C_1}{C_2} \frac{z^{-\frac{1}{2}}}{z^{\frac{1}{2}} - z^{-\frac{1}{2}}} V_i(z) - \frac{C_3}{C_2} \frac{z^{\frac{1}{2}}}{z^{\frac{1}{2}} - z^{-\frac{1}{2}}} V_o(z)$$

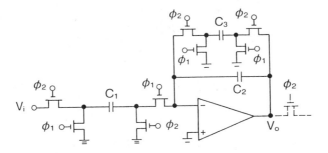

Fig. 6   Damped noninverting integrator

Collecting the terms in $V_o(z)$ results in the transfer function

$$\frac{V_o(z)}{V_i(z)} = \frac{C_1 z^{-\frac{1}{2}}}{C_2(z^{\frac{1}{2}} - z^{-\frac{1}{2}}) + C_3 z^{\frac{1}{2}}} \qquad (16)$$

For physical frequencies,

$$\frac{V_o}{V_i}(\omega) = \frac{C_1 e^{-j\omega T/2}}{j2C_2 \sin\left(\frac{\omega T}{2}\right) + C_3\left(\cos\frac{\omega T}{2} + j\sin\frac{\omega T}{2}\right)} \qquad (17)$$

Note that the damping term in the denominator is a function of frequency and, furthermore, includes an imaginary part. Thus, damping modifies the time constant of the SC integrator. This effect can, however, be accounted for in the design by simply combining the imaginary part of the damping term with the integrator time constant to obtain

$$\frac{V_o}{V_i}(\omega) = \frac{C_1 e^{-j\omega T/2}}{j2\left(C_2 + \frac{C_3}{2}\right)\sin\left(\frac{\omega T}{2}\right) + C_3 \cos\frac{\omega T}{2}} \qquad (18)$$

Now the damping term though real is still frequency dependent. For $(\omega T/2) \ll 1$,

$$\frac{V_o}{V_i}(\omega) \simeq \frac{C_1 e^{-j\omega T/2}}{j\omega T\left(C_2 + \dfrac{C_3}{2}\right) + C_3} \tag{19}$$

Thus the integrator time constant is $\dfrac{T}{C_1}\left(C_2 + \dfrac{C_3}{2}\right)$ and its 3 dB (corner) frequency is $\dfrac{1}{T}\dfrac{C_3}{C_2 + \dfrac{C_3}{2}}$.

The transfer function of the damped integrator (Eqn. (16)) can be expressed in terms of the LDI variable γ as follows. We define a new complex variable

$$\mu \equiv \frac{1}{2}(z^{\frac{1}{2}} + z^{-\frac{1}{2}}) \tag{20}$$

and express the $z^{\frac{1}{2}}$ term in the denominator of (16) as

$$z^{\frac{1}{2}} = \gamma + \mu$$

Thus (16) can be written as

$$\frac{V_o}{V_i} = \frac{C_1 z^{-\frac{1}{2}}}{2\gamma C_2 + (\gamma + \mu)C_3}$$

which can be manipulated to yield

$$\frac{V_o}{V_i} = \frac{C_1 z^{-\frac{1}{2}}}{\gamma(2C_2 + C_3) + \mu C_3} \tag{21}$$

This represents the transfer function of a damped integrator in the γ plane. Still an unusual feature of this integrator is the dependence of the damping term on frequency (through μ). Here we note that for physical frequencies, Eqn. (20) shows that μ is real and is given by

$$\mu = \cos(\omega T/2) \tag{22}$$

2.4 General form of the First-Order Building Block

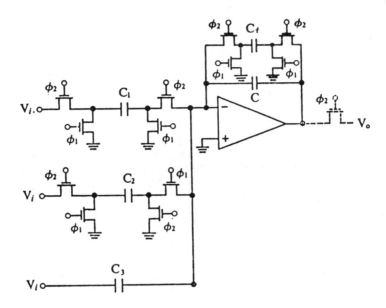

Fig. 7   General form of the strays-insensitive first-order building block.

Fig. 7 shows a general form of the strays-insensitive first-order building block. It has three feed-in branches: an inverting integrator branch through the switched capacitor C_1, a noninverting integrator branch through the switched capacitor C_2 and an inverting gain branch through the unswitched capacitor C_3. In addition, the circuit is damped via the switched capacitor C_f. To obtain the transfer function of the general block we express V_o as the sum

$$V_o = -\frac{C_1}{C}\frac{z^{\frac{1}{2}}}{z^{\frac{1}{2}}-z^{-\frac{1}{2}}}V_i + \frac{C_2}{C}\frac{z^{-\frac{1}{2}}}{z^{\frac{1}{2}}-z^{-\frac{1}{2}}}V_i - \frac{C_3}{C}V_i - \frac{C_f}{C}\frac{z^{\frac{1}{2}}}{z^{\frac{1}{2}}-z^{-\frac{1}{2}}}V_o$$

This leads to the transfer function

$$\frac{V_o}{V_i} = \frac{-C_1 z^{\frac{1}{2}} + C_2 z^{-\frac{1}{2}} - C_3(z^{\frac{1}{2}}-z^{-\frac{1}{2}})}{C(z^{\frac{1}{2}}-z^{-\frac{1}{2}}) + C_f z^{\frac{1}{2}}} \tag{23}$$

which can be expressed in terms of γ and μ as

$$\frac{V_o}{V_i} = \frac{-C_1 z^{\frac{1}{2}} + C_2 z^{-\frac{1}{2}} - 2\gamma C_3}{\gamma(2C + C_f) + \mu C_f} \tag{24}$$

With a bit of practice one can write these transfer functions by inspection.

3. BIQUADRATIC SECTIONS

In this section the first-order building blocks will be employed in the design of SC biquads. These biquad circuits will be generated by simply replacing the resistors in active-RC biquads with their SC equivalent branches. Thus the response of the resulting SC biquads will approach that of their active-RC counterparts only if the clocking frequency is much greater than the biquad pole frequency.

3.1 Circuit Generation

Fig. 8a shows an active-RC biquad based on the two-integrator-loop topology [8]. Note that the noninverting integrator is implemented using a Miller circuit with a negative feed-in resistor $(-R_3)$. This biquad realizes a bandpass function at the output of OA1 and a low-pass function at the output of OA2. The transfer functions are

$$\frac{V_{o1}(s)}{V_i(s)} = \frac{-s\dfrac{1}{C_2 R_1}}{s^2 + s\dfrac{1}{C_2 R_6} + \dfrac{1}{C_2 C_4 R_3 R_5}} \tag{25}$$

$$\frac{V_{o2}(s)}{V_i(s)} = \frac{-1/C_2 C_4 R_1 R_3}{s^2 + s\dfrac{1}{C_2 R_6} + \dfrac{1}{C_2 C_4 R_3 R_5}} \tag{26}$$

Thus the pole frequency is given by

$$\omega_0 = 1/\sqrt{C_2 C_4 R_3 R_5}, \tag{27}$$

the pole-Q is given by

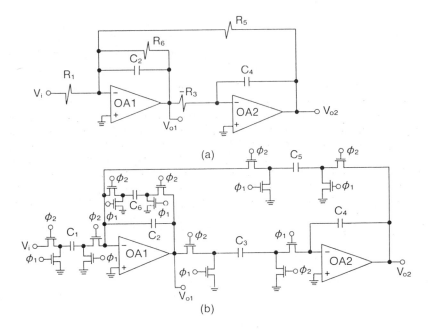

Fig. 8 (a) A two-integrator-loop active-RC biquad. (b) The equivalent SC biquad obtained by replacing each resistor R_i with a switched capacitor $C_i = T/R_i$.

$$Q = \sqrt{\frac{C_2}{C_4}} \frac{R_6}{\sqrt{R_3 R_5}}, \tag{28}$$

and the center-frequency gain of the bandpass function is given by

$$G_0 = -\frac{R_6}{R_1} \tag{29}$$

Usually the biquad is designed to have equal time-constants for the two integrators as follows

$$C_2 = C_4 = C \ , \ R_3 = R_5 = R$$

Thus,
$$CR = \frac{1}{\omega_0}$$

The damping resistor R_6 is given by $R_6 = QR$ and the feed-in resistor R_1 determines the gain.

The equivalent SC biquad in Fig. 8b is obtained by replacing each resistor with a switched capacitor. R_1 is replaced with $C_1 = T/R_1$, $-R_3$ is replaced with $C_3 = T/R_3$, and so on. Note that positive resistors are replaced with inverting SC branches while the negative resistor is replaced with the noninverting SC branch. If the clocking frequency is much greater than the pole frequency, the SC biquad of Fig. 8b realizes a second-order bandpass function at V_{o1} and a second-order low-pass function at V_{o2}. The pole frequency is given by

$$\omega_0 = \frac{1}{T} \sqrt{\frac{C_5}{C_2} \frac{C_3}{C_4}} \tag{30}$$

and the pole-Q is given by

$$Q = \sqrt{\frac{C_2}{C_4}} \frac{\sqrt{C_3 C_5}}{C_6} \tag{31}$$

Examination of the two-integrator loop in Fig. 8b confirms our earlier expectation that the excess phase ($z^{\pm \frac{1}{2}}$ numerator terms) of the two integrators cancel out. Note that there is a delay of one clock period around the loop.

Another important feature of this SC biquad circuit is that it is tolerant of the effects of the finite settling time of the op amps [20]. This comes about because the clock phasing is such that the output of each integrator is not applied to the succeeding integrator during the clock phase in which the output is changing. To be specific, consider OA1. Its output will be changing during ϕ_2. Simultaneously, the output is used to charge C_3. It will not be applied, however, to OA2 until the next half cycle (ϕ_1). Thus the output of OA1 is allowed to settle before it is read by OA2. Similarly, the output of OA2 which changes during ϕ_1 is allowed to settle before it is read during the next half cycle (ϕ_2). In designing SC filters one should always attempt to achieve this "decoupling" property. Fortunately, it will be automatically obtained in all the design methods studied in this chapter.

3.2 Exact Transfer Function

The exact transfer function of the SC biquad in Fig. 8b can be derived directly in the z domain. For the two output voltages we write by inspection

$$V_{o1}(z) = -\frac{C_1}{C_2}\frac{z^{\frac{1}{2}}}{z^{\frac{1}{2}}-z^{-\frac{1}{2}}}V_i(z) - \frac{C_6}{C_2}\frac{z^{\frac{1}{2}}}{z^{\frac{1}{2}}-z^{-\frac{1}{2}}}V_{o1}(z)$$

$$-\frac{C_5}{C_2}\frac{z^{\frac{1}{2}}}{z^{\frac{1}{2}}-z^{-\frac{1}{2}}}V_{o2}(z)$$

$$V_{o2}(z) = \frac{C_3}{C_4}\frac{z^{-\frac{1}{2}}}{z^{\frac{1}{2}}-z^{-\frac{1}{2}}}V_{o1}(z)$$

These two equations can be combined to obtain the two transfer functions $\dfrac{V_{o1}}{V_i}$ and $\dfrac{V_{o2}}{V_i}$. As an example, for the first transfer function we obtain

$$\frac{V_{o1}(z)}{V_i(z)} = \frac{-\dfrac{C_1}{C_2}z^{\frac{1}{2}}(z^{\frac{1}{2}}-z^{-\frac{1}{2}})}{(z^{\frac{1}{2}}-z^{-\frac{1}{2}})^2 + \dfrac{C_6}{C_2}z^{\frac{1}{2}}(z^{\frac{1}{2}}-z^{-\frac{1}{2}}) + \dfrac{C_5}{C_2}\dfrac{C_3}{C_4}} \tag{32}$$

For physical frequencies we have

$$\frac{V_{o1}}{V_i}(\omega) = \frac{-j\dfrac{C_1}{C_2}2\sin\left(\dfrac{\omega T}{2}\right)e^{j\omega T/2}}{\left[\dfrac{C_5}{C_2}\dfrac{C_3}{C_4} - 4\sin^2\left(\dfrac{\omega T}{2}\right)\left(1+\dfrac{C_6}{2C_2}\right)\right] + j\dfrac{C_6}{C_2}\sin\omega T} \tag{33}$$

which for $\omega T \ll 1$ reduces to

$$\frac{V_{o1}}{V_i}(\omega) \simeq \frac{-j\omega T\dfrac{C_1}{C_2}e^{j\omega T/2}}{\left[\dfrac{C_5}{C_2}\dfrac{C_3}{C_4} - \omega^2 T^2\left(1+\dfrac{C_6}{2C_2}\right)\right] + j\left(\dfrac{C_6}{C_2}\right)\omega T} \tag{34}$$

Apart from the unimportant numerator term $e^{j\omega T/2}$, this equation is in the form of the second-order bandpass function

$$T(\omega) = \frac{-j\omega k}{(\omega_0^2 - \omega^2) + j\frac{\omega\omega_0}{Q}}$$

It follows, that

$$\omega_0 = \frac{1}{T}\sqrt{\frac{C_5}{C_2}\frac{C_3}{C_4}\Big/\left(1 + \frac{C_6}{2C_2}\right)} \tag{35}$$

and,

$$Q = \sqrt{\frac{C_2}{C_4}}\frac{\sqrt{C_3 C_5}}{C_6}\sqrt{1 + \frac{C_6}{2C_2}} \tag{36}$$

These values are slightly different from those given by Eqns. (30) and (31). The difference is due to the assumption, made in deriving (30) and (31), that damping does not affect the integrator time constant. That this is not the case for SC integrators has been demonstrated in the previous section. Thus Eqns. (35) and (36) are more accurate than (30) and (31) and should be the ones used in design. Nevertheless, the resulting design will still be approximate since it is based on the assumption that $\omega T \ll 1$. Better results can be obtained if design is based on the exact transfer function in Eqn. (33). For instance, ω_0 can be found from

$$4 \sin^2\frac{\omega_0 T}{2} = \left(\frac{C_3}{C_4}\frac{C_5}{C_2}\right)\Big/\left(1 + \frac{C_6}{2C_2}\right) \tag{37}$$

More details on this approach for designing SC biquads can be found in [5].

From the exact transfer function in Eqn. (32) it can be seen that the SC biquad of Fig. 8b has one transmission zero at $z = 1$ and the other at $z = 0$. While the first zero corresponds to $\omega = 0$ the second has no corresponding physical frequency. Therefore it contributes little to the selectivity of the bandpass filter response. In other words, while the original continuous-time active-RC circuit has a zero at $\omega = 0$ and another at $\omega = \infty$, both contributing to selectivity, the discrete-time SC filter has only one physical-frequency zero (at $\omega = 0$). A more selective bandpass response can be obtained by placing a transmission zero at half the clock frequency ($\omega = \frac{\pi}{T}$). Such a design will be shown in Section 5.

3.3 Capacitive Damping

The active-RC circuit of Fig. 8a is damped by placing the resistor R_6 across the integrator capacitor C_2. The damping resistor can of course be placed across the other integrator capacitor, C_4. Alternatively, damping can be achieved by placing a *capacitor* across $-R_3$ or R_5. This capacitive damping is seldom used in active-RC filters because in that technology one attempts to use resistors in favor of reducing the number of capacitors. Capacitive damping, however, can be attractive in SC filters. As an example, it can be implemented in the circuit of Fig. 8b by connecting an unswitched capacitor between the output of OA2 and the inverting input of OA1 (while, of course, removing the C_6 branch).

Both forms of damping biquads have been used in the literature [6,9,10]. Depending on the application one form might lead to lower total capacitance and hence becomes preferred.

3.4 Switch Sharing

In SC circuits, it is usually possible to share a set of switches between two or more paths, thus reducing the total number of switches required. To illustrate, consider the circuit in Fig. 8b. The left-hand-side plates of C_5 and C_6 are switched to ground during ϕ_1 and to the inverting input of OA1 during ϕ_2. Obviously, only one set of switches is sufficient to accomplish this task. Similarly, the two switches connecting to the right-hand-side plate of C_6 can be eliminated and that plate of C_6 connected to the left-hand-side plate of C_3. Although switch sharing reduces the number of switches and thus silicon area it can cause signal coupling problems that limit the performance of high-frequency SC filters.

4. LADDER FILTERS

Doubly-terminated LC ladder networks that are designed to effect maximum power transfer from source to load over the filter passband feature very low sensitivities to variations in their component values [11]. This fact has over the years spurred considerable interest in finding active-RC, digital and switched-capacitor filter structures which simulate the internal workings of LC ladder prototype networks. Essentially these filter structures consist of a connection of first-order blocks that implement the I-V integral relationship of the L and C elements in the prototype. That is, the basic building blocks are integrators.

The straightforward approach to the design of SC filters based on the simulation of LC ladder prototype, to be referred to as SC ladder filters, is to first design a suitable active-RC filter and then replace the integrators with the SC blocks of Fig. 1. This approach works well only if the clock frequency is much higher than the signal

289

frequencies. In this section we study a more refined approach to the design of SC ladder filters. Our approach takes into account the fact that the transfer functions of the SC building blocks are inversely proportional to $\sin(\omega T/2)$ and not ω. It also takes into account the fact that damping affects the time constants of SC integrators. The resulting designs, therefore, have frequency responses that are closer to the desired ideal than those obtained by replacing the resistors in an active-RC filter. Nevertheless, the resulting designs are still not exact because the method does not take into account the frequency dependence inherent in the damping of SC integrators. Exact design methods will be studied in Section 5.

4.1 Circuit Generation

The method proposed here makes use of the observation made in Section 2 that the SC blocks behave as perfect integrators in the γ plane. Therefore, we shall perform our design in that plane. The starting point in the design is a set of attenuation specifications posed versus frequency ω, such as those shown in Fig. 9a for a low-pass filter. The specifications can be cast versus the γ-plane frequency by transforming the horizontal axis points according to $\text{Im}(\gamma) = \sin(\omega T/2)$. This process, known as *prewarping*, results in the attenuation specifications shown in Fig. 9b. Note that half the clock frequency maps to $\text{Im}(\gamma) = 1$.

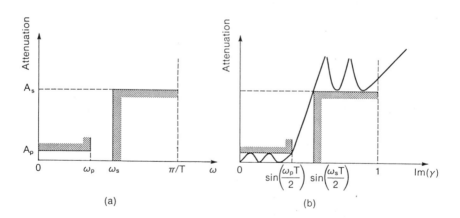

Fig. 9   (a) Attenuation specifications for a low-pass SC filter. (b) The specifications in (a) versus $\text{Im}(\gamma) = \sin(\omega T/2)$.

Next, filter design tables or computer programs are used to find a suitable transfer function and/or LC ladder network that meets the specifications in Fig. 9b. For illustration, we shall assume that a fifth-order elliptic filter meets the given specifications. The attenuation response for such a filter is shown in Fig. 9b, and Fig. 10a shows its LC ladder network realization. Because this network is designed to meet specifications posed versus $\mathrm{Im}(\gamma)$, its inductances have impedances γl_i and its capacitances have impedances $1/\gamma c_i$. (This of course has no bearing on the process of finding the element values of the LC network.)

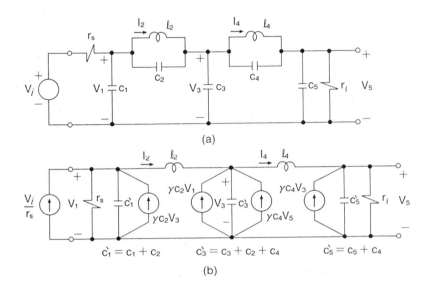

(a)

(b)

Fig. 10 (a) γ-plane LC ladder network realization of a fifth-order elliptic filter that meets the specifications in Fig. 9b. (b) Norton's theorem is used to replace c_2 and c_4 with voltage-controlled current sources.

The next step in the design involves the use of Norton's theorem to replace the bridging capacitors c_2 and c_4 by voltage-controlled current sources, resulting in the modified network shown in Fig. 10b. For this network we can write five equations that describe the operation of its five reactive components, as follows:

$$V_1 = \frac{(V_i/r_s) - I_2 + \gamma c_2 V_3}{\gamma c_1' + (1/r_s)} \tag{38}$$

291

$$I_2 = \frac{V_1 - V_3}{\gamma l_2} \tag{39}$$

$$V_3 = \frac{I_2 - I_4 + \gamma c_2 V_1 + \gamma c_4 V_5}{\gamma c_3'} \tag{40}$$

$$I_4 = \frac{V_3 - V_5}{\gamma l_4} \tag{41}$$

$$V_5 = \frac{I_4 + \gamma c_4 V_3}{\gamma c_5' + (1/r_l)} \tag{42}$$

These equations are in the form of the transfer function of the general first-order block of Fig. 7 (Eqn. (24)). From this analogy we can directly sketch the SC circuit realization shown in Fig. 11. The transfer functions of the five blocks in Fig. 11 can be written by inspection as

$$V_1' = \frac{-C_1 z^{\frac{1}{2}} V_i' - C_2 z^{\frac{1}{2}} V_2' - \gamma 2 C_{16} V_3'}{\gamma (2C_3 + C_s) + \mu C_s} \tag{43}$$

$$V_2' = \frac{C_4 z^{-\frac{1}{2}} V_1' + C_5 z^{-\frac{1}{2}} V_3'}{\gamma 2 C_6} \tag{44}$$

$$V_3' = \frac{-C_7 z^{\frac{1}{2}} V_2' - C_8 z^{\frac{1}{2}} V_4' - \gamma 2 C_{15} V_1' - \gamma 2 C_{17} V_5'}{\gamma 2 C_9} \tag{45}$$

$$V_4' = \frac{C_{10} z^{-\frac{1}{2}} V_3' + C_{11} z^{-\frac{1}{2}} V_5'}{\gamma 2 C_{12}} \tag{46}$$

$$V_5' = \frac{-C_{13} z^{\frac{1}{2}} V_4' - \gamma 2 C_{18} V_3'}{\gamma (2C_{14} + C_l) + \mu C_l} \tag{47}$$

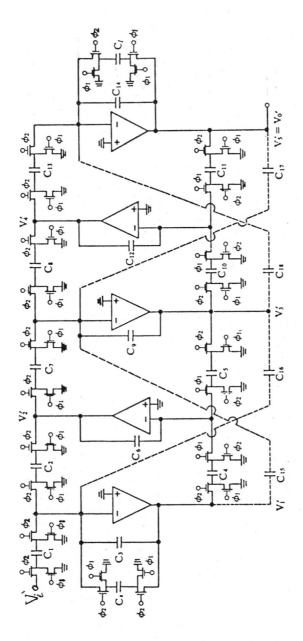

Fig. 11   Switched-capacitor circuit that simulates the operation of the LC network of Fig. 10b.

Comparing corresponding equations between the set (38)-(42) and the set (43)-(47) we obtain the following correspondences between the ladder currents and voltages (Fig. 10b) and the op amp voltages (Fig. 11):

$$(-V_i') \Leftrightarrow V_i \qquad (z^{-\frac{1}{2}}V_1') \Leftrightarrow V_1 \qquad V_2' \Leftrightarrow I_2$$

$$(-z^{-\frac{1}{2}}V_3') \Leftrightarrow V_3 \qquad (-V_4') \Leftrightarrow I_4 \qquad (z^{-\frac{1}{2}}V_5') \Leftrightarrow V_5$$

With these correspondences we see that the SC circuit in Fig. 11 exactly simulates the operation of the LC network of Fig. 10b except for the two μ terms in the denominators of Eqns. (43) and (47). For physical frequencies $\mu = \cos(\omega T/2)$ and thus the SC circuit simulates a ladder having frequency-dependent terminations. We should therefore expect the frequency response of the SC circuit to deviate from that of the ladder with the deviation being small at clock frequencies that are sufficiently high so that $\cos(\omega T/2) \simeq 1$. A simple formula that predicts the expected deviation in attenuation response has been given in [12].

4.2 Design

Initial values for the capacitors of the SC circuit in Fig. 11 are obtained by equating the coefficients of corresponding terms between the set (38)-(42) and the set (43)-(47). The result is

$$C_1 = 1/r_s \qquad C_2 = 1 \qquad C_s = 1/r_s \qquad C_3 = (c_1' - \frac{1}{r_s})/2$$

$$C_4 = C_5 = 1 \qquad C_6 = l_2/2 \qquad C_7 = C_8 = 1 \qquad C_9 = c_3'/2$$

$$C_{10} = C_{11} = 1 \qquad C_{12} = l_4/2 \qquad C_{13} = 1 \qquad C_l = 1/r_l$$

$$C_{14} = (c_5' - \frac{1}{r_l})/2 \qquad C_{15} = C_{16} = \frac{c_2}{2} \qquad C_{17} = C_{18} = \frac{c_4}{2}$$

Using these capacitor values gives the SC filter a dc gain equal to that of the LC ladder. If the gain of the SC filter is to be a factor K greater than that of the LC ladder then the value of C_1 must be changed to KC_1.

The initial capacitor values above must then be scaled so that the SC filter has the widest possible dynamic range. Dynamic range scaling is usually based on equalizing the peaks obtained at the outputs of the op amps as the frequency of an input sinusoid is swept over a desired band. To perform scaling one needs the value of the signal peaks, denoted $\hat{V}_1'$, $\hat{V}_2'$, $\hat{V}_3'$, $\hat{V}_4'$ and $\hat{V}_5'$. These can be found from a simulation of the SC circuit [13] with the capacitor values given above. Alternatively, the peak values can be obtained from an analysis of the LC ladder network.

Once the spectral peaks have been determined the capacitor values can be scaled as follows: capacitor C_{ij} which is connected between the output of op amp i and the input of op amp j is changed to $C_{ij}(\hat{V}_i'/\hat{V}_j')$. As an example, capacitor C_2 changes from the initial value of unity to $(\hat{V}_2'/\hat{V}_1')$. Note that this scaling process preserves the magnitude of loop transmission of every two-integrator loop, and hence the filter transfer function (except for a gain constant) remains unchanged. To keep the overall gain unchanged, C_1 must be scaled by the factor $(\hat{V}_5'/\hat{V}_1')$.

The next and final step in the design involves scaling the capacitor values to minimize the total capacitance. This is achieved by considering the five blocks one at a time. For block #i, assume that the smallest capacitance connected to the virtual ground input of the op amp is of value $\check{C}_i$. We wish to scale so that this capacitor is equal to the minimum capacitor value possible with the given technology; call it C_{min} (this may be 0.1 pF or so). It follows that for block #i the scaling factor $k_i = C_{min}/\check{C}_i$. Every capacitor of block #i is multiplied by k_i. The transfer function of block #i obviously remains unchanged. The process is then repeated for the other blocks.

This completes the design procedure for SC ladder filters. As noted earlier, this design is not exact because of the frequency-dependent terminations realized. If the clock frequency is lowered to the point where the deviation in response is unacceptably high then one must consider using the exact design method of the next section. Another shortcoming of the method above is that transmission zeros at $\gamma = \infty$ do not map to physical frequencies and thus their effect on selectivity is lost.

5. EXACT DESIGN METHODS

In this section we present methods for the *exact* synthesis of SC filters. These design methods result in an SC filter whose transfer function is identical to the desired one. Each of the methods presented forms a complete design process: beginning with the filter specifications and ending with an SC circuit realization.

5.1 The Bilinear Transformation

All exact design methods [9,17,21] make use of the bilinear transformation [14,8]

$$\lambda = \frac{z-1}{z+1} \tag{48}$$

which maps the unit circle in the z-plane (which is the contour of physical frequency points) onto the entire imaginary axis of the λ plane. The relationship between the frequency ω and the corresponding point Ω on the imaginary axis of the λ plane is given by

$$\Omega = \tan(\omega T/2) \qquad (49)$$

We note that $\omega = 0$ maps to $\Omega = 0$ and $\omega = \dfrac{\pi}{T}$ (half the clocking frequency) maps to $\Omega = \infty$. This is a one-to-one mapping with half the unit circle ($\omega T = 0$ to $\omega T = \pi$) being mapped onto the positive half of the $j\Omega$ axis ($\Omega = 0$ to $\Omega = \infty$). In the continuous-time analog filter literature there exists a wealth of design data and synthesis methods for filters whose responses are specified along the imaginary axis of a complex plane. These tools can be applied to the design of discrete-time filters (of which SC filters is a special case) via the bilinear transformation.

5.2 Approximation and LC Ladder Synthesis

Synthesis of SC filters using the bilinear transformation proceeds as follows. The given attenuation specifications versus frequency ω are recast versus Ω. This step, known as *prewarping*, uses the relationhip in (49) and results in modifying the critical frequency points along the horizontal axis without affecting the vertical (attenuation) axis. For example, the low-pass attenuation specifications of Fig. 9a transform to those shown in Fig. 12.

We note that the attenuation specifications versus Ω look identical to those of continuous-time analog filters. Thus filter design tables and/or computer programs [15] can be used to obtain a transfer function $T(\lambda)$ and/or an LC ladder network whose attenuation function meets the prewarped specifications. If $T(\lambda)$ has a transmission zero at $\lambda = \infty$, the corresponding SC filter will have a zero at $\omega = \pi/T$. Thus, unlike the γ-plane design method of Section 4, transmission zeros at ∞ are not lost and their full effect on filter selectivity is realized.

Having performed filter approximation, and possibly also obtained an LC realization, it now remains to obtain an SC circuit implementation. In the following we present a number of methods for obtaining SC realizations starting from a transfer function $T(\lambda)$ or from an LC ladder realization of $T(\lambda)$.

5.3 Cascade Design

The SC filter can be realized as a cascade of biquads and, in the case of odd-order filters, a first-order section. Cascade design is performed directly on the function $T(\lambda)$ obtained from filter approximation. First, $T(\lambda)$ is factored into the product of second-order (and possibly one first-order) functions. In performing factorization the designer faces the questions of pole-zero pairing, cascading sequence, and gain assignment. These problems are identical to those encountered in the design of

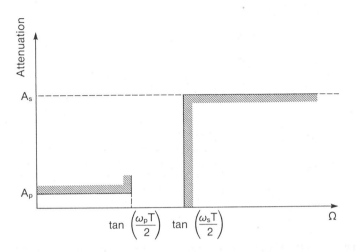

cascade active-RC filters [8] and will not be considered here.

Second-order transfer functions obtained in the factorization of $T(\lambda)$ will be special cases of the general biquadratic function

$$t(\lambda)=\pm \frac{a_2\lambda^2 + a_1\lambda + a_0}{\lambda^2 + b_1\lambda + b_0} \qquad (50)$$

SC biquad circuits capable of realizing this general transfer function have been given in [6,9,10]. As an example, we show one such circuit in Fig. 13. Analysis of this circuit, following the method of Section 3, results in the z-plane transfer function

$$\frac{V_o(z)}{V_i(z)} = -\frac{K_3z^2 + (-2K_3+K_1K_5+K_2K_5)z + (K_3-K_2K_5)}{z^2 + (-2+K_4K_5+K_5K_6)z + (1-K_5K_6)} \qquad (51)$$

The λ-plane transfer function is obtained by substituting

$$z = \frac{1+\lambda}{1-\lambda} \qquad (52)$$

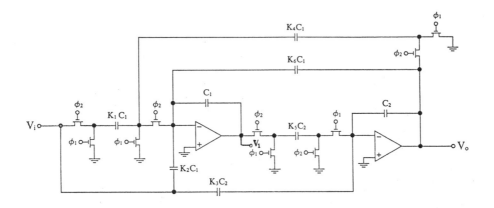

Fig. 13   An SC biquad circuit capable of realizing the general bil-
inearly transformed function in Eqn. (50).

which is the inverse of the transformation in (48). Design equations for the SC
biquad can be then obtained by equating the coefficients in the resulting transfer
function to the corresponding coefficients in (50). Equating corresponding denominator coefficients results in two equations in the three capacitor ratios that determine the biquad poles,

$$K_4 K_5 = 4mb_0 \tag{53}$$

$$K_5 K_6 = 2mb_1 \tag{54}$$

where $m = 1/(1 + b_1 + b_0)$ (55)

Because we have only two equations in the three unknowns K_4, K_5 and K_6, one of
the capacitor ratios can be chosen arbitrarily. This choice is usually done with a
view to maximizing the biquad dynamic range. For high-Q biquads, this is approximately obtained when the two loop time constants are made equal, that is, $K_4 = K_5$.

Design equations for the feed-in capacitor ratios, which determine the biquad
transmission zeros, are obtained by equating coefficients of corresponding numerator
terms. The result for some special cases are:

(a) Low-Pass $(a_1 = a_2 = 0)$

$$K_2 = 0 \quad K_1 K_5 = 4ma_0 \quad K_3 = ma_0$$

(b) Bandpass $(a_0 = a_2 = 0)$

$$K_1 = 0 \quad K_2 K_5 = 2ma_1 \quad K_3 = ma_1$$

(c) High-Pass $(a_0 = a_1 = 0)$

$$K_1 = K_2 = 0 \quad K_3 = ma_2$$

(d) Notch $(a_1 = 0)$

$$K_2 = 0 \quad K_1 K_5 = 4ma_0 \quad K_3 = m(a_0 + a_2)$$

The circuit of Fig. 13 realizes a negative bandpass function. As will be seen below, a positive bandpass is frequently needed. It can be obtained by taking the output at V_1. While the pole design equations remain unchanged (Eqns. (53) and (54)) the zeros' equations become

$$K_1 = 0 \qquad K_3 K_4 = 2ma_1 \qquad K_2 = ma_1 \left(1 + \frac{b_1}{b_0} \right)$$

The reader is urged to note that low-pass and bandpass biquad circuits designed using the bilinear transformation (such as the circuit in Fig. 13) have frequency responses that are more selective than those studied in Section (3). This is a result of the transmission zero placed at $\omega = \pi/T$ (which is the mapping of $\lambda = \infty$).

5.4 Coupled-Biquad Designs

Though simple to obtain, cascade realizations are usually quite sensitive, especially in the filter passband, to the inevitable tolerances in the values of their components. Less sensitive realizations can be obtained by coupling biquad circuits in a feedback structure. A number of such structures, including the leap-frog and the follow-the-leader feedback, have been proposed for active-RC filters [16,8]. These coupled-biquad designs can be applied to the SC case as follows. Use active-RC design methods [8] to obtain a block-diagram realization of the transfer function $T(\lambda)$. Such a block diagram will consist of biquadratic blocks coupled together with feedback and feedforward paths. Then implement each of the biquadratic blocks with an SC circuit such as that in Fig. 13. Coupling the biquads can be achieved by using multiple feed-in components, with the summing performed at the virtual grounds of the op amps. We note that this method applies only for even-order filters. We shall illustrate this design approach by two different and important structures.

5.4.1 Design of Geometrically-Symmetric Bandpass Filters

A bandpass filter whose prewarped specifications (versus Ω) can be made geometrically symmetric around the center frequency can be designed using the low-pass to bandpass transformation. Consider as an example a sixth-order elliptic bandpass filter. Fig. 14a shows the LC ladder realization of the third-order elliptic low-pass prototype. Replacing the bridging capacitor c_2 with two voltage-controlled current sources, and the input signal source with its Norton's equivalent, results in the network shown in Fig. 14b. A leapfrog block diagram realization of this network is given in Fig. 14c. Here the summing coefficients are included for the purpose of maximizing the dynamic range of the resulting realization. Dynamic range optimization is based on equalizing the spectral peaks obtained at the op amp outputs as the input signal is swept in frequency. The values of the summing coefficients are obtained from an analysis of the LC ladder of Fig. 14a with a swept-frequency input and determining the peaks of the spectra of V_1, I_2 and V_3 (denoted $\hat{V}_1$, $\hat{I}_2$ and $\hat{V}_3$). The factor k in the input summing coefficient is the ratio of the desired center-frequency gain to the dc gain of the LC ladder.

Applying the low-pass to bandpass transformation $\lambda = \dfrac{\lambda^2 + \Omega_0^2}{\lambda B}$ to the block diagram realization in Fig. 14c results in the coupled-biquad realization of the bandpass filter, shown in Fig. 14d. Note that Ω_0 is the center-frequency and B is the ripple bandwidth (in the λ plane; i.e. versus Ω).

The realization in Fig. 14d consists of three biquads. As an example, consider biquad #1. It has three separate inputs: two realizing bandpass functions and one realizing a notch function. The realization of different filtering functions (same poles, different zeros) within the same biquad circuit is achieved by using different sets of feed-in branches.

The final step in the design is to implement each biquad in the block diagram realization with an SC circuit, such as the circuit of Fig. 13.

5.4.2 Design Based on Simulating the Node Voltages of an LC Ladder Prototype

Our second coupled-biquad design method applies to general-parameter bandpass filters, that is, those that are not necessarily derived from a low-pass prototype. First an LC ladder realization of the prewarped specifications is obtained. This ladder must, however, be in a special form. To illustrate, we show in Fig. 15a a suitable LC ladder realization of an eighth-order bandpass filter. Note that each branch in the ladder consists of either a single component or a parallel connection of single components. The reader will shortly appreciate the reason for imposing this restriction.

The voltage at each node in the ladder circuit can be expressed in terms of the voltages of the preceding and succeeding nodes. For the ladder in Fig. 15a we can write

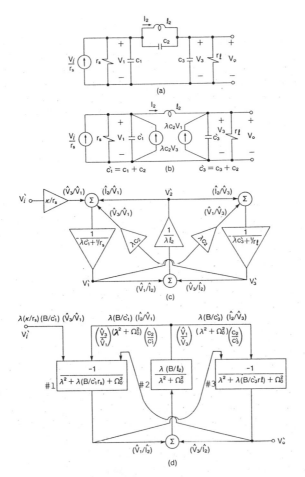

Fig. 14 Illustrating the exact design of a sixth-order geometrically-symmetric bandpass filter. (a) LC realization of the low-pass prototype. (b) The bridging capacitor c_2 is eliminated by the application of Norton's theorem. (c) Leapfrog simulation of the operation of the network in (b). (d) Coupled-biquad realization obtained by applying the low-pass-to-bandpass transformation to the block diagram in (c).

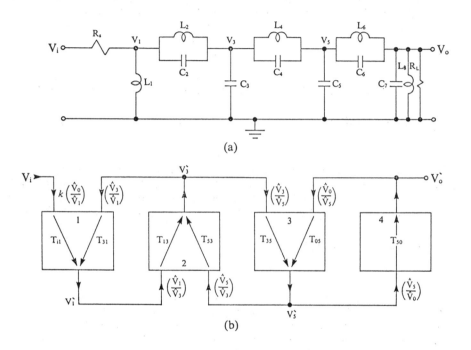

Fig. 15 (a) LC ladder realization of an eighth-order general-parameter bandpass filter. (b) Block diagram of a realization that simulates the node voltages of the LC ladder.

$$V_1 = T_{i1}V_i + T_{31}V_3 \qquad\qquad (56)$$

$$V_3 = T_{13}V_1 + T_{53}V_5 \qquad\qquad (57)$$

$$V_5 = T_{35}V_3 + T_{o5}V_o \qquad\qquad (58)$$

$$V_o = T_{5o}V_5 \qquad\qquad (59)$$

where the transfer functions are defined as

$$T_{i1} = \left.\frac{V_1}{V_i}\right|_{V_3=0} \qquad\qquad T_{31} = \left.\frac{V_1}{V_3}\right|_{V_i=0}$$

$$T_{13} = \left.\frac{V_3}{V_1}\right|_{V_5=0} \qquad\qquad T_{53} = \left.\frac{V_3}{V_5}\right|_{V_1=0}$$

$$T_{35} = \left.\frac{V_5}{V_3}\right|_{V_o=0} \qquad\qquad T_{o5} = \left.\frac{V_5}{V_o}\right|_{V_3=0}$$

$$T_{5o} = \frac{V_o}{V_5}$$

Each pair of transfer functions associated with a particular node (e.g. T_{i1} and T_{31}) have the same poles but different zeros. Therefore such a pair of transfer functions can be realized in one biquad circuit. Fig. 15b shows a block diagram realization of the eighth-order bandpass filter under discussion. This realization is obtained directly from Eqns. (56)-(59). The summing coefficients shown are for the purpose of maximizing dynamic range. The various transfer functions shown can be obtained from the LC ladder using the definitions above. As an example, we have

Biquad #1

$$T_{i1} = \frac{\lambda\dfrac{1}{C_2R_s}}{\lambda^2 + \lambda\dfrac{1}{C_2R_s} + \dfrac{1}{C_2(L_1 \| L_2)}}$$

303

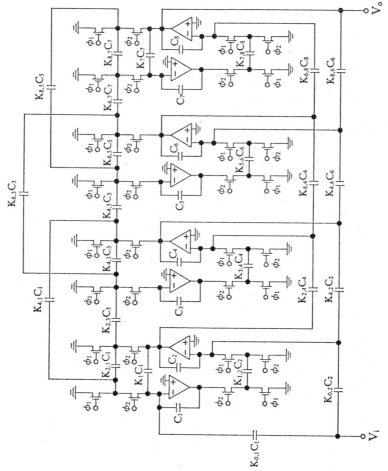

Fig. 16   SC realization of the eighth-order bandpass filter whose block diagram is given in Fig. 15b.

$$T_{31} = \frac{\lambda^2 + \dfrac{1}{L_2 C_2}}{\lambda^2 + \lambda \dfrac{1}{C_2 R_s} + \dfrac{1}{C_2(L_1 \| L_2)}}$$

Biquad #2

$$T_{13} = \frac{C_2}{C_2 + C_3 + C_4} \cdot \frac{\lambda^2 + \dfrac{1}{L_2 C_2}}{\lambda^2 + \dfrac{1}{(L_2 \| L_4)(C_2 + C_3 + C_4)}}$$

$$T_{53} = \frac{C_4}{C_2 + C_3 + C_4} \cdot \frac{\lambda^2 + \dfrac{1}{L_4 C_4}}{\lambda^2 + \dfrac{1}{(L_2 \| L_4)(C_2 + C_3 + C_4)}}$$

and so on for the remaining functions. Each biquad can be realized using the circuit of Fig. 13. The complete realization of the eighth-order bandpass filter is shown in Fig. 16. Further details on this particular design can be found in [9]. Fig. 17 shows the response measured on a discrete prototype of this design.

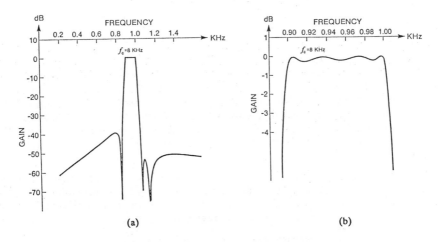

(a) (b)

Fig. 17 Measured frequency response of the circuit in Fig. 16 — with an expanded view of the passband shown in (b).

5.5 A General Exact Design Method

The coupled-biquad synthesis methods above apply only to even-order filters. We now present a general method [17] for the synthesis of SC filters based on the exact simulation of LC ladder prototypes. Although the method will be demonstrated for the case of odd-order elliptic low-pass filters, it has been extended to other filter functions [18], to high-pass filters [19] and to bandpass filters. As will be seen shortly, the method presented removes all the shortcomings of the approximate method studied in Section 4.

The general method employs the three complex variables γ, μ and λ. From the definitions of these variables above we can write

$$\gamma = \tfrac{1}{2}(z^{\frac{1}{2}}-z^{-\frac{1}{2}}) = \tfrac{1}{2}(e^{sT/2}-e^{-sT/2}) = \sinh(sT/2) \tag{60}$$

$$\mu = \tfrac{1}{2}(z^{\frac{1}{2}}+z^{-\frac{1}{2}}) = \tfrac{1}{2}(e^{sT/2}+e^{-sT/2}) = \cosh(sT/2) \tag{61}$$

$$\lambda = \frac{z-1}{z+1} = \frac{z^{\frac{1}{2}}-z^{-\frac{1}{2}}}{z^{\frac{1}{2}}-z^{-\frac{1}{2}}} = \tanh(sT/2) \tag{62}$$

Two important relationships that we will find useful are:

$$\lambda = \gamma/\mu \tag{63}$$

$$\text{and} \quad \mu^2 = 1 + \gamma^2 \tag{64}$$

To illustrate the method consider as an example low-pass filter specifications (Fig. 9a) whose prewarped version (Fig. 12) can be met by a fifth-order elliptic filter. Fig. 18a shows the LC ladder realization of such a filter, which is designed in the λ plane. If we attempt to directly simulate the operation of this network we will need SC circuits having transfer functions of the form $1/\lambda$. Unfortunately, these λ-plane or *bilinear* integrators cannot be realized in a single-op amp strays-insensitive circuit. We must therefore seek an alternative to the direct simulation of the λ-plane ladder of Fig. 18a.

The alternative approach involves scaling the impedances of the ladder in Fig. 18a by the factor μ. Specifically, if we divide all impedances by μ, the voltage transfer function remains unchanged and we obtain the ladder network in Fib. 18b. Here we note that the termination resistors have changed to r_s/μ and r_l/μ. A capacitor C in the ladder of Fig. 18a gives rise to an element with impedance $\dfrac{1}{\mu}\left(\dfrac{1}{\lambda C}\right)$ which using Eqn. (63) is simply $1/\gamma C$. Thus a capacitor in the λ-plane ladder of Fig. 18a transforms into a capacitor of equal value in the γ-plane ladder of Fig. 18b.

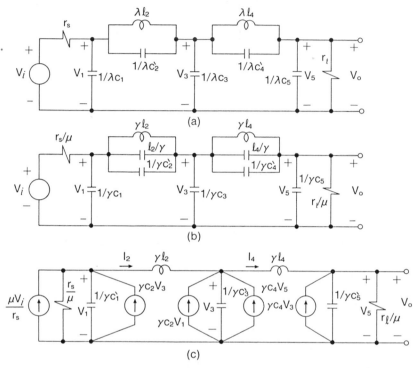

Fig. 18   (a) LC ladder realization of a fifth-order elliptic low-pass filter that meets the prewarped specification of Fig. 12. (b) The network of (a) impedances scaled by μ. (c) The network of (a) after eliminating the bridging capacitors using Norton's theorem.

An inductance l in the ladder of Fig. 18a transforms to a branch with impedance $\lambda l/\mu$ which using Eqn. (64) can be shown to be the parallel equivalent of the two impedances γl and l/γ. In terms of the variable γ, this is an inductance l in parallel with a capacitance $1/l$.

Consider now the transformed network in Fig. 18b. The parallel capacitors in each of the series arms of the ladder can be combined as follows

$$c_2 = c_2' + (1/l_2)$$

$$c_4 = c_4' + (1/l_4)$$

These bridging capacitors can be replaced with voltage-controlled current sources as shown in Fig. 18c where

$$c_1' = c_1 + c_2' + (1/l_2)$$

$$c_3' = c_3 + c_2' + (1/l_2) + c_4' + (1/l_4)$$

$$c_5' = c_5 + c_4' + (1/l_4)$$

Now the network of Fig. 18c is identical to that of Fig. 10b except that here the terminations are frequency dependent. This is a most welcome result because we know that strays-insensitive damped SC integrators, which have to be used to simulate the end elements, do in fact realize frequency-dependent terminations. To be specific, the operation of the ladder in Fig. 18c is fully described by the following five equations:

$$V_1 = \frac{\mu(V_i/r_s) - I_2 + \gamma c_2 V_3}{\gamma c_1' + (\mu/r_s)} \tag{65}$$

$$I_2 = \frac{V_1 - V_3}{\gamma l_2} \tag{66}$$

$$V_3 = \frac{I_2 - I_4 + \gamma c_2 V_1 + \gamma c_4 V_5}{\gamma c_3'} \tag{67}$$

$$I_4 = \frac{V_3 - V_5}{\gamma l_4} \tag{68}$$

$$V_5 = \frac{I_4 + \gamma c_4 V_3}{\gamma c_5' + (\mu/r_l)} \tag{69}$$

These equations are identical to the set (38) - (42) except for the μ terms in the denominators of (65) and (69) and the μ term that multiplies V_i in the numerator of (65). The denominator μ terms are automatically realized in the damped SC

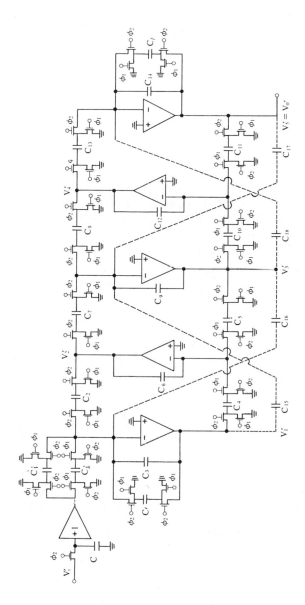

Fig. 19   SC realization of a fifth-order elliptic low-pass filter designed
using the general exact synthesis method of Section 5.5.

309

integrators. To multiply V_i by μ we need to feed V_i through the input branch shown in Fig. 19. Except for this modified feed-in branch the SC circuit configuration in Fig. 19 is identical to that of Fig. 11. Thus the resulting circuit is described by the set of five equations (43)-(47) except that (43) is replaced with

$$V_1' = \frac{-2C_i\mu V_i' - C_2 z^{\frac12} V_2' - \gamma 2C_{16} V_3'}{\gamma(2C_3 + C_s) + \mu C_s} \tag{70}$$

The remainder of the design procedure and indeed the equations giving the component values are identical to those in Section 4 with C_1 replaced by $(2C_i)$.

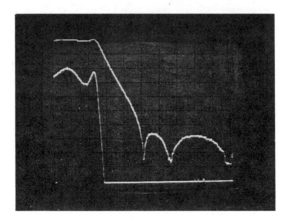

Fig. 20   Measured response of a fifth-order elliptic low-pass filter designed using the exact design method of Section 5.5.

Fig. 20 shows the measured response of a fifth-order elliptic low-pass filter designed using the exact method above. This filter has the following specifications:

Passband Edge	1	kHz
Max. Passband Ripple	1	dB
Stopband Edge	1.5	kHz
Min. Stopband Att.	60	dB

As a final remark, we note that not only is this method exact, thus enabling one to use a clock to passband-edge ratio of as low as 5 to 1, but it also preserves the transmission zero at ∞, locating it at $\omega = \pi/T$. This is a result of performing filter approximation and ladder synthesis in terms of the bilinear (λ) variable.

6. CONCLUDING REMARKS

Beginning with the pair of strays-insensitive integrators we have presented a number of methods, approximate and exact, for the synthesis of SC filters. The general exact design method discussed above seems to yield the best results in most applications. Very-high-selectivity bandpass filters might, however, be an exception. We are currently investigating this case.

Another topic that is the subject of vigorous research at the present time is the application of SC filters at high-frequencies (hundreds of kHz and possibly few MHz). Although initial results [20] have indicated that SC filters are more tolerant of the finite settling time of the op amp than active-RC filters, there are still many factors that limit the high-frequency operation of SC filters. These will be discussed by Paul Gray in the next chapter.

7. REFERENCES

[1]   J.T. CAVES, M.A. COPELAND, C.F. RAHIM, and S.D. ROSENBAUM, "Sampled analog filtering using switched capacitors as resistor equivalents", *IEEE J. Solid-State Circuits*, vol. SC-12, pp. 592-600, Dec. 1977.

[2]   B.J. HOSTICKA, R.W. BRODERSEN, and P.R. GRAY, "MOS sampled-data recursive filters using switched-capacitor integrators", *IEEE J. Solid-State Circuits*, vol. SC-12, pp. 600-608, Dec. 1977.

[3]   R.W. BRODERSEN, P.R. GRAY, and D.A. HODGES, "MOS switched-capacitor filters", *Proc. IEEE*, vol. 67, no. 1, pp. 61-75, Jan. 1979.

[4]   G.C. TEMES and Y. TSIVIDIS, eds., "Special Section on Switched-Capacitor Circuits", *Proc. IEEE*, vol. 71, no. 8, pp. 926-1005, Aug. 1983. (Contains four excellent papers on various aspects of SC circuit analysis and design by recognized leaders in the field.)

[5]   K. MARTIN, "Improved circuits for the realization of switched-capacitor filters", *IEEE Trans. Circuits Syst.*, vol. CAS-27, pp. 237-244, Apr. 1980.

[6] K. MARTIN and A.S. SEDRA, "Strays-insensitive switched-capacitor filters based on the bilinear z transform", *Electron. Lett.*, vol. 19, pp. 365-366, June 1979.

[7] L.T. BRUTON, "Low-sensitivity digital ladder filters", *IEEE Trans. Circuits Syst.*, vol. CAS-22, pp. 168-176, Mar. 1975.

[8] A.S. SEDRA and P.O. BRACKETT, *Filter Theory and Design: Active and Passive*, Matrix Publishers, Portland, Oregon, 1982.

[9] K. MARTIN and A.S. SEDRA, "Exact design of switched-capacitor bandpass filters using coupled-biquad structures", *IEEE Trans. Circuits Syst.*, vol. CAS-27, pp. 469-475, June 1980.

[10] P.E. FLEISCHER and K.R. LAKER, "A family of active switched-capacitor biquad building blocks", *Bell Syst. Tech. J.*, vol. 58, pp. 2235-2269, Dec. 1979.

[11] H.J. ORCHARD, "Inductorless filters", *Electron. Lett.*, vol. 2, pp. 224-225, June 1966.

[12] K. MARTIN and A.S. SEDRA, "Transfer function deviations due to resistor-SC equivalence assumption in switched-capacitor simulation of LC ladders", *Electron. Lett.*, vol. 16, no. 10, pp. 387-389, May 1980.

[13] S.C. FANG, Y. TSIVIDIS, and O. WING, "SWITCAP: A switched-capacitor network analysis program", Parts I and II, *IEEE Circuits and Systems Magazine*, Sept. and Dec. 1983.

[14] A.V. OPPENHEIM and R.W. SCHAFER, *Digital Signal Processing*, Prentice-Hall, Englewood Cliffs, N.J., 1975.

[15] W.M. SNELGROVE, "FILTOR-2: A computer aided filter design program", Dept. of Elec. Engrg., Univ. of Toronto, Toronto, Canada, 1981.

[16] K. LAKER, R. SCHAUMANN, and M.S. GHAUSI, "Multiple-loop feedback topologies for the design of low-sensitivity active filters", *IEEE Trans. Circuits and Syst.*, vol. CAS-26, pp. 1-21, Jan. 1979.

[17] R.B. DATAR and A.S. SEDRA, "Exact design of strays-insensitive switched-capacitor ladder filters", *IEEE Trans. Circuits Syst.*, vol. CAS-30, pp. 888-898, Dec. 1983.

[18] R.B. DATAR, "Exact design of strays-insensitive switched-capacitor ladder filter", Ph.D. Thesis, Dept. of Elec. Engrg., Univ. of Toronto, Oct. 1983.

[19] R.B. DATAR and A.S. SEDRA, "Exact design of strays-insensitive switched-capacitor high-pass ladder filters", *Electron. Lett.*, vol. 19, no. 29, pp. 1010-1012, 24th Nov. 1983.

[20] K. MARTIN and A.S. SEDRA, "Effects of the operational amplifier gain and bandwidth on the performance of switched-capacitor filters", *IEEE Trans. Circuits Syst.*, vol. CAS-78, pp. 822-829, Aug. 1981.

[21] M.S. LEE, G.C. TEMES, C. CHANG, and M.G. GHADERI, "Bilinear switched-capacitor ladder filters", *IEEE Trans. Circuits Syst.*, vol. CAS-28, no. 8, pp. 811-822, Aug. 1981.

[22] P.E. ALLEN and E. SANCHEZ-SINENCIO, *Switched-Capacitor Circuits*, Van Nostrand Reinhold, New York, 1984.

10

PERFORMANCE LIMITATIONS IN SWITCHED-CAPACITOR FILTERS

PAUL R. GRAY and RINALDO CASTELLO
Department of Electrical Engineering
and Computer Sciences
University of California
Berkeley, California

1. INTRODUCTION

The performance of switched capacitor filters has improved steadily since the first commercial application of such devices in 1977. These improvements have come about largely from improvements in techniques for the synthesis and design of the filters themselves, improvements in operational amplifier performance, and better understanding of the factors limiting dynamic range and power supply rejection. This chapter is devoted to a discussion of the performance limitations in such filters associated with the circuit level design of the operational amplifiers, switches, and so forth. Filter synthesis and architectural issues are discussed in a separate chapter.

Section 2 addresses factors limiting the accuracy of the frequency response attainable in monolithic switched capacitor filters. These include capacitor ratio errors, capacitive parasitics, switch charge injection, and finite operational amplifier gain. Section 3 addresses factors limiting the maximum bandedge frequency and clock frequency in the filter, and section 4 addresses the problem of power supply rejection ratio. Section 5 addresses the factors limiting dynamic range in the filter, and further explores the ultimate limits on the achievable dynamic range as a function of power dissipation and area.

2. FACTORS LIMITING ACCURACY OF INTEGRATOR FREQUENCY RESPONSE

Depending on the clocking scheme used, switched capacitor integrators can implement LDI, DDI, or bilinear integrator transfer functions. A typical switched capacitor integrator is shown in fig. 1. In this section, the "ideal" integrator response will be taken to be the response that the integrator would have if the operational amplifier were ideal (noiseless, infinite gain, zero settling time, zero offset), the switches were ideal, and the capacitors have no nonlinearity and have perfectly linear with precisely the nominal ratio. Because the filter is actually implemented with non-ideal circuit elements, integrator performance will deviate from this ideal in important ways. Several of the more important of these deviations are discussed in this chapter.

314

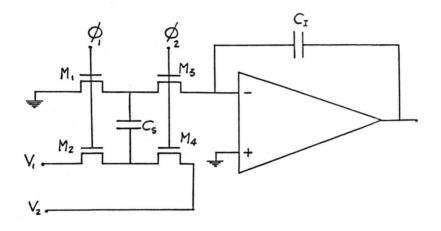

1. Typical Switched Capacitor Integrator

2.1 Capacitor Ratio Errors

Switched capacitor integrators have frequency behavior which is primarily determined by the ratio of the sampling capacitor to the integrating capacitor and the clock frequency. While the capacitors are defined with a nominal ratio by the mask geometries used, the actual ratio achieved in practice is a random variable with a mean at or near the desired ratio and a standard deviation which typically lies in the range of 0.1% to 5% of the mean ratio. This variation occurs because of a number of factors, including long-range process gradients, short-range oxide thickness variations, global variations in lithography edge location (undercutting)and short-range variations in edge location due to the limited resolution of the lithographic process used. (1) A number of layout techniques for improving the matching of capacitor arrays have been described in the literature.(2) These include the use of common-centroid geometries to alleviate the effects of long range gradients, replication of unit element geometries to achieve insensitivity to undercutting, and the use of large geometries for reducing the effects of random edge variations. For large geometries, oxide thickness variations tend to dominate, while for small geometries, edge location variations dominates. In perhaps the most extensive published study of the effect of geometry on capacitor matching, McCreary (1) concluded, at least for the process he examined, that for arrays incorporating element replication and common centroid geometry, oxide variations began to dominate for capacitor sizes in the 25 microns range and above.

An important aspect of switched capacitor filter design stems from the fact that when large ratios of capacitors must be realized, one of the capacitors is required to have a small size if reasonable total area is to be consumed. In this situation the ratio accuracy is limited by variations in the smaller capacitor. Thus the realization of large ratios usually results in an increase in the standard deviation of the ratio error distribution as a fraction of the ratio value itself.

An important practical effect of capacitor ratio variations is to limit the useful range of selectivity in bandpass switched capacitor filters. In such filters the center frequency is related to the clock frequency by capacitor ratios. If the ratios display 0.5% variation, for example, the center frequency will display the same variation. Most applications for highly selective bandpass filters also require that the center frequency be controlled with a tolerance small compared to the selectivity. Thus filter bandwidth to center frequency ratios smaller than about 1% become very problematical.

2.2 Frequency Response Errors Due to Charge Injection and Parasitic Capacitance

In the switched capacitor integrator shown in fig. 1, the signal is applied only to the "bottom plate" of the sampling capacitor and the top plate is connected only to ground and to the operational amplifier summing node. This configuration is referred to as "parasitic free" since parasitic capacitances from either of the plates of either of the two capacitors to ground has no effect on the integrator frequency response assuming the operational amplifier gain is high enough that the virtual ground is indeed ground.

The injection of channel charge from the four transistor switches as they turn off can result in input-referred dc offsets in the integrator, clock feedthrough to the output of the integrator, an error in the integrator time constant, and an input-referred nonlinearity. The channels of the top plate switches M_1 and M_2 always reside at ground potential when the switches are turned on, so that their charge injection offset is not signal dependent. Thus charge injection from these switches produces an input-referred dc offset and a clock feedthrough to the output which is proportional to the clock voltage swing and the gate capacitance of the switches. The channels of

316

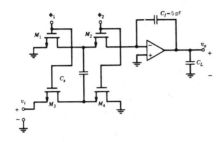

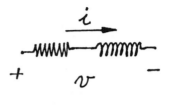

$$\frac{v_o}{v_i} \simeq \frac{1}{s\tau + \frac{1}{a}}$$

$$i = \frac{v}{sL + R}$$

$$\tau = \frac{C_I}{f_s \, C_s}$$

2. Illustration of the Effect of Finite Operational Amplifier Gain

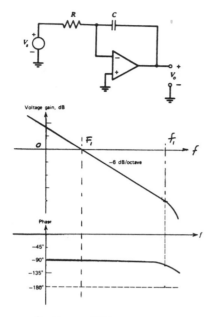

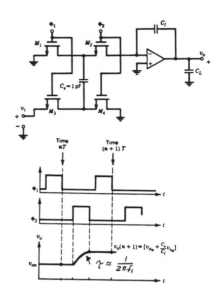

a. Continuous RC Active Filter

b. Switched Capacitor Filter

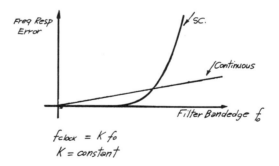

$$f_{clock} = K f_o$$
$$K = constant$$

c. Comparison of error vs filter bandedge frequency for continuous and switched capacitor filters.

3. Effect of Operational Amplifier Bandwidth on Filter Response

318

M_3 and M_4, however, reside at a potential equal to the input voltage, and as a result the channel charge varies with input voltage. This in turn results in a signal-dependent charge-injection offset on the sampling capacitor when transistor M3 turns off. This charge is integrated and appears as a shift in the effective integrator unity-gain frequency. The analysis of this effect is further complicated by the fact that the effect does not occur if M1 turns off before M3, which is the case for input signals less than zero, and by the fact that the magnitude of the charge injected is strongly dependent on the clock waveform and the source impedence from which the integrator is driven. For the values of sampling capacitor and switch size normally used in voiceband filters, the effect is negligible. However, in high frequency filters where large switches must be used, the effect can be large. It can be shown that the gain error portion of the injection can be eliminated by adding a clock phase to insure that transistors M_1 and M_2 are turned off prior to the other two.
(3)

2.3 Operational Amplifier Amplifier Gain and Bandwidth

MOS operational amplifiers desplay many deviations from ideal behavior which can strongly affect filter performance. In this section we discuss the effect of the limited gain and bandwidth of the amplifier.

2.3.1 Finite Operational Amplifier Gain: MOS operational amplifiers typically display far lower values of open loop gain than their bipolar counterparts. This occurs primarily because of the lower transconductance of the MOS transistors at a given value of drain current. The effect of this finite gain on the frequency response of a switched capacitor integrator is similar to the effect in a continuous-time R-C integrator. Because the dc gain of the integrator is finite, the pole at the origin is moved a small distance away from the origin and lies at a frequency equal to the integrator unity-gain frequency divided by the open loop gain of the operational amplifier. It is easily demonstrated that this effect is directly analogous to a resistive loss in a passive energy storage element as illustrated in fig. 2.

In a filter made with such integrators, the effect is to shift the poles of the filter transfer function to the left by an amount proportional to the inverse of the gain. The effect that this has on the frequency response depends on how close the original poles were to the axis. This problem is normally not significant for lowpass filters in the voiceband, since the pole Qs are typically less than 10 and operational amplifier gains are over 1000. However, the effect can be very significant in high-Q bandpass filters, particularly when implemented at high frequency. The high pole Q in such filters causes the addition of loss to have a more pronounced effect, and the higher bias currents used in high-frequency operational amplifiers tends to result in lower gain in such amplifiers.

2.3.2 Finite Operational Amplifier Bandwidth: The effect of operational amplifier bandwidth on the performance of switched capacitor filters has been studied analytically by several authors(4,5). The results of their analysis will be described qualitatively here.

In continuous RC active filters the effect of the finite bandwidth of the operational amplifiers is to cause additional phase shift in the transfer function of various circuit blocks which in turn distorts the filter transfer function. For example, the continuous RC integrator shown in fig. 3a would have exactly 90 degrees of phase shift at the unity gain frequency if the operational amplifier were ideal. However, if the integrator has a unity gain frequency F_1 and is implemented with an operational amplifier with unity gain frequency f_1 and a single-pole open loop response, the integrator displays an excess phase shift of $\dfrac{(F_1)}{(f_1)}$ radians at the unity gain frequency. For an integrator with a 10khz unity gain frequency using an op amp with 1Mhz unity gain frequency, this results in approximately 0.5 degrees of phase error at 10khz. The phase error is an approximately linearly decreasing function of the ratio of amplifier bandwidth to the filter bandedge. It should be pointed out that many techniques have been developed for alleviating this problem in RC active filters, such as active compensation or more complex op amp compensation networks.

The effect of finite operational amplifier bandwidth in a switched capacitor integrator is quite different, as illustrated in fig. 3b. The important aspect here is the transient response of the operational amplifier in the integrator configuration. The amplifier must be fast enough so that the charging transient goes to completion during one clock half-period. Actual monolithic operational amplifiers have complex transient behavior, but some insight can be gained by approximating the amplifier as having a single-pole response with no slew rate limiting. In this case the residual error as a function of time at the summing node of the operational amplifier following a clock transition has the form $e^{-(\frac{T}{2\tau})}$ where T is the clock period and τ is the time constant of the operational amplifier in the integrator configuration. If the sampling capacitor is small compared to the integrating capacitor, and if no large summing capacitors are connected to the summing node, this time constant is approximately the inverse of the amplifier unity-gain frequency in radians.

There are two important points to be observed from this simple analysis. The first is that in contrast to the continuous case, the finite bandwidth has virtually no effect for voiceband filters using operational amplifiers with bandwidths in the megahertz range. Again consider the case of a 10khz unity gain frequency integrator, with operational amplifier of 1mhz bandwidth, and assume it is clocked at 256khz. The value of τ is approximately 160ns while the clock half-period is 2 microseconds. The residual error at the amplifier summing node settles to a value equal to $e^{-(12.5)}$ of the initial value or about 3 parts per million. This will have no significant effect on the frequency response, in contrast to the continuous case.

The reason for this difference is the fact that in the switched capacitor case a fixed delay has been introduced in the integrator transfer function by virtue of it's sampled data operation. The output of the integrator is not sampled until the end of the transfer clock period, independent of how long it takes the operational amplifier to reach a correct final value. The clock delay in the integrator can be easily accomodated in the filter design, since it is a fixed, known delay, by using LDI clocking or other techniques. In the continuous case, the delay represented by the op amp excess phase shift added directly to the phase shift in the integrator thereby affecting the transfer function. The dependence of the frequency response error on the ratio of filter bandedge frequency to op amp unity-gain bandwidth is illustrated in fig. 3c. Because of the exponential nature of the switched capacitor error voltage, the filter response tends to degrade very rapidly when the value of the clock half-period approaches a value on the order of 5 or 6 τ. It has been shown (3,4) that the incomplete charge transfer that results in this case causes gain and excess phase errors in the integrator which distort the transfer function and increase pole Qs.

Another important consideration is that in many filters continuous paths are implemented by connecting capacitors from the outputs of operational amplifiers directly to the summing nodes of others. This tends to lengthen the effective settling time since the operational amplifiers are connected together in a complex continuous-time feedback network.

3. MAXIMUM BANDEDGE FREQUENCY

As pointed out in the last section, operational amplifier settling time is the principle factor limiting the maximum usable clock rate in switched capacitor filter. The time constant contributed by the on resistance of the various switches and the sampling capacitances can also be important, but in principal at least larger switches can be used to alleviate this problem provided

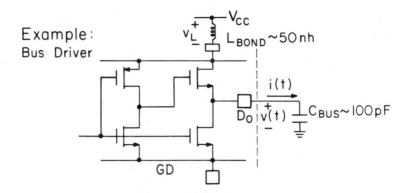

Example:
Bus Driver

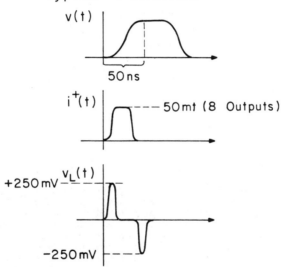

Typical Waveforms:

50 ns

$i^+(t)$ — — — 50 mt (8 Outputs)

+250 mV $v_L(t)$

−250 mV

4. Power supply noise generated by an 8-bit bus driver.

appropriate measures are taken to limit the charge injection gain and offset errors.

The most important factors in determining the maximum usable bandedge frequency for a switched capacitor filter in a given technology are:

1. The minimum ratio of clock frequency to filter band edge. While in principle it is possible to operate the filter at a clock rate as low as twice the filter bandedge frequency, practical antialiasing consideration usually dictate that the clock frequency be at least on the order of eight time the filter bandedge.

2. Filter pole Qs. The higher the selectivity of the filter, the smaller the allowable residual error at the end of the clock period for a given response degradation. A low-order lowpass filter might require that the clock interval last 6 time constants while a high-Q, high-order filter might require as much as 10 time constants.

3. The unity-gain frequency of the operational amplifier. Present circuit approaches to the implementation of CMOS operational amplifiers tend to be limited in unity-gain frequency to a value on the order of a device transconductance divided by the amplifier load capacitance. In single stage folded cascode amplifiers for example (6) the amplifier transconductance is that of the input stage. For typical values of load capacitance this has resulted in operational amplifier unity-gain frequencies which are lower than the inherent transistor f_t at the bias points used by a factor of from 10 to 30.

These three constraints taken together would dictate that the filter bandedge frequency be approximately a minimum of a factor of 25 smaller than the operational amplifier unity-gain frequency, assuming that 10 time constants are allowed for settling. Thus for example the implementation of a video lowpass filter with 4Mhz bandedge would require an operational amplifier with approximately 100Mhz bandwidth. Such a filter has recently been reported.(7) Further progress will come from improved operational amplifier design which allows the unity-gain frequency of a loaded amplifier to more closely approach the f_t of the devices used.

4. POWER SUPPLY NOISE COUPLING

A key technological advantage of switched capacitor filter over other filter implementations is the ability to incorporate the filter on the same chip with complex digital and analog-digital conversion functions. This fact implies, however, that the filters are likely to reside in a noisy environment, particularly with regard to the substrate and the power supplies. This problem is illustrated by the example shown in fig. 4. Here, an push-pull digital output driver is connected through a pad to a typical 8-bit bus, which can present a capacitance of as much as 100pF. The bus must be driven with a rise time of 50ns, which when the eight drivers are taken together gives rise to peak currents of approximately 25mA. This current flows through the package leads which display and inductance on the order of 50nH. The approximate current wave form shown would generate power supply spikes with magnitudes on the order of 250mV. The frequency of this noise is in the Mhz range. While separate digital and analog supplies can be used, the noise still couples into the substrate and other parts of the circuit.

This type of noise causes particular problems in sampled data analog circuits because the signal is resampled at each stage of the filter. It can thus be aliased into the passband at many different points and degrade the system dynamic range. Good power supply and substrate rejection at high frequencies are essential to maintain good system dynamic range.

Circuit design techniques for improving dynamic range have been extensively discussed in the literature. They include shielding of capacitive nodes in the circuit from the substrated vol-

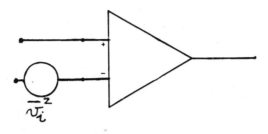

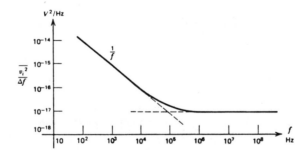

5. Equivalent input noise of an MOS operational amplifier

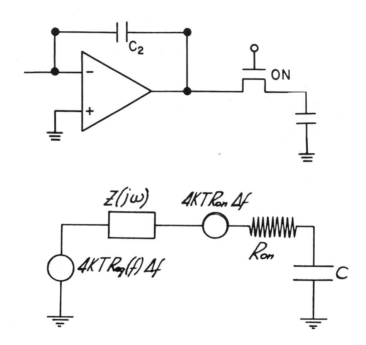

6. Sampling of an integrator output by the following stage.

tage with well diffusions(8), proper design of the operational amplifiers to limit internal capacitive coupling to the summing node(8,9), limiting of clock swings to a fixed reference to prevent a power supply dependence to the clock swing and resulting feedthrough,(9) and many others.

An alternative approach to the improvement of supply rejection is the use of a completely differential signal path(10,11). Here the signals inside the filter are represented by two voltages balanced with respect to ground so that the supply variations add to the signal as a common-mode component. This technique results in excellent rejection but also results in a considerable increase in the complexity of the operational amplifier, since it must contain common-mode feedback. It appears likely that such an approach will be most applicable, for example, in wide dynamic range filters which must be implemented on a 5 volt supply on the same chip with a digital signal processor. An alternative approach is to use a single ended supply and reference all internal signals to a potential which is a fixed offset from the substrate potential(12)

5. DYNAMIC RANGE LIMITATIONS IN SWITCHED CAPACITOR FILTERS

Because of the inherently high 1/f noise of MOS transistors, the achievement of large dynamic range in switched capacitor filters has been a difficult task. Steady progress has been made through improved operational amplifier design, improved process technology, and better techniques for filter design. Currently, PCM filters operating from a 10 volt power supply with effective dynamic range in the 95 dB range are routinely manufactured. In this section we first discuss the effects of operational amplifier noise, and then consider the more fundamental limitation of kT/C noise.

5.1 Operational Amplifier Noise

The noise contributed by the operational amplifier can be represented as an equivalent input noise generator as shown in fig. 5. The noise contributed by the operational amplifier can be divided into two parts- the 1/f or flicker noise and the thermal noise. The latter is usually restricted to frequencies far below the sampling rate and can therefore be analyzed as if the filter were a continuous one. The output noise of the filter resulting from this source can be found by simply multiplying each noise source by the transfer function from that point in the filter to the output and summing the resulting noise powers. Many currently manufactured filters have output noise which is dominated by 1/f noise. It can best be minimized by optimizing the operational amplifier design, as covered in other chapters. This noise can be further reduced by chopper stabilization in critical applications (11).

The analysis of operational amplifier broadband thermal noise is considerably more complex because it extends to frequencies far beyond the sampling rate and the higher frequency components can be aliased into the passband. That portion of the thermal noise residing in the filter passband can be analyzed directly as in the case of 1/f noise. The in-band thermal noise is a direct function of the transconductance of the operational amplifier input transistors and is usually in the range of 4 to 20 ($\frac{nV}{\sqrt{hz}}$) for non-micropower operational amplifiers.

The additional noise contributed by the aliasing of broadband noise has been considiered by several authors (11, 13, 14). The results will be illustrated qualitatively here by considering two

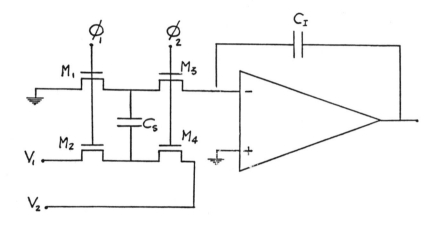

7. Simplified switched capacitor integrator

important specific cases. Consider the circuit shown in fig. 6. Here a sampling capacitor and switch are sampling the output of a previous integrator. The integrator circuit consisting of the operational amplifier together with the feedback capacitor can be represented as a Thevinin source having some effective output impedence vs frequency and some equivalent noise resistance which is also a function of frequency. One can identify one limiting case in which the output impedence is a pure resistance, and the noise equivalent resistance is equal to the output resistance. In this case the bandwidth of the sampling circuit is reduced by the same factor that the noise power spectral density is increased, and one obtains the result that the op amp noise adds NO noise over and above the fundamental kT/C noise contributed by the switch on resistance, to be discussed in the next section. This limiting case is approximated by a wideband transconductance operational amplifier with no output stage.

A second important case is one in which a second stege or an output stage is present in the operational amplifier, which would give rise to a low output resistance at high frequencies but might also give a high noise equivalent resistance, depending on the internal design of the amplifier. The operational amplifier, viewed as a Thevinin source, displays excess noise at high frequency in the sense that it's equivalent noise resistance is higher than its output resistance. This excess noise would be aliased into the passband by the sampling process. For this reason, output stages are generally avoided in internal switched capacitor filter amplifiers.

5.2 kT/C Noise

The most fundamental source of noise in switched capacitor filters comes from the noise in the MOS transistor switches. In fig. 7, when switch M_1 is in the on state, it contributes an equivalent noise resistance of $4kTR_{on}\Delta f$ where R_{on} is the switch on resistance. This white noise is bandlimited by the RC circuit formed by the on resistance of the switch and the sampling capacitor. When this bandlimited white noise is sampled by the opening of the switch, a random component is added to the sample of the signal which has a mean of zero and a variance equal to kT/C. The expected value of the stored energy associated with this random component is kT/2.

The effect of this random component on the noise of the filter can be analyzed by regarding the filter as a discrete time system in which such a random component is added to the samples at the input of each integrator in the system. The variance of the filter output samples is thus given by:

$$n_i{}^2 = \frac{kT}{C_s} \sum_{m=0}^{\infty} h^2(m) \tag{1}$$

where $h(m)$ is the discrete time impulse response from the noise source to the output. Using Parseval's theorem the variance may be expressed in terms of the frequency response of the filter as:

$$n_i{}^2 = \frac{kT}{C_s} \frac{1}{2\pi} \int_{-\pi}^{\pi} H(e^{j\omega}) H(e^{-j\omega}) d\omega \tag{2}$$

where $H(e^{j\omega})$ is the z transform of $h(m)$ evaluated on the unit circle. By introducing the the following definition

$$B_o = \frac{f_{clock}}{2\pi} \int_{-\pi}^{\pi} H(e^{j\omega}) H(e^{-j\omega}) d\omega \tag{3}$$

327

and making use of the relation between C_s, C_i, the sampling rate, and the unity gain frequency of the integrator, this can be written as:

$$n_i^2 = \frac{1}{2\pi} \frac{kT}{C_i} \frac{B_o}{f_{unity}} \qquad (4)$$

The quantity B_o is the effective noise bandwidth from the input of the switched capacitor integrator to the output of the filter. It is the integral of the magnitude squared of the frequency response of the sampled data filter from the integrator input to the filter output, taken around the unit circle. For lowpass and bandpass filters where the clock rate is far above the passband, this is equivalent to the integral over the passband of the transfer function from the integrator input to the filter output in the continuous equivalent filter. The quantity f_{unity} is the unity-gain frequency of the integrator.

The noise contributed by the right hand side switch is also sampled by C_s. However, in this case, the resulting signal cannot rigorously be considered as a first order low pass filtered noise. The reason is that the circuit through which the white noise of the switch is sampled does not have a single pole roll-off since it contains also the op. amp. The amount of noise transferred to the output is, to first order, proportional to the ratio between the op. amp. unity gain bandwidth and the bandwidth of the circuit formed by the switch resistance and the sampling capacitor. For simplicity it is assumed that the op amp is ideal. In this case the two switches behave in the same way and the total output noise, n^2, becomes

$$n^2 = \frac{1}{\pi} \frac{kT}{C_i} \frac{B_o}{f_{unity}} \qquad (4)$$

Assuming that the maximum undistorted output signal is approximately equal to the supply voltage V_s, i.e. $\frac{V_s}{\sqrt{2}}$ rms, the dynamic range of the integrator,(DR), becomes:

$$(DR)^2 = \frac{s^2}{n^2} = \frac{\pi}{2} \frac{V_s^2 C_i}{kT} \frac{f_{unity}}{B_o} .sp\, .5 \qquad (5)$$

Note that the numerator is proportional to the maximum energy stored on the integrating capacitor for a peak value of the signal, which we call E_{max}.

$$(DR)^2 = \frac{\pi}{4} \frac{E_{max}}{kT} \frac{f_{unity}}{B_o} \qquad (5)$$

Eq. (5) implies that the square of the dynamic range is given by the ratio between the maximum energy stored in the integrator and the unit of thermal energy kT, modified by the ratio between the noise bandwidth to the output and the unity gain bandwidth of the integrator.

As an example consider the unity gain feedback circuit shown in Fig. 8. This is the simplest configuration in which the S.C. integrator can be operated. It corresponds to a first order low pass filter whose z domain transfer function from C_s to the output is given by

$$H (z) = \frac{\dfrac{C_s}{C_i}}{1 - z^{-1} + \dfrac{C_s}{C_i}} \qquad (6)$$

$H (e^{j\omega T})$ is shown in Fig. 5b. In this simple case B_o can be easily computed since it si just

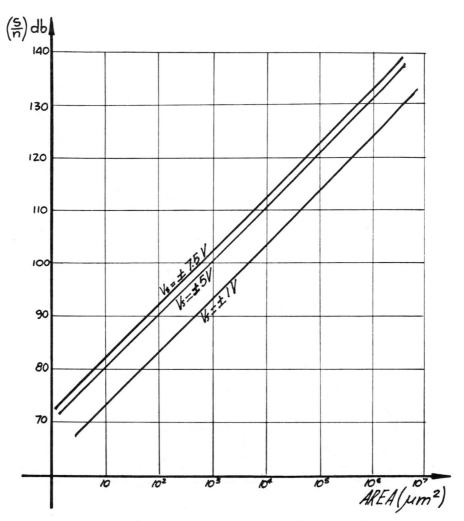

8. Maximum achievable dynamic range as a function of capacitor area for several power supply voltages

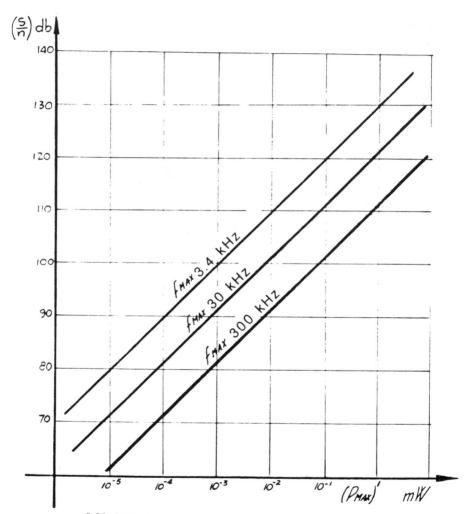

9. Maximum dynamic range as a function of power dissipation.

the noise bandwidth of a single-time constant lowpass filter. Assuming that the integrating capacitor is much larger than the sampling capacitor, this becomes:

$$B_o = \frac{f_{clock}}{2} \frac{C_s}{C_i} = \pi f_{unity} \tag{7}$$

the circuit dynamic range is therefore

$$(DR)^2 = \frac{V_s^2 C_i}{2 kT} \tag{8}$$

which is a particularly simple result. Note that this ratio is simply the maximum energy stored on the integrating capacitor divided by kT. This result has strong implications for the ultimate limit on the ability to scale switched capacitor filters with technological feature size. In effect, silicon dioxide can only store a certain amount of energy per unit volume as dictated by the maximum field strength of silicon. For a given oxide thickness and power supply voltage this dictates a maximum energy storage per unit area, which dictates a minimum area for a given dynamic range and power supply voltage. If one makes the assumption that the oxide thickness is scaled so that the field in the oxide has a value E_{max} when the power supply voltage V_s is applied across it, then this dictates that the minimum integrating capacitor area for a given desired dynamic range and power supply voltage is given by:

$$(DR)^2 = \frac{\pi}{4} \frac{V_s \epsilon_{diel} E_{max} AREA}{kT} \frac{f_{unity}}{B_o} \tag{9}$$

This indicates that the ultimately achievable dynamic range is proportional to the square root of the product of the power supply voltage and the area.

It is also possible to extract a minimum achievable level of power supply dissipation based on the necessity for charging and discharging the integrating capacitor as the signal passes through the filter. This relationship is derived elsewhere to be:.

$$(DR)^2 = \frac{\pi}{16} \frac{P}{kT B_o} \frac{f_{unity}}{f} \tag{10}$$

Thus the dynamic range is proportional to the square root of the minimum power dissipation necessary to charge and discharge the sampling and integrating capacitors from the power supply. Here f is the signal frequency. Eq. (10) is only valid for $f \leq f_{unity}$ since outside this range the gain of the integrator is less than 1 and therefore it is not possible to have $V_o = V_s$ for an input signal v_i smaller than the supply voltage. The absolute maximum for P (P_{max}), when both V_i and V_o are not allowed to exceed the supply voltage, corresponds to $f = f_{unity}$. In this case Eq. (10) becomes

$$(DR)^2 = \frac{\pi}{16} \frac{P_{max}}{kT B_o} \tag{11}$$

It can be shown that these results are valid for both single ended and fully differential integrators. These two relationships are plotted in figs 9 and 10. for the special case of a single-pole lowpass filter. It is interesting to note that the ultimately achievable minimum power dissipation and area are orders of magnitude below current filter sizes. While these represent only the area of the integrating capacitor, in principle the rest of the circuitry can be directly scaled, assuming that the 1/f noise problem can be solved by some technique such as chopper stabilization. The minimum power dissipation of course does not include power associated with the biasing of the operational amplifier. Subthreshold effects in the operational amplifiers will result in other limitations on the minimum achievable power at the low-power end of the performance spectrum. However, these results provide a strong indication that additional savings and power and area are possible in switched capacitor filters through effective use of scaled technologies.

REFERENCES

1. J. MC CREARY, "Matching Properties, and Voltage and Temperature Dependence of MOS Capacitors" , IEEE Journal of Solid-State Circuits, vol SC-16, December, 81

2. D.J. ALLSTOT, and W.C. BLACK, "Technological Design Considerations for Monolithic MOS Switched-Capacitor Filtering Systems" ' Proceedings of IEEE, vol. 71, August, 83

3. P. W. LI, M. CHIN, P. R. GRAY, and R. CASTELLO, " A Ratio-Independent Algorithmic Analog-Digital Conversion Technique, Digest of Technical Papers, 1984 International Solid-State Circuits Conference, February, 1984

4. K. MARTIN, and A.S. SEDRA, " Effects of the op amp Finite Gain and Bandwidth on the Performance of Switched-Capacitor Filters" ' IEEE trans. Circ. Syst. vol. CAS-28, Aug. 81.

5. G.C. TEMES, " Finite Amplifier Gain and Bandwidth Effects in Switched Capacitor Filters" , IEEE Journal of Solid-State Circuits, vol SC-15, June, 80.

6. T.C. CHOI, R.T. KANESHIRO, R.W. BRODERSEN, P.R. GRAY, W.B. JETT, and M. Wilcox, " High-Frequency CMOS Switched-Capacitor Filters for Communications Applications IEEE Journal of Solid-State Circuits, vol SC-18, December, 83.

7. S. MASUDA, Y. KITAMURA, S. OHYA, and M. KIKUCHI, " CMOS Sampled Differential Push-Pull Cascode Operational Amplifier Proc 1984 Int. Symp. Circuit and Systems, May, 84.

8. W.C. BLACK, Jr., D.J. ALLSTOT, R.A. Reed, "A High Performance Low Power CMOS Channal Filter," IEEE J. Solid-State Circuits, vol SC-15, December, 80.

9. H. OHARA, P.R. GRAY, W.M. BAXTER, C.F. RAHIM, and J.L. MC CREARY " A Precision Low-Power PCM Channel Filter with on Chip Power Supply Regulation," IEEE J. Solid-State Circuits, vol SC-15, December, 80.

10. D. SENDROWICZ, S.F. DREYERr, J.H. HUGGINS, C.F. RAHIM, and C.A. LABER, "A Family of Differential NMOS Analog Circuits for a PCM Codec Filter Chip" IEEE J. Solid-State Circuits, vol. SC-17, pp. 1014-1023, Dec. 1982. .lp 11. K.C. HSIEH, and P.R. GRAY, D. SENDERWOICZ, D. MESSERSCHMITT "A Low-Noise Chopper-Stabilized Differential Switched Capacitor Filtering Technique," IEEE J. Solid-State Circuits, vol. SC-16, Dec. 1981.

12. J.R. IRELAND, B. TERRY, and D.J. ALLSTOT, " CMOS Analog Cells for General Purpose A/D Conversion in Custom and Semi-Custom Applications Proc 1984 Int. Symp. Circuit and Systems, May, 84.

13. C.A. GOBET, and A. KNOB, "Noise Analysis of Switched Capacitor Networks," IEEE Trans. Circuits and Syst. vol. CAS-30, pp. 37-43, Jen. 1983.

14. J.H. FISCHER, " Noise Sources and Calculations Techniques for Switched-Capacitor Filters IEEE J. Solid-State Circuits, vol. SC-17, Aug. 1982.

15. R. CASTELLO, and P.R. GRAY, to be published

11

CONTINUOUS-TIME FILTERS

YANNIS TSIVIDIS
DEPARTMENT OF ELECTRICAL ENGINEERING
COLUMBIA UNIVERSITY
NEW YORK, NY 10027, USA

1.   INTRODUCTION

As is apparent in several chapters of this book, VLSI telecommunication circuits must often incorporate both analog and digital functions. In this context, the need for continuous-time filtering arises often; three special cases are illustrated in Fig. 1. In (a), the most straightforward use of a continuous-time filter is shown; here a continuous-time signal is processed directly by this filter, with no sampled-data processing involved. In (b) and (c) switched-capacitor and digital filters are respectively used in the processing. In these two cases a continuous-time filter is used at the input to attenuate the high-frequency components of the incoming signal, so as to prevent the "aliasing" of these components by the sampling process; at the output, a "smoothing" filter is used to smooth out the staircase waveform provided by the sampled-data filters, and also to reject high frequency noise and switch feedthrough components. In some cases one or the other of the two continuous-time filters shown in (b) or (c) is not used. For example, in a PCM encoder the sampled-data filter is followed directly by an A/D converter, with no smoothing filter used; similarly, the output of a PCM decoder is fed directly into a switched-capacitor filter with no antialiasing filter in between. The cases in (b) and (c) will be used as an example; all three cases in Fig. 1 have continuous input and output, and thus it will be possible to meaningfully compare them.

The case of Fig. 1(a) will be considered first. Special-purpose continuous-time filters for the cases in Figs. 1(b) and 1(c) will be discussed at the end of this chapter. In accordance with the general spirit of this book, only techniques of proven high performance will be considered in detail. For the case of Fig. 1(a), this needs some qualification. The only

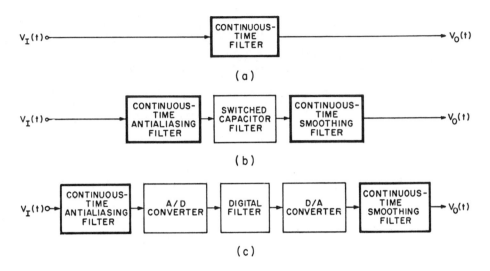

Fig. 1.   Three ways for processing a continuous-
          time signal.

high performance techniques available for this case have been
proposed very recently, and thus they haven't yet had the
opportunity to stand the "test of time", as would be the case
with switched-capacitor or digital filters. However, given that
there are no other high performance techniques for fully
integrated continuous-time filters, and given the wide-ranging
promise of these recently proposed techniques, it was felt that
they deserve a detailed description in this book.

2.   PRECISION CONTINUOUS-TIME FILTERS

2.1. <u>Advantages of precision continuous-time filtering</u>

     In cases where a continuous-time signal must be processed,
one is faced with a choice of one of the techniques illustrated
in Fig. 1. The popular techniques shown in (b) and (c) are not
without problems. First, they need antialising and smoothing
filters, whose purpose is not to provide useful frequency
shaping but rather to get around limitations inherent to
sampled-data signal processing. Unfortunately, although the

input antialiasing filter reduces signal aliasing it can do nothing to reduce aliasing of noise generated at its output and beyond. In the case of switched-capacitor filters this can become a serious problem. High frequency noise coming from switches, op amps, and power supply lines is sampled by the switches and is aliased into the baseband. Some MOS op amps (especially NMOS designs) are known to have large amounts of output high frequency noise, in fact well above their unity-gain frequency; the aliasing of this noise has plagued switched-capacitor filters for some time. Power supply rejection in switched-capacitor filters can be easily claimed to be, say, 40 dB when it is measured by injecting a signal into the power supply line with a frequency in the baseband. In a real situation, however, a switched capacitor filter usually coexists with a multitude of other switching circuits, including digital circuits, on the same chip, and "noise" in power supply lines due to switching (e.g., coupled to other circuits through a common bonding wire inductance) can exist at frequencies well above the Nyquist rate; at these high frequencies the power supply rejection of the op amps can be low, and in addition there can be coupling directly to switches and capacitors through parasitic capacitances. This would not represent a severe problem if the noise at these high frequencies were not aliased into the baseband due to the sampling process. Fully differential switched-capacitor techniques have been devised to avoid such effects, but in these not only the signal paths must be fully balanced but also the parasitics to the power supply; this does not represent a trivial circuit design and layout problem. Clock feedthrough, especially at high sampling rates, is a well-known problem in switched capacitor filters. Switch charge injection depends on signal level; also, the order of opening certain switches, although it makes no difference "in principle", actually does make a difference in practice due to switch nonidealities; this has even prompted the use of three-phase clocks in some cases. Needless to say, predicting noise and distortion due to these and other nonidealitieis in switched-capacitor circuits is a very difficult task. The presence of switching waveforms is also responsible for "mixing products" between the input frequencies and the switching frequency along with its harmonics; this is a potential problem for example in modems, as discussed in the chapter on switched-capacitor modems by Kerry Hanson. Finally, an observation can be made on the performance of these circuits at high frequencies. Switched-capacitor filters are known to exhibit virtually unchanged performance as the operating speed is increased, up to a certain point; beyond that, an abrupt deterioration is observed. Among the nonidealities responsible for such behavior is switch nonzero resistance, as well as op amp slewing and settling. The latter can be a complicated phenomenon in circuits with continuous-path loops, such as in certain leapfrog filters, and op amp nondominant poles and zeros can play an important role in such behavior; the prediction of these effects requires extensive transient analyses and is difficult. Also, the abrupt appearance of such effects as one attempts to increase the operating frequency makes the use of frequency response "predistortion" techniques impractical.

Due to the above problems, although it may be relatively easy to understand and design a switched-capacitor filter on paper "in principle" using ideal switches and noiseless op amps fed from noiseless power supplies, it is quite a different story to try to make a working chip of professional quality, especially if one is attempting to advance the state of the art. As a result, the design time can be long and can require several iterations; when a final chip emerges, it incorporates solutions that may not work for a different fabrication process. What is more, such solutions are often not publicized by individual companies, as they actually make the difference between an academic "paper" design and a working product. Professional switched-capacitor chip design is a complex task, and those who have mastered it are few and hard to find.

Coming to digital filters, one can of course hardly deny their suitability in cases where the signal is already in digital form, where programmability is of prime importance, and when very long chains of filters must be used; however, in cases where analog signals must be processed they have the drawback of requiring A/D and D/A converters (Fig. 1(c)). In addition to the ensuing drawbacks in terms of chip area and power dissipation, these converters employ op amps and analog switches, and thus some of the problems mentioned above for switched-capacitor filters are, to some extent, present here too. For the digital filters themselves power dissipation and chip area have been reduced considerably over the last few years, but are still too large for many applications especially when high speed operation is required. The often-claimed advantage of physical dimension "scaling" may hold for the digital logic, but does not hold for the A/D converters as these are partly analog circuits, and reducing capacitances implies increased noise and possibly reduced accuracy. One should also note here that some of the techniques devised to improve the performance of A/D and D/A converters lead directly to techniques that improve the performance of analog filters; this should be taken into account when making projections for the future performance of the structure in Fig. 1(c) as compared to the one in Fig. 1(b) and 1(a). In evaluating digital filters one should also consider their quantization "noise", which has been found to be more objectionable to humans than true noise of comparable power. This can be traced to the fact that what is called quantization "noise" is actually a signal-dependent extraneous product with components at frequencies which are not harmonically related to the input frequency; an effect quite different from the harmonic distortion found in continuous-time systems, and one that requires special techniques for its reduction. Finally, an observation can be made on the bit accuracy required for certain tasks. In applications such as receiver IF filtering, the undesired signals that must be rejected are often orders of magnitude larger than the desired signal, although they may be very close to it in frequency. The filter must be designed with a sufficient number of bits in order to process the undesired large signals adequately, plus extra bits in order to produce at its output the desired signal with low quantization noise. The two requirements combined can mean a large number of bits, with ensuing complications in the design of the A/D converter. Note that oversampling techniques, while they may help aleviate some of the above problems to a degree, are certainly not a total cure and have potential drawbacks of their own, as discussed in the chapter on A/D conversion techniques by Paul Gray and David Hodges.

The advantage of continuous-time filters is precisely the absence of the problems mentioned above for their switched-capacitor and digital counterparts. The prospects for straighforward design, ease of computer simulation, low power, small area, quiet operation, and high frequency capability are behind the recent interest in continous-time filters. New techniques have made possible such filters with a precision comparable to that of switched-capacitor circuits. The rest of this section considers precision continous-time techniques in detail.

2.2 The MOSFET as a voltage-controlled resistor.

Precision RC filters cannot be implemented using capacitors and non-tunable resistors as passive elements, since the resulting RC products vary widely with fabrication process and

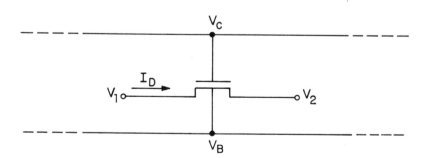

Fig. 2.   A MOS transistor with all terminal voltages refered to ground. V_C and V_B are DC biases; V_1 and V_2 are signals.

temperature variations (e.g., by 50-100%). In the techniques to be described in the bulk of this chapter [1,2], MOS transistors are used as voltage-controlled resistors which are automatically adjusted to provide precision RC products by an on-chip control system. The operation of transistors in this mode is considered in this section. All such transistors will be assumed to have their gate connected to a common "control voltage" bus and their substrate to a common substrate voltage bus, as shown in Fig. 2. Only operation in the <u>nonsaturation</u> region will be considered. For a careful investigation of distortion, the commonly used "square law" transistor model is inadequate; a more precise model will instead be used [3,4]. According to this model, if V_1 and V_2 are the source and drain voltages with respect to ground we have, assuming a n-channel device:

$$I_D = 2K \{(V_C - V_B - V_{FB} - \phi_B)(V_1 - V_2)$$

$$- \frac{1}{2} [(V_1 - V_B)^2 - (V_2 - V_B)^2]$$

$$- \frac{2}{3} \gamma [(V_1 - V_B + \phi_B)^{3/2} - (V_2 - V_B + \phi_B)^{3/2}] \} \tag{1}$$

with

$$\gamma = \frac{1}{C'_{ox}} (2q N_A \varepsilon_S)^{1/2} \tag{2}$$

$$K = \frac{1}{2} \mu C'_{ox} \frac{W}{L} \tag{3}$$

where I_D is the drain current in the nonsaturation region, V_C, V_B, V_1, V_2 are the gate, substrate, drain, and source potentials with respect to ground, W and L are the channel width and length, μ is the carrier effective mobility in the channel, V_{FB} is the flat-band voltage, N_A is the substrate doping concentration, C'_{ox} is the gate oxide capacitance per unit area, ε_S is the silicon dielectric constant, q is the electron charge and ϕ_B is the approximate surface potential in strong inversion for zero backgate bias. It is assumed that the source and the drain voltages V_1 and V_2 never become too low to forward bias the drain and the source junctions, and never become too high to drive the device into saturation. The ground potential is defined such that V_1 and V_2 vary around zero. In that case the substrate voltage should be negative in order to keep the drain and the source junctions reversed biased (such a definition for ground potential is convenient when two power supplies of opposite values are present).

If the 3/2 power terms in (1) are expanded in a Taylor series with respect to V_1 and V_2, one obtains [1,5]:

$$I_D = K[a_1(V_1-V_2)+a_2(V_1^2-V_2^2)+a_3(V_1^3-V_2^3)+\ldots] \qquad (4)$$

with:

$$a_1 = 2(V_C-V_T) \qquad (5a)$$

$$a_2 = -(1 + \frac{\gamma}{2 - V_B + \phi_B}) \qquad (5b)$$

$$a_n = - A(n)(-V_B+\phi_B)^{-\frac{2n-3}{2}}, \qquad n \geq 3 \qquad (5c)$$

where $A(3) = -1/12$, $A(4) = + 1/32$, $A(5) = -1/64$ etc., and V_T is the "threshold voltage" of the transistor corresponding to a "backgate bias"of $- V_B$ [3,4]:

$$V_T = V_{FB} + \phi_B + \gamma\sqrt{-V_B + \phi_B} \qquad (6)$$

The inverse of (Ka_1) in (4) is the small-signal resistance of the transistor R; thus from (5a) we have:

$$R = \frac{1}{2K(V_C-V_T)} \qquad (7)$$

The value of R may be varied with V_C (hence the name "control voltage"); therefore for small signals the MOSFET behaves as a voltage-controlled linear resistor. For large signals the nonlinear terms in (4) must be considered. Fig. 3 shows the ratios $|a_2/a_1|$, $|a_3/a_1|$, etc., for common process parameters, as computed from (5) [1]. A typical practical situation is illustrated by taking $V_2 = 0$ V, $V_1 = 1$ V, $V_C-V_T = 2$ V, $\mu C_{ox}'(W/L) = 10$ $\mu A/V^2$ and $V_B = -5$ V (for power supplies of $\pm$ 5 V the n-channel transistor substrate is considered connected to the minimum available potential). Then, the first term in the right-hand side of (4) is 20 μA, the second term is -6 μA, the third term is 3×10^{-2} μA, the fourth one is $- 2 \times 10^{-3}$ μA, etc. It is seen that the dominant deviation from linearity comes from the second-order term.

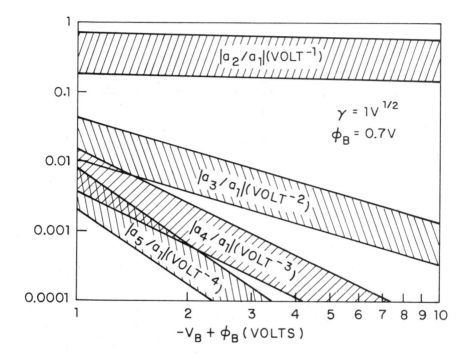

Fig. 3.   Coefficients of the nonlinear terms
in Eq. 4 normalized to a_1 for different
values of substrate bias and different
control voltage values.   Each band
corresponds to $1V < V_C - V_T < 4$ V, with
the bottom of the band corresponding to
4 V [1] (Copyright © 1983 by IEEE).

Based on (4) and (7), the MOS transistor in nonsaturation
can be represented by the model of Fig. 4.   The resistance value
R is given by (7); the current source represents the
nonlinearities and is given by:

$$I_{NL} = K \left[-a_2(V_1^2 - V_2^2) - a_3(V_1^3 - V_3^3) - \ldots \right]. \tag{8}$$

From the results derived above it follows that for $-V_B$ of
at least a few volts (a condition assumed in the rest of the

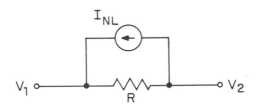

Fig. 4. A model for the device of Fig. 2, showing linear and nonlinear parts.

chapter) we can write:

$$I_{NL} \simeq - Ka_2 (V_1^2 - V_2^2) \qquad (9)$$

It will be seen later that the terms neglected in writing (9) have indeed a negligible contribution to the distortion of the proposed filters, and that this is so even for realistic devices with an effective mobility dependent on the gate field.

2.3 Cancellation of nonlinearities

MOS resistors can be used in lieu of linear resistors, while still maintaining linear circuit operation, if the nonlinearity terms I_{NL} are somehow cancelled. Several possibilities are shown [2] in Fig. 5, where the connection to $-V_B$ is not shown for simplicity. The cancellation of the nonlinearities can be accomplished in a single device (Fig. 5(a)) or among more than one devices (Figs. 5(b) and 5(c)). In these three cases, the results shown directly on the figures can be easily verified by using the model of Fig. 4. In Fig. 5(a), I_{NL} is forced to a zero value (assuming (9) is valid) by operating the device with antisymmetrical terminal voltages. In Fig. 5(b) $I_{NL} \neq 0$, but is the same for both devices and cancels out in the difference; similarly for Fig. 5(c). The cases shown in Figs. 5(d) and 5(e) do not provide complete cancellation of the square terms in I_{NL}, there being a residual part corresponding to the last term in eq. (5b); thus, these two cases may be satisfactory only if the body effect coefficient is

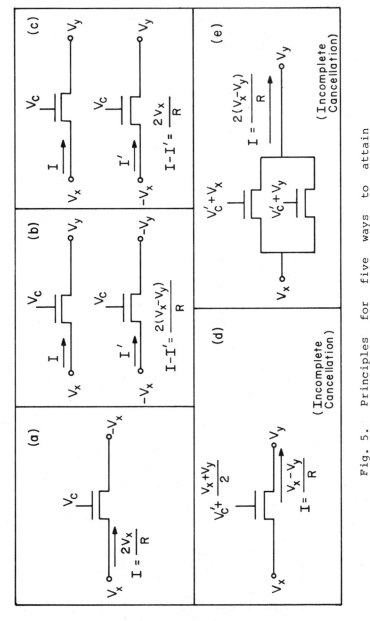

Fig. 5. Principles for five ways to attain linear resistor effects using MOS transistors [2]. All substrates are connected to V_B. (a), (b), and (c) achieve in principle complete cancellation of all even-order nonlinearities. (d) and (e) achieve partial cancellation.

small and/or $-V_B$ is large.* Various circuit possibilities exist
based on the various cases illustrated in Fig. 5. The
techniques described in this chapter are based on the principle
of Fig. 5(c); this technique has proven to be the most practical
sofar. It provides cancellation of all even-order terms in I_{NL}.

2.4 MOSFET-capacitor filters based on balanced structures

2.4.1. Principle. Throughout this section we will use topologies
resulting from classical filter structures using resistors,
capacitors, and op amps. The ideas to be discussed, however,
are also applicable to filters with other active elements. A key
component in the structures to be presented is a balanced output
op amp, whose symbol is shown in Fig. 6. This op amp will be

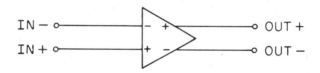

Fig. 6.   A balanced-output op amp.

assumed to have the magnitudes of its two outputs matched. This
is ilustrated schematically in Fig. 7. While this figure might
suggest an obvious way to make a balanced output op amp, one is
of course not limited to this approach.

---

*Several techniques based on partial cancellation of
nonlinearities can be used in cases where high linearity is not
of prime importance. A floating voltage-controlled resistor
based on the principle of Fig. 5(e) is discussed elsewhere [6],
and can be implemented in CMOS technology. In the same
technology, some designers use a "transmission gate" type of
roughly linear resistor, connecting in parallel the channels of
a n-type and a p-type devices; to make this combination work
well, though, one would have to match the K factors of the two
devices, which is not practical. In technologies with
depletion-mode devices one can use the circuit of Fig. 5(e)
with $V_C = 0$, connecting the gates to the corresponding sources;
the resulting resistor will be floating and relatively linear,
but it will not be voltage-controllable.

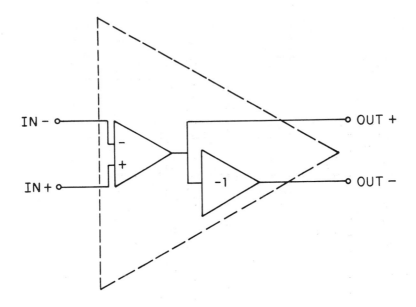

Fig. 7.   Equivalent circuit for a balanced-out-
            put op amp.

The balanced output op amp is shown with other devices in
Fig. 8.   For the top output one has:

$$V_{out}(t) = V_z - \frac{1}{C}\int_{-\infty}^{t} I_z(\tau)d\tau \qquad (10)$$

where V_z is the voltage at the input terminals with respect to
ground (assuming zero input offset).   For the bottom output one
has:

$$-V_{out}(t) = V_z - \frac{1}{C}\int_{-\infty}^{t} I_z'(\tau)d\tau \qquad (11)$$

Subtracting (11) from (10) and dividing by 2 one obtains:

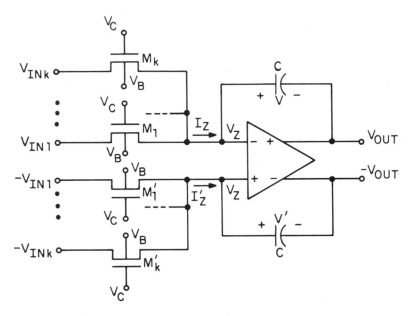

Fig. 8.   A balanced summing integrator using MOS transistors in lieu of resistors.

$$V_{out}(t) = -\frac{1}{2C}\int_{-\infty}^{t}[I_z(\tau) - I_z'(\tau)]\,d\tau \tag{12}$$

The case where M1 is matched to M1', M2 is matched to M2', and so on, is now considered. In the difference I_z-I_z' the nonlinearities of the devices within a matched pair will cancel out, as suggested by Fig. 5(c). Applying for each pair of matched devices the result shown directly on that figure, one obtains:

$$V_{out}(t) = -\frac{1}{R_1C}\int_{-\infty}^{t}V_1(\tau)\,d\tau - \frac{1}{R_2C}\int_{-\infty}^{t}V_2(\tau)\,d\tau - \ldots \tag{13}$$

i.e. a linear summing inverting integrator has been obtained.

This integrator is input-output equivalent to the circuit of
Fig. 9. It is emphasized here that the behavior of the circuit
in Fig. 8 is linear from input to output, but not within the
circuit; for example, the device currents are not linear
functions of the corresponding voltages. Also, V_Z at the op amp
inputs is in general nonzero. This value is continuously
adjusted by feedback action so that V_Z - V and V_Z - V' are equal
in magnitude and opposite in sign, as required by the balanced
output op amp (Fig. 7). The resulting values for V_Z are small
(typically no more than 20% of the input peak value); hence the
input common mode range requirements on the op amp are very
modest.

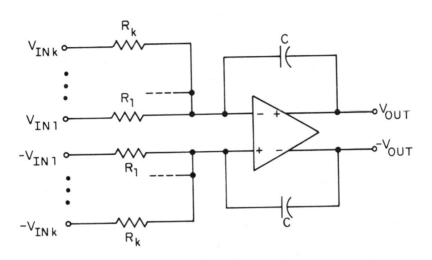

Fig. 9.   A   circuit   which   is   input-output
           equivalent to that in Fig.   8.

2.4.2. Filter synthesis. Since the circuit of Fig. 9 is input-
output replaceable with that in Fig. 8, a number of well-known
filter structures can easily be converted to use MOSFET's,
capacitors and op amps. An example will be given with the help
of Fig. 10. In (a), a Tow-Thomas biquad is shown [7,8]. The
top op amp is only used for inversion. If balanced-output op
amps were available, one would implement this filter as shown in
(b). The circuit can be easily "doubled-up" to provide the
structure in (c), which is simply a balanced version of (b) (it
is easy to show that the op amp gain should be halved in going

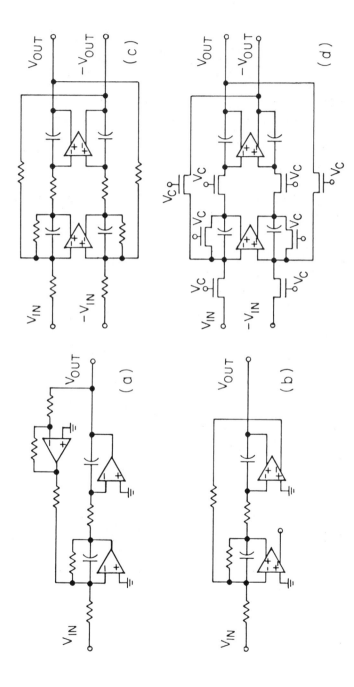

Fig. 10. An example of filter synthesis using MOStransistors in lieu of resistors. (a) A Tow-Thomas biquad. (b) The circuit in (a) using balanced-output op amps. (c) The fully balanced version of the circuit in (b). (d) The circuit in (c) with resistors replaced by MOS transistors.

from (b) to (c), if exact correspondence is to be maintained in the presence of finite gain effects). The structure in (c) can be seen to use building blocks of the form shown in Fig. 9. Hence, all resistors in (c) can be replaced by MOSFET's to produce the final circuit shown in (d). The resistance values for each MOSFET will be given by (7), and for a given V_C can be set by choosing the W/L ratios appropriately. More on the design process will be discussed later.

Proceeding in a similar manner one can obtain a fifth-order all-pole filter as shown in Fig. 11. The resulting circuit has been fabricated using CMOS technology, and is discussed in detail elsewhere [1]. A chip microphotograph is shown in Fig. 12. It is important to note here that the chip exhibited the high performance expected from computer simulation, the first time it was fabricated. No second iteration was needed and no unforseen effects were observed, although this was the very first time the proposed principle was tested in integrated form. This stems from the simplicity of the principle, the inherently "clean" continuous-time operation and the readily available simulation tools for circuits in such operation. The addition of eight capacitors to the circuit of Fig. 11 converts the structure to one capable of implementing a fifth-order elliptic response [9]. Again, this can be derived in a straightforward manner form the corresponding signle-ended version; the latter can be derived by using capacitive feedthrough to implement the zeros, as discussed elsewhere [10].

2.4.3. Accuracy. Capacitors in MOS technologies are known to match within 0.1% "with extra care". However, this is true for composite structures, each composed of many "unit" capacitors. The matching of two unit capacitors, or of one unit capacitor to an array of unit capacitors, is worse [11]; a typical working limit is around 0.5% [12]. Long-channel transistors have been known to match to 0.1%-0.3% for some time [13, 14, 15]. This often comes as a surprise, due to the transistor's being viewed as a more complex device than a capacitor. The two devices, though, undergo different processing steps and the mechanisms contributing to edge effects are different; direct comparisons are not easy. On the continuous-time chips reported [1,9], a matching of about 0.2% has been measured for transistors next to each other. Although good matching is observed even for 4μ wide devices, such devices can show "narrow width effects" (see the chapter on modelling by F.M. Klassen); these effects increase the threshold voltage and are often not well modelled. Preferably somewhat larger widths should be used, unless the technology employed is well characterized. Standard techniques can be used to obtain good matching, such as laying out the devices in close proximity and using cross-coupled structures, just as is usually done to achieve low input offset in operational amplifiers. This approach is evident in the

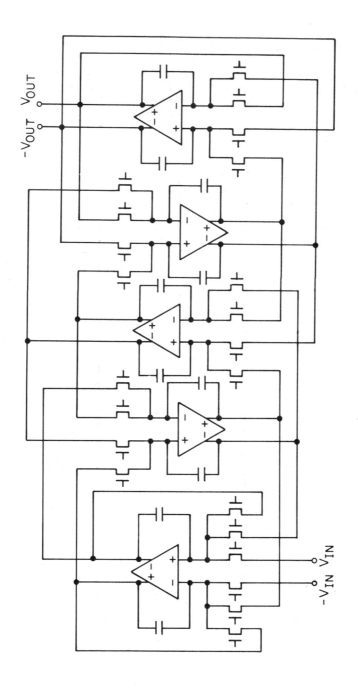

Fig. 11. A balanced fifth-order all-pole filter [1] (Copyright © 1983 by IEEE).

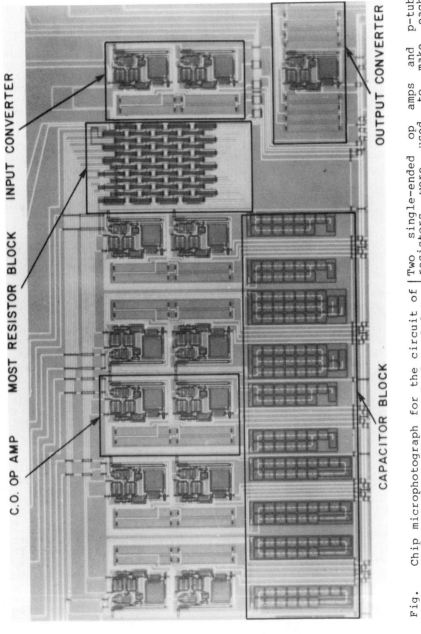

C.O. OP AMP   MOST RESISTOR BLOCK   INPUT CONVERTER

CAPACITOR BLOCK

OUTPUT CONVERTER

Fig. 12. Chip microphotograph for the circuit of Fig. 11, an input single-ended-to-balanced converter, and an output balanced-to-single-ended converter [11]. Two single-ended op amps and p-tub resistors were used to make each balanced-output op amp, for simplicity (Copyright © 1983 by IEEE).

351

microphotograph of Fig. 12. However, in a more recent design [9] cross-coupling was not used and the matching was still be found to be good. MOSFET "resistor" matching can be easily measured with a small drain-source voltage, along with appropriate gate-source and substrate-source bias. Such measurements are preferable to measurements in the saturation region, since the latter is characterized by mechanisms different from those in nonsaturation (where the above "resistors" operate); one can thus expect that the achievable degree of matching may also be different in the two regions. This is one of the reasons that op amp input offset numbers cannot be used to infer "resistor" matching, other reasons being that the offsets are determined by several devices, and that some of these devices have relatively short channels.

The high degree of MOSFET resistor matching can also be infered indirectly from the reliability with which precision specs can be realized on chips (e.g., a fifth-order elliptic filter with 0.06 dB ripple [9]).

2.4.4. Distortion and dynamic range. Equation (9) gives the main nonlinearity in a MOST "resistor"; this is in principle cancelled out in the integrators described, along with all other even-order terms. Odd-order terms are not cancelled; however, they can easily be made negligible by applying a few volts of substrate bias (see Fig. 3). It has been found that in this way the odd-order distortion becomes small even in the presence of nonideality effects such as mobility dependence on the gate field. Rather, distortion in fully balanced schemes such as the one in Fig. 8 typically comes mostly from mismatches, which cause less than perfect cancellation of even-order distortion. Both experiment [1] and calculations [5] show that in practice most of the distortion is of second order. For the circuit of Fig. 8, assuming a single pair of inputs, a conservative estimate of the second-order distortion in the presence of mismatches can be derived from results presented elsewhere [5], and is given by:

$$D \simeq \frac{V_{i,\,peak}}{16(V_C - V_T)} \times (|\frac{\Delta R}{R}| + |\frac{\Delta C}{C}| + 2|\frac{\Delta V_i}{V_i}|) \qquad (14)$$

where $\Delta R/R$ and $\Delta C/C$ is the resistance and capacitance mismatches, respectively, $\Delta V_i/V_i$ is the mismatch between the absolute values of the two (supposedly balanced) inputs, and V_i,peak is the peak value for each of the two balanced-input sinusoids (the corresponding differentially-measured peak-to-peak input would be four times as large). The control voltage V_C is automatically adjusted to keep the filter tuned (Sec.

2.5); the worst-case distortion corresponds to the minimum value of V_C needed over all temperature and process variations. As an example, let $V_{C, min}$ = 2.5 V, V_T = 1 V, $\Delta R/R$ = $\Delta C/C$ = 0.003, and $\Delta V_i/V_i$ = 0.01. Then for a 1 V peak input the distortion is 0.1%. It should be emphasized that the above estimate is for a single integrator, and cannot be used in an obvious manner to predict distortion in a complete filter; in fact, our experience sofar has been that the above numbers are typical for the behavior of a filter as a whole. For example, the filter in Figs. 11 and 12 (including the input single-ended-to-balanced and the output balanced-to-single-ended converters) could handle signals of 0.75 V to 1 V peak (depending on the value of V_C) before the total harmonic distortion rose to 0.1%. This distortion was indeed found to be mostly second-order, as claimed above, indicating device mismatch as the main cause. The fact that a complete filter can have a distortion as low, or even lower than, a single integrator is due to additional nonlinearity cancellations that take place in the filter structure as a whole and which come more or less as an unexpected bonus to the designer. Such cancellations can eliminate both even and odd-order nonlinearities in certain cases; an explanation of such effects, along with examples, is under preparation. Due to the low distortion and low noise, dynamic ranges of 95 dB for complete filters are easily achieved as demonstrated experimentally [1]; this number is based on a 1% THD overload level, which occured at peak amplitudes of 2V-3.5V (depending on the value of V_C).

For low distortion, the chip internal signal level should be kept low (e.g., 1 V peak for ± 5 V supplies). In a SC filter, higher swings are possible (e.g., 3 V peak for ± 5 V supplies). To make handling a large external signal possible, a filter of the type described above can be placed between an on-chip attenuator and an on-chip amplifier, thus maintaining internal signal levels sufficiently low, as shown in Fig. 13.

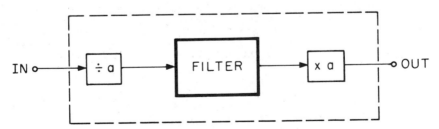

Fig. 13. A method for increasing the signal handling capability of a filter chip. For the filters described, the ensuing noise penalty is small.

Since attenuation and amplification factors are small (e.g. 3), this approach will not cause a large noise increase.

The use of "depletion-mode" MOSTs can increase the allowable internal signal levels, since for such devices $V_T < 0$ and the approximate worst-case swing $V_{C,min} - V_T$ is large. However, for such devices (1) is known not to hold very accurately; thus, whether the remaining odd-order nonlinearity is negligible or not is not obvious. Signal levels could also be increased by using on-chip voltage multipliers to raise the values of V_C and $-V_B$, especially in cases of limited power supply voltages [16, 17]. However, this would work against the simplicity of the technique, and is not likely to be needed in view of the possibility of using the approach illustrated in Fig. 13.

Although a balanced structure is inherently resistant to noise common to balanced paths, for critical applications it is desirable to provide control and substrate voltages (V_C and V_B) as free from extraneous signals as possible, to prevent such signals from varying the resistance values and creating modulation products with the input signal. The possibility for such products exists, for example, if the extraneous signals contain frequency components in a range where the filter's frequency response is steep; then they could modulate the gain of the filter. Whether this effect is any serious for a given application can be checked through computer simulation. If desired, the V_C and V_B lines can be filtered easily since they do not carry DC current, or can be generated from reference voltages. One should note that in CMOS technologies the devices for which a "clean" V_B should be provided are especially the ones inside wells, since for them the body effect coefficient is larger; fortunately, the very presence of the well makes the task easy. On-chip crude filtering to rid the substrate of extraneous signals has also been used in switched-capacitor circuits [18].

2.4.5. Parasitic capacitances. The parasitic capacitances of a MOSFET resistor are shown in the small-signal equivalent circuit of Fig. 14. Here C_S and C_d are the junction capacitances of the source and the drain; C_p is the distributed parasitic capacitance of the channel to the gate and to the substrate. We have:

$$C_p = C_{ox} + C_b \qquad (15)$$

where C_{ox} is the total oxide capacitance:

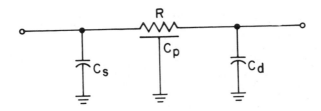

Fig. 14. A small-signal model for a resistor
          implemented by a MOS transistor.

$$C_{ox} = C'_{ox} \; WL \tag{16}$$

with C'_{ox} the oxide capacitance per unit area, and C_b is the channel-to-substrate capacitance given by [4]:

$$C_b = \frac{\gamma}{2\sqrt{-V_B + \phi_B}} \; C'_{ox} \; WL \tag{17}$$

where we note that the body effect coefficient γ is inversely proportional to C'_{ox}, so C'_{ox} actually cancells out in (17); the form of this equation, though, is useful for comparing C_b to C_{ox} in terms of the widely used coefficient γ.

The effects of C_s and C_d are not significant in the structures presented, at frequencies where the op amp is practically ideal. These capacitances are either voltage driven, or are connected to one of the op amp input terminals in Fig. 8. Due to the matched transistor pairs in this figure, the total junction capacitance between either op amp input terminal and ground is the same. As V_Z varies with time, these capacitances add equal parasitic currents to I_Z and I'_Z, and thus cancel out in the difference in (12).

The effect of the distributed capacitance C_p can be very important, and can distort the frequency response if not taken into account at the design stage. A simple empirical technique that has proven very successful is the following [1]. A filter is first designed assuming ideal resistors with no distributed capacitance. Next, distributed effects are added and the filter is computer-simulated to determine whether such effects have significantly affected the frequency response. The distributed effects can be included in two ways. One can replace each transistor by the model of Fig. 14, where the middle part is simulated using the transmission line model found today in popular CAD programs. Alternatively, one can use the approach illustrated in Fig. 15. A transistor is split into several sections, with lengths adding up to the total transistor length. This results in a many-segment lumped approximation to the transistor. Since the individual devices in Fig. 15 are fictitious and no source-drain junctions are meant to exist at the intermediate points, the model for these devices should of course not include any junction capacitances at these points, neither should it include short channel effects. Computer simulation including distributed effects usually predicts a certain effect on the frequency response (typically peaking at the band edge for a low-pass filter). Provided this peaking is relatively small (e.g., a fraction of 1 dB), it can be eliminated by empirically modifying slightly the value of one or two capacitor pairs in the filter, using computer simulation as a guide. The capacitors to which the response appears most sensitive are preferable. This process is fast, and has resulted in successful chips from the first run; design iterations were found unnecessary, as the measured response had

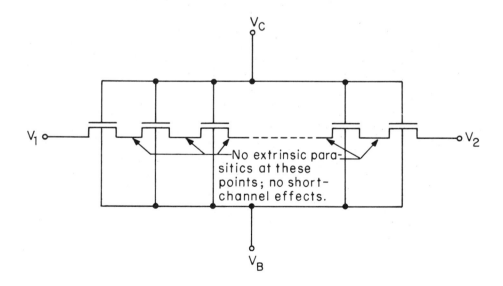

Fig. 15. A lumped approximation to a MOS transistor with distributed effects.

been very well predicted during the above simulation stage. If
the peaking produced by the distributed capacitance is large,
the above procedure will not work well since elimination of the
peaking might cause distortion of the response at other points
(e.g., an increase of the passband ripple). In such cases, the
filter should be first modified by increasing all capacitor
sizes and decreasing all resistor sizes proportionately, thus
making the effect of the distributed capacitances less
pronounced; then the above empirical adjustment should be
carried out. That this approach will decrease the effect of the
distributed capacitance can be seen as follows. With R the
value of a MOSFET resistor and C an integrator capacitor one
has, from (7) and (3):

$$RC = \frac{LC}{W\mu C'_{ox}(V_C-V_T)} \tag{18}$$

It will be assumed for simplicity that C_b in (17) is
negligible (it is typically 0.1 of C_{ox}). Then, from (15), (16)
and (7) one obtains:

$$RC_p = \frac{L^2}{\mu(V_C-V_T)} \tag{19}$$

If C in (18) is increased by a factor α, L must be
decreased by α to maintain the proper value of RC. Thus, from
(19) the product RC_p will be decreased by α^2. The ratio of the
two products in (19) and (18) is a measure of the relative
influence of the parasitic distributed capacitance C_p. Hence,
there is good reason to use as large a capacitance C as is
convenient, to make L as small as possible. Fortunately,
increasing C at the expense of R makes possible lower thermal
noise at the same time.

Another observation concerns the choice of p- or n-channel
devices. If μ is decreased by a factor β, L should also be
decreased by β as seen from (18), other things being equal.
From (19), this will decrease the product RC_p by β, thus making
parasitic effects less important. It follows that p-channel
devices, with their lower mobility, are normally preferable to
n-channel devices for implementing the filter resistors.
Equation (19) suggests the possibility of using MOS transistors
as voltage-controlled distributed RC structures [19], which can
be made very narrow since W is not a factor in (19) and hence

will not affect matching. Two problems with this approach, though, are that all capacitance must be at AC ground (since they are connected to V_C) and that the nonlinear terms are no longer given by (9) and may be difficult to cancel.

2.4.6. A warning concerning computer simulation [20]. In CAD programs, model parameters are often chosen to provide a "best overall fit" of device I-V characteristics to measurements, and in the process many important details are lost. Since models with thirty parameters or more are in use, the optimization routines used can assign to these parameters quite fictitious values. One is sometimes much better off working by hand, using simple expressions. Thus, for example, it has been found that equation (1) results in relatively accurate distortion predictions. In contrast to this, CAD programs with semiempirical models can give distortion figures which are in error by an order of magnitude (either lower or higher), especially if the distortion is low in the first place. This is not surprising; a model which provides the drain current with an accuracy, say, of 5% for digital applications cannot be assumed to preserve the nuances necessary to predict distortion of the order of 0.1%. If a choice is available, one should choose a CAD model corresponding to (1) as close as possible. Another problem with some models is in the prediction of thermal noise. The latter is represented in some programs by a drain current noise spectral density proportional to the device transconductance. This representation is in fact correct only in the saturation region; in nonsaturation with zero drain-source voltage the transconductance is zero and the above models predict zero thermal noise, an erroneous result since then the device operates as a linear resistor [20]. Finally, it was recently found that for one of three models available in a popular simulator, the drain-source small-signal resistance predicted changed by a factor of three as the drain-source bias voltage was changed from OV to any minute value, e.g. 1 μV! This is obviously artificial, but potentially dangerous since it can result in very wrong predictions of frequency response. The model used in computer simulations of the filters described in this chapter should be checked against such errors.

In view of the above problems with simulation, one would rightly question an advantage claimed earlier for continuous-time filters as opposed to switched capacitor filters, namely ease of computer simulation. Such worries can be answered as follows. Some effects in switched-capacitor filters, such as noise and clock feedthrough are inherently complicated, and theories leading to their computationally efficient prediction have yet to be developed. In contrast to this, expressions for transistor thermal noise or the behavior of the current at zero drain-source voltage have been known since the sixties; if some

CAD programs are having difficulty predicting such effects, it is only due to errors in model implementation in these programs. The resulting bugs went undetected for a long time in the course of digital circuit simulation. With the advent of analog MOS circuits, though, such bugs are being detected, and their removal is trivial. In two facilities familiar to this author it took a knowledgeable person less than one afternoon to locate the problems mentioned and correct them, and chances are such corrections have already taken place in most places with an active analog MOS design group.

2.5 On-chip tuning schemes

The general approach to automatic tuning of the filters presented is a variation of schemes proposed elsewhere [21-25] and is illustrated in Fig. 16. The reference circuit can be an oscillator made by the same basic structures as are used in the filter itself. A phase comparator continuously compares the output of the oscillator to an external clock signal, and adjusts V_C until the oscillator tracks the clock. At that point, the RC products within the oscillator attain the desired value. Since all MOSFET resistors and capacitors in the main filter are ratio-matched, respectively, to the resistors and capacitors of the oscillator, and since a common V_C is used for all MOSFET resistors, the various RC products in the filter attain the desired value, and frequency response becomes stabilized. Measurements on complete chips show practically no frequency variation of the -3dB point over the commercial temperature range, and a worst-case passband variation of only 0.04 dB. The clock feedthrough at the output of the filter is measured at 110 μV rms [9].

An alternate scheme uses a filter as a reference circuit, again made of structures matched to those in the main filter. The clock signal is passed through the reference filter, and the phase of the resulting signal is compared to the original clock signal phase. Yet another scheme uses no clock signal at all, the external reference being a temperature-insensitive resistor [25]. The reference circuit is now simply a resistor or group of resistors, and the comparison circuit compares resistance values. The external resistance must be initially adjusted until the prescribed frequency response is achieved. However, this type of control loop cannot tune out any subsequent variations in the capacitors due to temperature. A "self-tuning" scheme, in which the filter itself is being periodically tuned (as opposed to being locked to a reference circuit), is described elsewhere [26]. While the filter is being tuned, another filter is placed in the signal path through a suitable switching arrangement that avoids transients. This scheme may prove attractive for high-Q applications.

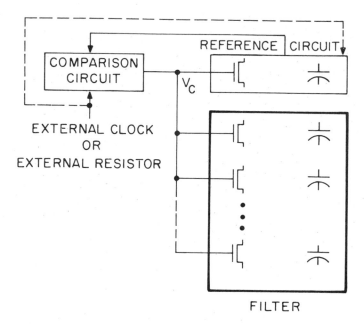

Fig. 16. On-chip automatic tuning scheme for the filters described.

2.6 Worst-case design

In the automatic tuning schemes described, the value of V_C will change with fabrication process and with temperature, to keep the RC products stable at a desired value. Since drain and source voltages must be kept at least a threshold voltage below V_C to maintain operation in the nonsaturation region, one desires a V_C as high as possible to allow for large signal swings. A worst-case design strategy can be summarized as follows, considering a single RC product for simplicity. First, all process tolerances and temperature are combined in the worst-case direction resulting in a required V_C as high as possible for a given value of τ = RC. Curves of the required V_C versus τ might look as in Fig. 17(a), where each curve is for a different nominal value of channel length for the MOSFET resistor (assuming L/W > 1 and that W has been set to the minimum acceptable value). If the maximum allowed value of V_C is set at the power supply voltage, V_{DD}, then by choosing L = L_2 from the figure we are guaranteeing that V_C will never need be

360

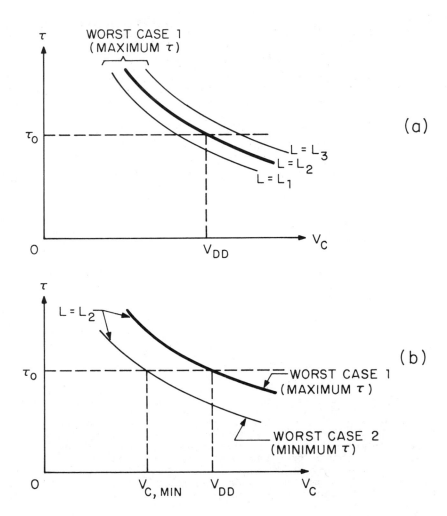

Fig. 17. Illustration of worst-case design precedure. (a) Choosing channel length for a maximum allowable V_C and worst case corresponding to large RC. (b) Determining the minimum V_C required for the choice in (a).

higher than that value to adjust τ to the required value τ_o; and it will only be equal to V_{DD} in the improbable case where all process tolerances and temperature happen to combine in a worst-case sense to require maximum V_C. Keeping now L at the chosen nominal value L_2, all process tolerances and temperature are combined in the _opposite_ worst-case direction, resulting in the curve shown in Fig. 17(b) where the middle curve from Fig. 17(a) is also reproduced for comparison. From the new curve, the minimum V_C value that will be required is found. The corresponding minimum guaranteed peak swing will then be approximately one threshold voltage below $V_{C,min}$. A detailed numerical example is considered elsewhere [5]. With + 5 V power supplies, minimum guaranteed swings of 1 V to 2 $\overline{V}$ peak are typical for modern processes. As mentioned before, these can be effectively increased using the scheme of Fig. 13 while still maintaining a dynamic range of over 90 dB.

2.7 High frequency operation

As mentioned earlier, continuous-time filters exhibit a "graceful degradation" of performance as the speed of operation is increased. Hence frequency response predistortion techniques are feasible. In addition, several known techniques for compensating the op amp finite gain-bandwidth effects can be employed [7]. We note here that passive compensation in fully integrated filters is much more feasible than in active filters with discrete elements. This is because tracking is much better in the former. The MOSFET's inside the op amp can be made to track the MOSFET resistors, all being made simultaneously on the same chip; similarly, a capacitor used for passive compensation tracks the op amp compensation capacitor(s) for the same reason. MOS op amps with unity-gain frequencies of 50-100 MHz are already in use. High frequency performance will improve if double-pole, simple-zero compensation can be employed [8]. Special consideration should be given to the op amp output impedance at high frequencies, since the filter resistance levels are then low. If that output impedance cannot be made very small, at least one should carefully take into account its effect. Preliminary work indicates that filters in the MHz range are feasible [27].

Transconductance-capacitance structures have been proposed at the time of this writing for implementing high frequency filters [28]. The basic integrator in these structures is shown in Fig. 18. M1 and M2 are the input transistors of a differential pair, with M7 and M8 as loads; M5 and M6 provide common-mode feedback for output quiescent point stabilization. M3 and M4 provide resistive termination when needed, by connecting their drains to the drains of M1 and M2, respectively. M10 and M11 are current sources; their gate

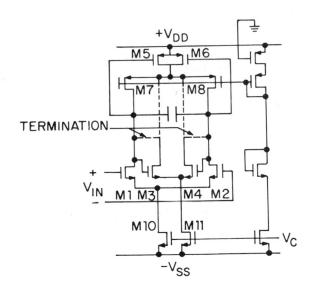

Fig. 18. A transconductance-capacitance integrator [28] (Copyright ©1984 by IEEE).

voltage is used to adjust the integrator unity-gain frequency $g_{m1,2}/2C$, through on-chip tuning techniques similar to those described in Sec. 2.5. This integrator structure can be designed for high frequency operation (a 500 kHz filter with a Q and 5 has been reported [28]), but its drawback is the nonlinearity of the input devices, which limits the dynamic range to about 60 dB.

3. ANTIALIASING AND SMOOTHING FILTERS FOR SAMPLED-DATA SIGNAL PROCESSORS

This chapter is concluded with a discussion of the special-purpose continuous-time filters used in the cases of Figs. 1(b) and 1(c). Such filters are usually kept simple; their only purpose is to attenuate undesired high-frequency components, whereas baseband filtering is reserved for the switched-capacitor or digital filters in the system. This discussion will concentrate on antialiasing filters, the considerations for output smoothing filters being of a similar nature.

The requirements imposed on an antialiasing filter will be illustrated with the help of Fig. 19. In (a) the magnitude frequency response of the sampled-data filter is shown; let us denote by f_C the maximum frequency of interest in the baseband. If f_S is the sampling frequency, an extraneous input component at frequency f_S-f_C will be viewed by the sampling process as indistinguishable from a baseband component at frequency f_C. To reduce this "aliasing" effect to an acceptable level, the frequency response $G(f)$ of the input "antialiasing" filter, shown in Fig. 19(b), must satisfy a constraint of the form:

$$| G(f_S - f_C) | \leq \beta \qquad\qquad (20)$$

Antialiasing filters have traditionally been implemented using nontunable resistors and capacitors which can vary greatly

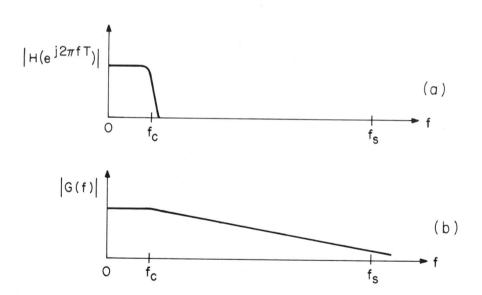

Fig. 19. (a) Frequency response of a sampled-datafilter. (b) Frequency response of an associated antialiasing filter.

with fabrication process and temperature (note that the technique described in Sec. 2 was not proposed until very recently). Focusing for the present on such nontunable structures, it is clear that G(f) is not known pecisely and cannot be considered as part of the overall baseband filtering action. Thus one is limited to antialiasing filter responses which are sufficiently flat in the baseband over all element variations. Assuming $|G(0)| = 1$ and a monotonically decreasing response in the baseband, one must then impose a constraint of the form:

$$| G(f_c)| \geq \alpha \qquad\qquad (21)$$

Since in most technologies the ratio accuracy for resistors and capacitors is good, all resistances and capacitances in the filter will be represented as multiples of R and C respectively, where R and C are arbitrarily chosen values. Then working with the transfer function for a given RC filter structure and response type one can determine constraints corresponding to (20) and (21), respectively, in the following form:

$$R_{min}\ C_{min} \geq \tau_2 \qquad\qquad (22)$$

$$R_{max}\ C_{max} \leq \tau_1 \qquad\qquad (23)$$

The above constraints are illustrated in Fig. 20. For given R_{max}/R_{min} and C_{max}/C_{min}, and for given α and β, the above constraints impose a minimum possible ratio f_s/f_c. If this ratio is not acceptable, a higher order antialiasing filter would have to be used. One of the reasons for using large f_s/f_c in sampled-data filters has been to keep the antialiasing and smoothing filters simple. Often second or first-order active or passive structures are employed. The Sallen-Key structure shown in Fig. 21 is typically used [7,8]. This structure has the attractive feature that G(0) = 1 (within the accuracy of the unity gain buffer) no matter what the resistor and capacitor values are, since at DC the capacitors become open circuits and the input is effectively fed directly to the unity gain buffer.

Other structures that have been used are the biquad of Fig. 22 [7,8] often refered to as the Rauch structure [29, 30], and the Tow-Thomas biquad [7, 8] shown in Fig. 10(a). These structures do not have a guaranteed DC gain of 1, and thus will be useful mostly in cases where the overall gain of a system is

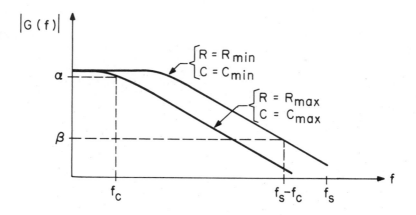

Fig. 20.   Illustrating the constraints on the
frequency response of an antialiasing
filter.

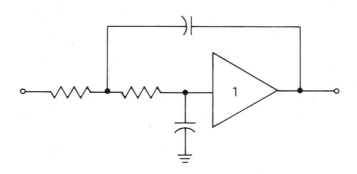

Fig. 21.   A Sallen-Key low-pass filter.

trimmed (e.g., in a switched-capacitor filter-codec chip). The
structures have the advantage that the op amp input terminals
are at practically zero potential, and thus the common mode
input range requirements for the op amps are relaxed.   In
addition, the Tow-Thomas structure is implementable using
MOSFET's in lieu of resistors as shown in Fig. 10(d). All gates
can be connected to the power supply voltage for a non-tunable

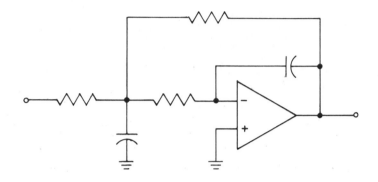

Fig. 22. A Rauch low-pass filter.

implementation [31]; however, if an automatically tuned structure as described in Sec. 2.5 is used, a smaller f_S/f_C is possible since the effect of element variations is eliminated. Thus one can take better advantage of the recent "exact" switched capacitor design techniques, which in principle work for f_S as low as twice the maximum input frequency (such techniques are described by Adel Sedra in the chapter on switched-capacitor filter synthesis). Now it is even possible to allow for a larger baseband response variation and, since G(f) is stabilized and reliable, compensate for this variation by modifying the response of the sampled-data filter accordingly. Of course, at this point one may want to consider eliminating the sampled-data filter alltogether, and performing the whole filtering in continuous time as described in Sec. 2. Such an approach might in fact be preferable for several applications, as already discussed in Sec. 2.1.

For the commonly used nontunable filter structure of Fig. 21, the resistors are typically realized using polysilicon material (in CMOS or NMOS technologies) or p-well material (in standard CMOS; n-well in "inverted CMOS" technology). Typical sheet resistances are $100\,\Omega/\square$ for poly and $4k\,\Omega/\square$ for p-well [32]. P- or n-well realizations are thus much more economical in terms of chip area, but are not always preferable for reasons to be discussed shortly.

The voltage power spectral density of the thermal noise in a resistor is given by 4kTR, where k is Boltzmann's constant and T the absolute temperature. Hence resistor thermal noise can be reduced by decreasing R and increasing C in the filters considered. However, usually the noise performance of these filters is dominated by power supply noise, which is coupled to the structure though distributed parasitic capacitances. In typical antialiasing applications most of the noise can be filtered out by the sampled-data filter, provided it is concentrated at frequencies below half the sampling rate; noise at frequencies higher than that, though, can be aliased into the baseband. In the smoothing filter, all of the noise will appear at the output since no further filtering action takes place. Thus, depending on the type of application and the noise spectrum it may be very desirable to use special shields for resistors and capacitors in order to reduce power supply noise coupling [32,33]. Poly resistors can be shielded from the substrate (which is connected to a power supply voltage) by using a p-well or n-well in CMOS technologies, or a "depletion implant" in NMOS technologies. Similar shields are also possible for capacitors. In double-poly technologies, one can implement the resistors with the second level poly and use the first level poly as a shield. In all cases the shield can be made more effective by contacting it throughout its periphery, rather than at a single point [33]. Unfortunately, p-well or n-well resistors in CMOS technologies cannot normally be shielded; in addition their large voltage coefficient [32] can cause baseband distortion, especially in the Rauch and Tow-Thomas structures. Hence the attractive chip area savings possible with such resistors cannot always be taken advantage of. An interesing alternative is the use of individually linear ized MOS transistor combinations in lieu of resistors; several of these were described in Sec. 2.3.

Antialiasing and smoothing filters implemented using the techniques presented in Sec. 2.4 are inherently resistant to power supply noise effects by virtue of their balanced structure, as already mentioned.

4. CONCLUSIONS

Recent work indicates that certain forms of continuous-time filters are capable of precision comparable to that of switched capacitor circuits. On-chip automatic tuning schemes are used to ensure filter accuracy over fabrication process tolerances and temperature variations. By using MOS resistors, capacitors, and op amps as basic building blocks the design is kept very simple, and a wealth of knowledge on active RC filters can be used to advantage. In addition, since the op amp is a generic element it continues to be improved for a variety of applications and is available as a building block in a typical design environment. The filter designer thus is not faced with the problem of designing special-purpose building blocks, and the design time is short. Thus fully integrated continuous-time filters need not be limited to simple antialiasing and smoothing functions. Instead, they can be added to the choices available to the designer of high performance integrated filters. Numerous tradeoffs must of course be considered in order to decide which technique is best for a situation at hand. For certain telecommunications applications, continuous-time filtering will be found to be the solution of choice.

5. REFERENCES

[1]   M. BANU and Y. TSIVIDIS, "Fully integrated active RC filters in MOS technology," IEEE J. Solid-State Circuits, vol. SC-18, no.6, pp. 644-651, Dec. 1983.

[2]   Y. TSIVIDIS and M. BANU, "Integrated nonswitched active RC filters with wide dynamic range," Proc. 1983 European Conf. Circ. Theory and Design, Stuttgart, pp. 111-113.

[3]   W.M. PENNEY and L. LAU (eds.), MOS Integrated Circuits, New York, Van Nostrand-Reinhold, 1972.

[4]   Y. TSIVIDIS, "The MOS Transistor," McGraw-Hill, New York, to be published in 1985.

[5]   M. BANU and Y. TSIVIDIS, "Detailed analysis of nonidealities in MOS fully integrated active RC filters based on balanced networks," IEE Proceedings, Part G (Electronic Circuits and Systems), vol. 131, no.5, pp. 190-196, Oct. 1984.

[6]   M. BANU and Y. TSIVIDIS, "Floating voltage-controlled resistors in CMOS technology," IEE Electron. Lett., vol. 18, no. 15, pp. 678-679, July 1982.

[7]   A.S. SEDRA and P.O. BRACKETT, Filter Theory and Design: Active and Passive, Matrix Publishers, Beaverton, 1978.

[8]   M.S. GHAUSI and K.R. LAKER, Modern Filter Design, Prentice-Hall, Englewood Cliffs, 1981.

[9]   M. BANU and Y. TSIVIDIS, "On-chip automatic tuning for a CMOS continuous-time filter", Digest, 1985 Int. Solid-State Circ. Conf., New York.

[10] D.J. ALLSTOT, R.W. BRODERSEN, and P.R. GRAY, "MOS switched capacitor ladder filters," IEEE J. Solid-St. Circuits, vol. SC-13, no.6, pp. 806-814, Dec. 1978.

[11] J.L. McCREARY, "Matching properties and voltage and temperature dependence of MOS capacitors", IEEE J. Solid-State Circuits, vol. SC-16, pp. 608-616, Dec. 1981.

[12] P.R. GRAY, University of California, Berkeley, Private communication.

[13] S. KELLY and D. ULMER, "A single-chip CMOS PCM Codec", IEEE J. Solid-State Circuits, vol. SC-14, no. 1, pp. 54-59, Feb. 1979.

[14] H.U. POST and K. WALDSCHMIDT, "A high-speed NMOS A/D converter with a current source array", IEEE J. Solid-State Circuits, vol. SC-15, no. 3, pp. 295-301, June 1980.

[15] R.J. APFEL, Advanced Micro Devices, private communication.

[16] F. KRUMENACHER, H. PINIER and A. GUILLAUME, "Higher sampling rates in SC circuits by on-chip clock-voltage multiplication", Digest, 1983 European Solid-State Circ. Conf., Lausanne, pp. 123-126.

[17] E.M. BLASER, W.M. CHU, and G. SONODA, "Substrate and load gate voltage compensation", Digest, 1976 Int. Solid-State Cir. Conf., pp. 56-57.

[18] H. OHARA, P.R. GRAY, W.M. BAXTER, C.F. RAHIM, and L.L. McCREARY, "A precision low-power PCM channel filter with on-chip power supply regulation", IEEE J. Solid-State Circuits, vol. SC-15, no.6, pp. 1005-1013, Dec. 1980.

[19] J. KHOURY, Y. TSIVIDIS, and M. BANU, "Use of MOS transistor as a tunable distributed RC filter element," Electronics Letters, vol. 20, no.4, pp. 187-188, 16 Feb. 1984.

[20] Y. TSIVIDIS and G. MASETTI, "Problems in precision modelling of the MOS transistor for analog applications," IEEE Trans. on Computer Aided Design, vol. CAD-3, no.1, pp. 72-79, Jan. 1984.

[21] J. R. CANNING and G. A. WILSON, "Frequency discriminator circuit arrangement," UK Patent No. 1 421 093, Jan. 1976.

[21a] K. RADHAKRISHNA RAO, V. SETHURAMAN, and P.K. NEELAKANTAN, "A novel 'follow the master' filter," IEEE Proceedings, vol. 65, no. 12, pp. 1725-1726, Dec. 1977.

[22] J.R. BRAND, R. SCHAUMANN, and E.M. SKEI, "Temperature stabilized active filters," Proc. 20th Midwest Symp. Circ. Syst., 1977, pp. 295-300.

[23] K.S. TAN and P.R. GRAY, "Fully integrated analog filters using bipolar-JFET technology," IEEE J. Solid-State Circuits, vol. SC-13, no. 6, pp. 814-821, Dec. 1978.

[24] K.W. MOULDING, J.R. QUARTLY, P.J. RANKIN, R.S. THOMPSON, and G.A. WILSON "Gyrator video filter IC with automatic tuning," IEEE J. Solid-State Circuits, vol. SC-15, no. 6, pp. 963-968, Dec. 1980.

[25] J.O. VOORMAN, W.H.A. BRULS, and P.J. BARTH, "Integration of analog filters in a bipolar process," IEEE J. Solid-State Circuits, vol. SC-17, pp. 713-722, Aug. 1982.

[26] Y. TSIVIDIS, "Self-tuned filters," IEE Electron. Lett., vol. 17, no. 12, pp. 406-407, 11 June 1981.

[27] J. KHOURY, B. X. SHI, and Y. TSIVIDIS, "Considerations in the design of high frequency fully-integrated continuous-time filters", Proceedings, 1985 Int. Symp. on Circuits and Systems, Kyoto.

[28] H. KHORRAMABADI and P.R. GRAY, "High frequency CMOS continuous time filters," Proc. 1984 Int. Symp. Circ. Systems, Montreal, pp. 1498-1501.

[29] W. HEINLEIN and H. HOLMES, Active Filters for Integrated Circuits, Prentice-Hall, 1974.

[30] B.K. AHUJA, "Implementation of active distributed RC anti-aliasing/smoothing filters", IEEE J. Solid-State Circuits, vol. SC-17, no. 6, pp. 1076-1080, Dec. 1982.

[31] C.F. RAHIM, private communication.

[32] D.J. ALLSTOT and W.C. BLACK, JR., "Technological design considerations for monolithic MOS switched-capacitor filtering systems," Proceedings of the IEEE, vol. 71, no.8, pp. 967-986, Aug. 1983.

[33] D.J. ALLSTOT and W.C. BLACK, JR., private communication.

12

NONLINEAR ANALOG MOS CIRCUITS

B.J. HOSTICKA
Lehrstuhl Bauelemente der Elektrotechnik
University of Dortmund, D-4600 Dortmund 50
Federal Republic of Germany

1 INTRODUCTION

This paper is devoted to the design of nonlinear analog MOS circuits. Variety of circuits will be discussed, such as comparators, Schmitt-triggers, peak detectors, waveform generators, multipliers, modulators, rectifiers, etc., with emphasis on switched-capacitor (SC) techniques. Three examples, an interpolative A/D converter, an FSK modulator, and an FSK demodulator using an SC phase-locked loop will be discussed and their integration in CMOS silicon-gate technology will be presented.

While the design of linear analog MOS circuits has greatly advanced in recent years, the design of nonlinear analog MOS circuits has received relatively little attention with the sole exception of A/D- and D/A-conversion [1-6]. Extending the operation of analog MOS circuits into nonlinear region greatly enhances MOS signal processing capabilities and offers interesting applications, such as clock recovery, waveform generation, adaptive filtering, rms averaging, frequency translation, and modulation. Classical RC-active nonlinear techniques require floating voltage limiting elements, e.g. pn- or Zener-diodes, beside resistors, capacitors, operational amplifiers, and comparators [7]. Adaptation of these techniques to MOS design is by no means always straightforward because some of these components are either not available in MOS technology or do not meet stringent requirements. Nonlinear MOS circuits relying on RC time constants necessarily suffer from imprecise RC products typical for MOS technologies [8]. The introduction of switched-capacitor (SC) principle into linear analog MOS design made it possible to use precise time constants derived from quartz oscillator frequencies and, in many instances, identical technique can be employed in nonlinear analog MOS circuits. It should be remembered, however, that the SC technique involves sampled-data operations and should be treated as such. A serious drawback of analog MOS circuits are rather high offset voltages that impede proper operation of many circuits, above all comparators. Offset cancellation schemes are well known but they are feasible only in sampled-data circuits. In some cases, a diode-connected

MOS transistor can replace the pn-diode; the MOS gate-to-source voltage, however, is too current dependent because the MOSFET transconductance is much lower than that of the bipolar device.

One convenient solution to voltage limitation is to use CMOS output stages with voltage sources as power supply rails, if the full voltage swing can serve as a voltage reference (see Fig. 1). As it will be seen, the precise value of the power supply voltage is often of no importance but the voltage must not change between consecutive samples. To warrant full voltage swing we have to ensure that one of the composite transistors always goes OFF [9]. This means capacitive loading only and a sufficient voltage gain in the preceding stage.

SC nonlinear technique does have a shortcoming: signal phase fluctuation in SC waveform generators. The effect arises from the fact that the clock frequency is not necessarily an integer multiple of the oscillation frequency. This might cause some phase jitter of the output signal of SC waveform generators [10]. Therefore it is important to keep the sampling ratio rather high to minimize this jitter.

In this work design of nonlinear analog MOS circuits, mostly of the SC type, will be presented. Three nonlinear SC circuits have been successfully integrated in CMOS silicon-gate technology and will be demonstrated.

2 BASIC NONLINEAR CIRCUITS

In SC circuits which are considered in this contribution only stray-insensitive SC-configurations are used. The two basic building blocks, the SC-amplifier and the SC-integrator [1], are depicted in Figs. 2 and 3, respectively. A 2-phase nonoverlapping clock is used with phases denoted by Φ and $\overline{\Phi}$. The clock phase notation is important as one has to take into account the clock timing between adjacent circuits. For example, assuming ideal circuit components, the output of the SC-amplifier in Fig. 2 is 0 V during the phase Φ and in the phase $\overline{\Phi}$ it is given by

$$v_{out}(nT) = C_A/C_F \left[v_{in1}((n-1/2)T) - v_{in2}(nT) \right] , \tag{1}$$

where T is the sampling period. As can be seen from Eq. 1, the component of the output voltage due to the noninverting input voltage v_{in1} is delayed by a half cycle while the component due to the inverting input voltage v_{in2} is not delayed at all. An analysis of the integrator circuit of Fig. 3 yields for the output voltage in the phase Φ

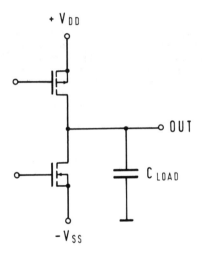

Fig. 1: CMOS output stage

$$\alpha_A = \frac{C_A}{C_F}$$

$$A = \frac{A_0}{(1 + s\tau_1)}$$

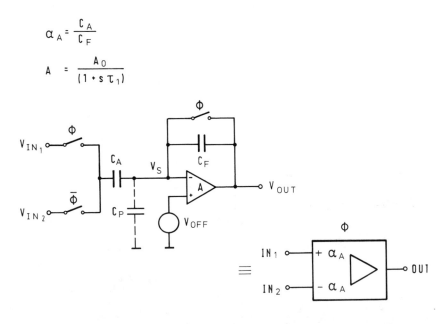

Fig. 2: Stray-insensitive amplifier

$$\alpha_A = \frac{C_A}{C_F}$$

$$A = \frac{A_0}{(1 + s\tau_1)}$$

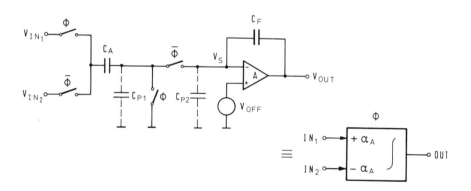

Fig. 3: Stray-insensitive integrator

$$v_{out}((n-1/2)T) = v_{out}((n-1)T) \tag{2}$$

and in the phase $\overline{\Phi}$

$$v_{out}(nT) = v_{out}((n-1)T) + C_A/C_F \left[v_{in1}((n-1/2)T) - v_{in2}(nT)\right] . \tag{3}$$

Eqs. 2 and 3 reveal that if the integrator output voltage is sampled during the clock phase Φ, the component contributed by the noninverting input voltage v_{in1} appears delayed by a full clock cycle at the output. This corresponds to Euler forward integration. On the other hand, sampling of the output voltage during the clock phase $\overline{\Phi}$ does not delay the component due to v_{in2} and thus realizes backward integration with respect to the inverting input.

A closer analysis of the SC-amplifier and integrator can be found in Appendix A. In the following, the clock notation Φ is used for the clocking as indicated in Figs. 2 and 3. If the clock notation is $\overline{\Phi}$ for a building block, then all clock phases in Figs. 2 and 3 have to be reversed. All unused inputs are assumed to be grounded. Naturally, the circuits could be easily extended to multiple input circuits by including more input capacitances in addition to C_A at the summing nodes. Due to the superposition principle, Eqs. I, II, and III in Appendix A remain valid.

The next building block can be readily obtained if the feedback capacitor C_F in the amplifier of Fig. 2 is omitted. In this way, the operational amplifier operates in the open-loop configuration during the clock phase $\overline{\Phi}$. Thus the "charge-balance" voltage comparator of Fig. 4 is obtained with offset cancellation provided by the input capacitors [1]. The input voltage of the operational amplifier during the phase Φ is (for $\tau_1 \to 0$)

$$v_s((n-1/2)T) = A_o/(A_o+1) \, V_{off} \tag{4a}$$

and in the phase $\overline{\Phi}$

$$v_s(nT) = C_A/(C_A+C_B+C_p+C_{pG}) \left[v_{in2}(nT) - v_{in1}((n-1/2)T)\right]$$
$$+ C_B/(C_A+C_B+C_p+C_{pG}) \left[v_{in4}(nT) - v_{in3}((n-1/2)T)\right]$$

377

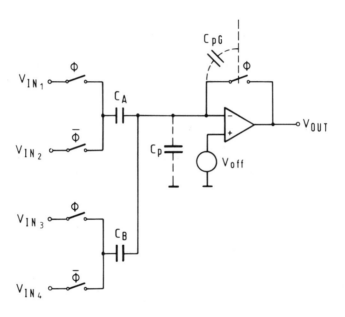

Fig. 4: Autozeroed charge-balance voltage comparator

$$+ \; C_{pG}/(C_A+C_B+C_p+C_{pG}) \; \Delta V_{clock} + A_o/(A_o+1) \; V_{off} \; .$$

$$(4b)$$

The corresponding output voltage in the phase $\overline{\Phi}$ is

$$v_{out}(nT) = (-V_{BATT}) \cdot sign \; \{v_s(nT) - A_o/(A_o+1) \; V_{off}\}$$

$$(5a)$$

ΔV_{clock} is the clock voltage swing and C_{pG} is the parasitic capacitance between the control electrode (gate) and the terminal (drain diffusion) of the feedback transistor switch, which is connected to the inverting input of the operational amplifier. The resolution of the comparator is limited by the injected charge through C_{pG}. While a constant clock feedthrough causes only a voltage offset, the superimposed noise and nonlinear clock feedthrough decrease the resolution. Thus the ratio of the network capacitors to the parasitic capacitances (C_{pG}) of the feedback switch should be as large as possible.

As indicated, the charge balance comparator can be modified to include more input capacitors. Their values can be chosen to weight individual voltages from each input by a factor

$$\beta_k = C_k / \sum_i C_i$$

$$(5b)$$

where C_k is the capacitor at the corresponding input and $\sum C_i$ is the sum of all input capacitors including parasitic capacitances at the inverting input node.

Fig. 5a shows an example of an SC Schmitt-trigger [11,12]. The circuit contains an open-loop amplifier used as a voltage comparator and an SC amplifier, which generates the switching threshold voltage $V_s = \pm\alpha_A V_{BATT}$ for the comparator. The amplifiers are connected in a positive feedback configuration to guarantee high loop gain and bistability. The nominal size of the hold capacitors C_H is determined solely by the clock feedthrough and noise requirements, and it does not affect the switching threshold which is defined by $\pm\alpha_A V_{BATT}$. The ideal switching threshold will be reduced by the parasitic effects. The differential input voltage between the noninverting and inverting input of the open loop amplifier v_i is then (for $\tau_1 \to 0$)

379

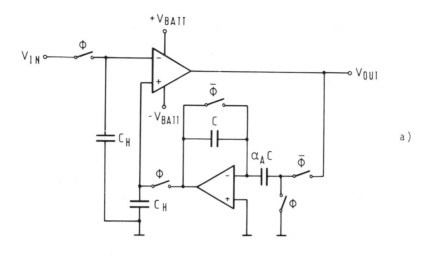

a)

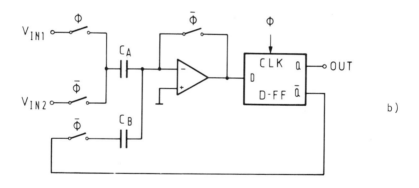

b)

Fig. 5: a) Schmitt-trigger,
      b) alternative Schmitt-trigger

$$v_i(nT) = V_{off1} + v_{in}(nT)$$

$$+ C_A/C_F \frac{A_{o2} \, v_{out}[(n-1)T]}{A_{o2} + 1 + (C_A + C_p)/C_F}$$

$$+ \frac{A_{o2} \left[1 + 1/(A_{o2}+1)(C_A+C_p)/C_F\right]}{A_{o2} + 1 + (C_A+C_p)/C_F} \, V_{off2} \, . \tag{6a}$$

The subscript 1 refers to the open-loop amplifier the subscript 2 denotes the op amp of the SC amplifier in the feedback path. The output voltage is

$$v_{out}(nT) = (-V_{BATT}) \cdot \text{sign}\{v_i(nT)\} \, . \tag{6b}$$

Another version of an SC Schmitt-trigger is shown in Fig. 5b. This circuit utilizes the "charge balance" comparator and a data latch with a positive feedback path. The threshold voltage is defined in the ideal case by $\pm V_{BATT} \, C_B/C_A$. The output voltage which is latched in the phase Φ is (for $\tau_1 \rightarrow 0$)

$$v_{out}(nT) = \begin{array}{l} V_{BATT} \text{ if } v_s(nT) \leqslant A_o/(A_o+1) \, V_{off} \\[8pt] -V_{BATT} \text{ if } v_s(nT) \geqslant A_o/(A_o+1) \, V_{off} \end{array} \tag{7}$$

where

$$v_s(nT) = C_A/(C_A+C_B+C_p+C_{pG}) \left[v_{in2}(nT)-v_{in1}((n-1/2)T)\right]$$

$$- C_B/(C_A+C_B+C_p+C_{pG}) \, v_{out}((n-1)T)$$

$$+ A_o/(A_o+1) \, V_{off} + C_{pG}/(C_A+C_B+C_p+C_{pG}) \, \Delta V_{clock} \, .$$

As in the case of the comparator of Fig. 4 the resolution is again limited by the clock feedthrough of the feedback switch.

Two examples of waveform generators are shown in Figs. 6a and 7. The square- and triangle-wave generator in Fig. 6a contains an integrator and a Schmitt-trigger [11,12]. As indicated, the circuit has two outputs. The operation of the circuit is as follows:
The integrator integrates the constant voltage delivered by the output of the Schmitt-trigger until the threshold voltage is reached. Then the sign of the Schmitt-trigger output voltage changes and the integrator integrates this voltage again until the other Schmitt-trigger threshold is reached. In order to obtain maximum dynamic range the threshold voltages of the Schmitt-trigger should be kept as near as possible to the supply voltages V_{BATT}. The limiting condition for this is that the integrator op amp must operate in the linear region to ensure high loop gain. The oscillation frequency is

$$f_0 = f_c \alpha_1/(4\alpha_2) \tag{8}$$

for a clock frequency f_c much higher than f_0 and stable power supply voltages $\pm V_{BATT}$ (see Fig. 5a). The parasitic capacitances and the finite open-loop gain of the SC-integrator reduces the integrator constant α_1, and this causes a somewhat lower oscillation frequency. The offset voltage of the integrator causes an offset of the triangle voltage which results in an unsymmetrical square voltage. The oscillation frequency is independent of the supply voltages, if the integrator op amp operates in the linear region.

The circuit was breadboarded with MC 14066 switches and LF 356 op amps. The output waveforms of the breadboarded oscillator are shown in Fig. 6b for $V_{BATT} = \pm 7.5$ V and 100 kHz sampling frequency. The oscillation frequency was chosen to be 1.5 kHz which yields $\alpha_1 = 1/20$ and $\alpha_2 = 5/6$. The measured oscillation frequency deviation was less than 1% of the desired one. The amplitude was less than V_{BATT} due to the limited output voltage swing of LF 356.

An alternative square-wave generator is illustrated in Fig. 7 [10,14]. It consists of an SC integrator with two inputs and a comparator. During the clock phase Φ the capacitor $\alpha_1 C$ is charged to the value of the comparator output voltage (e.g. $+V_{BATT}$) and in the next clock phase $\overline{\Phi}$ this charge is transferred from $\alpha_1 C$ into the integrator feedback capacitor C. As this operation is periodically repeated, the output voltage of the integrator rises and when it has reached the ground potential, the comparator switches and changes its output voltage to $-V_{BATT}$. The difference, the full supply voltage, will cause a charge packet $2\alpha_2 C V_{BATT}$ to be fed into C and the output voltage of the integrator becomes $2\alpha_2 V_{BATT}$. As the comparator output is now $-V_{BATT}$ the integration of this voltage produces decreasing voltage at the integrator output until the ground potential has been reached and the whole process is repeated. To reach the ground potential, $n = 2\alpha_2 V_{BATT}/(\alpha_1 V_{BATT})$ steps are needed, so that half the oscillation period

382

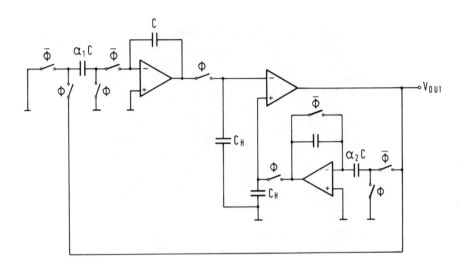

Fig. 6: a) Square- and triangle-wave generator

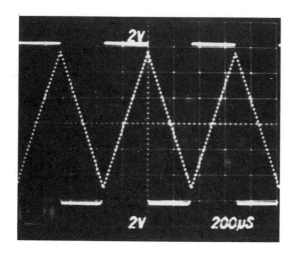

Fig. 6: b) Output waveform of the breadboarded square- and triangle-wave
generator

384

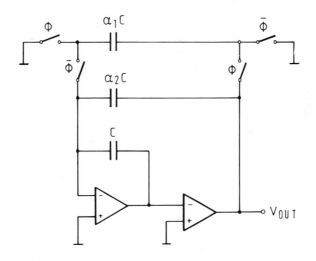

Fig. 7: Square-wave generator

can be calculated as $T_0/2 = nT_c$, where $f_c = 1/T_c$ is the clock frequency, and the oscillation frequency is given by $f_0 = f_c \alpha_1/(4\alpha_2)$. Unlike the previous generator, this one does not provide the triangle-wave signal.

Another class of nonlinear circuits are peak detectors. A peak detector is a circuit which stores the maximum peak value of an input signal. An example of a continuous-time peak detector is shown in Fig. 8a [13]. If $V_{IN} > V_{OUT}$, the capacitor C is charged to the new value of V_{IN}. Reset can be easily provided, e.g. by adding a parallel reset switch or a resistor to the capacitor C. A possible sampled-data realization of the peak detector is depicted in Fig. 8b [14]. Here, the comparator monitors the current value of the input signal and compares it with the stored peak value at C. Whenever the sampled-and-held input signal exceeds the stored peak value at C, the RS-latch is set HIGH during the phase Φ and then during the phase $\overline{\Phi}$ the peak value at C is updated through the input switch. Again, a reset of C can be easily implemented. If negative peak values are to be stored, the inputs of the comparator have to be reversed.

3 ADVANCED NONLINEAR CIRCUITS

In many analog systems, such as correlators, adaptive filters, dividers, voltage-controlled attenuators, etc., it is required to obtain an analog output which is a linear product of two signals. The circuit that performs such function is the analog multiplier. Besides multiplication, an analog multiplier can also be used as a modulator, phase comparator, and frequency translator. In many of these applications linearity or the four-quadrant capability are not required and thus the multiplier can be substantially simplified. There are several multiplication principles but from the variety of schemes only two will be presented here.

The first method is the time-division multiplier shown in Fig. 9a [15]. Essentially, it consists of a pulse-width modulator and a voltage-controlled current source that charges an output capacitor. The pulse-width modulator contains a comparator and a ramp generator. The ramp is generated by charging the capacitor C_1 using the constant current I_0. Simultaneously, the output capacitor C_2 is charged by the current V_{IN2}/R_2 from the current source controlled by the input voltage V_{IN2}. When the ramp voltage V_x has reached the value of the input voltage, the comparator opens the switch S_3. The charging time for C_2 is $T = C_1 V_{IN1}/I_0$ and the output voltage across C_2 is

$$V_{OUT} = V_{IN2}T/(R_2 C_2) = (C_1/C_2) \cdot [(V_{IN1}V_{IN2})/(R_2 I_0)] .\qquad (9)$$

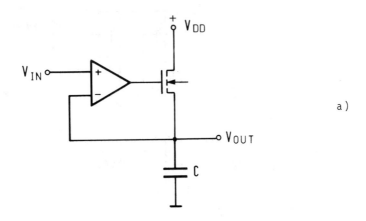

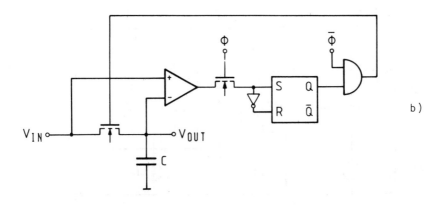

Fig. 8: a) Continuous peak detector,
        b) SC peak detector

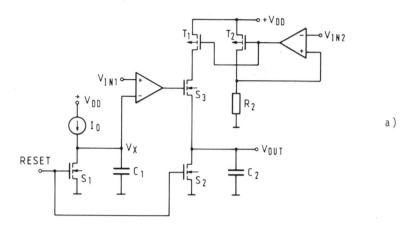

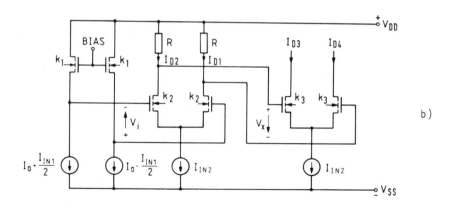

Fig. 9: a) SC time-division multiplier,
        b) NMOS multiplier

Simulated performance of this circuit indicated full-power bandwidth of 30 kHz, power dissipation of 7 mW (±5 V power supply voltages), and nonlinearity smaller than 2% within 3 V full-scale output.

The second multiplier relies on a principle commonly used in bipolar integrated circuits and termed "variable-transconductance" method (see Fig. 9b) [16]. The method has been adapted to make use of the square-law dependence of the MOSFET drain current on the gate-to-source voltage: $I_D = k(V_{GS}-V_T)^2$. It can be shown that the output current difference of a differential pair consisting of two perfectly matched devices is given by (see Fig. 9b)

$$I_{D1}-I_{D2} = k_2 V_i (2I_{IN2}/k_2-V_i^2)^{1/2} . \tag{10a}$$

If the differential voltage V_i is sufficiently small, the current $I_{D1}-I_{D2}$ is linearly dependent on V_i: $I_{D1}-I_{D2} \simeq k_2 V_i (2I_{IN2}/k_2)^{1/2}$. Cascading of two differential pairs provides an output current of

$$I_{OUT} = I_{D3}-I_{D4} = k_3 V_x (2I_{IN2}/k_3-V_x^2)^{1/2} \tag{10b}$$

and makes it possible to achieve linear dependence of I_{OUT} on I_{IN2}, because $V_x = R(I_{D1}-I_{D2})$ and, hence,

$$I_{OUT} = 2R(k_2 k_3)^{1/2} V_i I_{IN2} (1-R^2 k_2 k_3 V_i^2)^{1/2} . \tag{10c}$$

To obtain good linearity for large output swing, it pays off to predistort the input voltage for the first differential pair so that $I_{IN1} = k_1 V_i (2I_0/k_1-V_i^2)^{1/2}$ (see Fig. 9b). If the current I_0 is chosen so that the condition $I_0 = k_1/(2R^2 k_2 k_3)$ is met, the output current is given by

$$I_{OUT} = 2R^2 k_2 k_3 I_{IN1} I_{IN2}/k_1 \tag{10d}$$

and exhibits the desired linear dependence on the input currents I_{IN1} and

I_{IN2}.

In [16] the realization of an NMOS multiplier based on the principle shown above is described. The operation of the multiplier has been extended to four-quadrant service. It achieves small-signal bandwidth of 1.5 MHz, and the linearity is better than 0.3% at 75% of full-scale output swing, which is 0.66 V.

Unlike multipliers, modulators are required to exhibit linear response with respect to only one of the inputs. This input is usually called modulating input. The second input is driven by a carrier, typically a square-wave. Letting the subscripts m and c stand for modulating signal and carrier, respectively, it can be written

$$V_m(t) = V_m \cos(\omega_m t) \tag{11a}$$

for sine-wave modulating input and

$$V_c(t) = \sum_{n=1}^{\infty} A_n \cos(n\omega_c t), \quad A_n = [\sin(n\pi/2)]/(n\pi/4) \tag{11b}$$

for square-wave carrier. The output signal is given by

$$V_{OUT}(t) = K \sum_{n=1}^{\infty} A_n V_m \{\cos[(n\omega_c + \omega_m)t] + \cos[(n\omega_c - \omega_m)t]\} \ . \tag{11c}$$

The modulation generates lower and upper sideband components around the original carrier frequency but the carrier itself is suppressed. It should be noted that a carrier feedtrough can occur if the modulating input signal contains a DC component.

Main application of balanced modulators are frequency translation and phase detection. Fig. 10 shows a phase comparator which will find its use in the SC phase-locked loop described later in this work. The square-wave reference input multiplies the input signal either by +1 or -1. The output signal of the phase comparator is available during $\overline{\Phi}$:

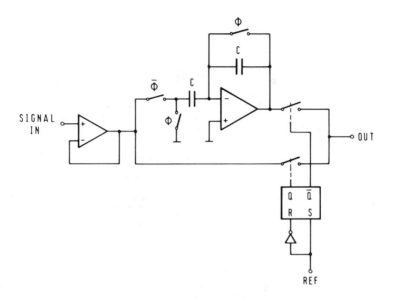

Fig. 10: Phase comparator

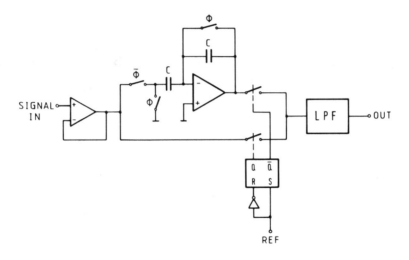

Fig. 11: Lock-in amplifier

$$v_{out}(nT) = REF \cdot v_{in}\left[(n-1/2)T\right], \qquad REF = 1;-1 \; . \qquad\qquad (12)$$

The RS-latch prevents the NMOS analog switches from being "ON" simultaneously. Two applications of the phase comparator are illustrated in Figs. 11 and 12. The lock-in amplifier of Fig. 11 contains an SC phase comparator which operates as a mixer, and an SC lowpass filter (LPF) [5]. If the input signal is not synchronized with the reference signal, as in the case of noise, the average level of the mixer output is zero. Otherwise the output voltage of the LPF is a function of the phase difference of the two signals. A full-wave rectifier can be easily designed using a phase and a voltage comparator (see Fig. 12). The output voltage of the full wave rectifier is

$$v_{out}(nT) = v_{in}\left[(n-1/2)T\right] \cdot sign\left\{v_{in}\left[(n-1/2)T\right]\right\} \; . \qquad\qquad (13)$$

Alternative approaches to rectification have been proposed in [12,14,17]. The phase comparator, when using a weighted capacitor array instead of fixed capacitors in the amplifying stages, could be easily extended to a modulator with a modulation frequency being an integer of the sampling frequency. For example a modulation of the input signal with a modulation frequency of a quarter of the sampling frequency could be easily achieved by multiplying the input signal with 0, 1, 0, and −1. The coefficients correspond to the sampled values of a sine wave at the times 0, $\pi/2$, π, and $3\pi/2$.

4 DESIGN EXAMPLES

4.1 Interpolative A/D-Converter

A/D- and D/A- converters using switched-capacitor arrays actually initiated interest in SC circuits [1]. In this section an A/D converter that is based on interpolative quantization will be demonstrated [18,19]. The interpolative converter consists of an integrator, a quantizer, a D-type flip-flop, and a binary up/down-counter (see Fig. 13a). The integrator integrates the difference between the analog input signal and the quantization-error feedback signal available at the output of the D-flip-flop. The result is then coarsely quantized. In this case, a 1-bit quantizer has been chosen, which was easily implemented using the charge-balance comparator of Fig. 4. Each quantizer decision is latched into the D-flip-flop, so that the offset cancellation of the comparator during $\overline{\Phi}$ does not affect the operation of the converter. Besides, the converter loop

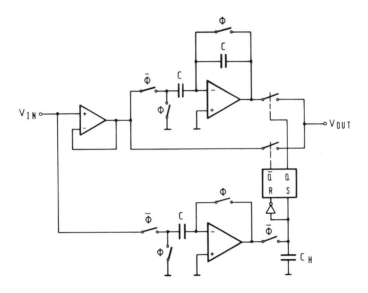

Fig. 12: Full-wave rectifier

394

gain is increased. The encoding loop operates at a clock frequency f_c. As the integrator tries to keep the integral of the quantized signal equal to the integral of the input signal due to the negative feedback action, the average value of N consecutive quantizer decisions y^* approximates the analog input signal x. If q is designated as the quantization error, it can be shown that the instantaneous value of $y^*(nT)$ at a n-th time slot is [12]:

$$y^*(nT) = \alpha x[(n-1/2)T] - (1-\alpha)y[(n-1)T] - \alpha q[(n-1)T] + q(nT) ,$$

(14a)

where $T = 1/f_c$. The averaging is carried out in the binary up/down-counter, which accumulates the values of $y^*(nT)$. The result is the sum of N values that is available each NT seconds which corresponds to the conversion time. The average value is

$$\frac{1}{N} \sum_{n=1}^{N} y^*(nT) = \frac{1}{N} \sum_{n=1}^{N} x[(n-1/2)T] - \frac{1}{N} [q(0)-q(NT)] + \frac{1-\alpha}{\alpha N} y(NT) .$$

(14b)

Larger N minimizes the error terms and thus provides finer resolution but requires a higher clock rate. The main problems of this converter are voltage offsets. Operational amplifiers and comparators integrated in MOS technology exhibit offsets in the range 1-10 mV. While in our case the offset of the comparator is reduced by the amplifier gain due to the feedback, the offset of the operational amplifier contributes fully to the DC offset of the converter [12]. This can be reduced, if an offset cancellation scheme is employed [1]. As the output voltage of the comparator serves as a voltage reference, it is necessary to keep the power supply voltages symmetrical, i.e. $|V_{DD}| = |V_{SS}|$. Any assymmetry will produce a DC offset at the output of the converter.

The circuit has been integrated in CMOS silicon-gate technology with 5 μm minimum feature size, 40 nm gate oxide, and 1.8×10^{14} cm^{-3} p-substrate. The doping of the n-wells was 1.8×10^{16} cm^{-3} [20]. The converter used dynamic operational amplifiers of which the performance data are given in Table I [21,22]. The power supplies were ±5 V , $\alpha = 0.5$, and $C_{min}= 5$ pF. All biasing capacitors were 15 pF. Fig. 13b shows the error plot for the clock rate $f_c = 100$ kHz and N = 256. The abscissa depicts the analog input voltage, while the ordinate shows the deviation of the converter output, reconstructed with DAC 1200, from the analog input ramp. The performance data of the conveter are summarized in Table II. The photomicrograph of the

395

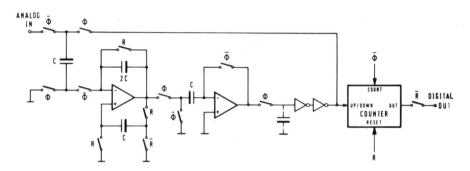

CLOCK FREQUENCY Φ : f_C
RESET FREQUENCY R : f_C / N

Fig. 13: a) Interpolative A/D converter

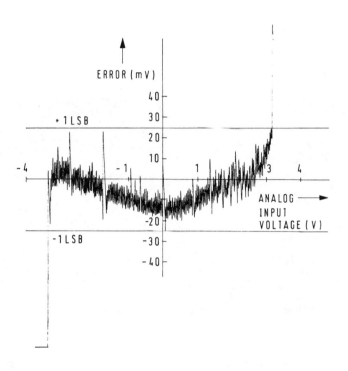

Fig. 13: b) Error plot of the interpolative A/D converter

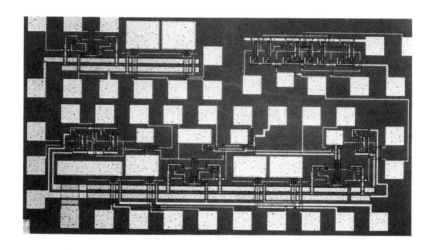

Fig. 13: c) Chip photomicrograph of the interpolative A/D converter

TABLE I

D Y N A M I C   O P E R A T I O N A L   A M P L I F I E R

POWER SUPPLY VOLTAGES	±5 V
OPEN-LOOP GAIN	66 dB
OUTPUT VOLTAGE SWING	±5 V
COMMON MODE INPUT RANGE	+4.5 V / −4.8 V
PSRR$^+$	60 dB
PSRR$^-$	57 dB
POWER DISSIPATION (f_c = 100 kHz)	500 μW
CHIP AREA	0.2 mm^2

TABLE II

I N T E R P O L A T I V E   A / D - C O N V E R T E R

POWER SUPPLY VOLTAGES	±5 V
CLOCK RATE	100 kHz
RESOLUTION	8 BIT
LINEARITY	±1/2 LSB
INPUT RANGE	±3.25 V
CONVERSION TIME	2.56 ms
POWER DISSIPATION	1.4 mW
CHIP AREA	1 mm^2

chip containing the integrator, the D-flip-flop, and the comparator is shown in Fig. 13c. The chip area is 1 mm^2 without bonding pads. The counter was built using two synchronous 4-bit binary up-down counters MC 14193. It deserves notice, that the interpolative A/D converter does not require any precision components. For the same resolution, N can be substantially decreased, if an M-bit quantizer is used but then an M-bit D/A converter has to convert the quantizer output into an analog signal for the integrator input. Note that the error characteristic of the interpolative A/D converter does not exhibit the sawtooth waveform typical for linearity test of A/D converters because this converter interpolates values of the input signal by oscillating between the two values of the coarse quantizer. Still, the total error lies within ±1 LSB band as required for ±1/2 LSB linearity.

4.2 FSK Modulator

A convenient way of sending narrow-band digital data up to 1800 baud over telephone lines is to use frequency-shift keying (FSK) (see Fig. 14). For this purpose, an FSK modulator encodes the digital data into two discrete audio frequencies. The higher frequency, called SPACE, corresponds to a binary ZERO, while the lower frequency, called MARK, corresponds to a binary ONE. An FSK demodulator demodulates the received FSK signal and restores the original data in binary format. In general, additional bandpass filters are used to suppress the outband noise and to reduce the harmonics of the FSK transmitter.

Since MOS modem filters tend to be realized in switched-capacitor technique, it should be obvious to implement the modulator and demodulator compatible with the technology of these filters. In addition, CMOS technology is also well suited for realization of nonlinear analog SC circuits since the capacitively loaded CMOS output stages of the operational amplifiers ensure full voltage swing and can therefore be used as voltage limiting elements with the power supplies as voltage references, as it was described above [11,12].

The FSK modulator has been designed using an SC harmonic oscillator and an optional crossover detector to avoid phase shifts if phase-coherent FSK is desired. In this case the frequency transitions are supposed to occur only at zero crossovers to minimize isochronous distortion. But in general, the oscillation frequency can be changed at any time. The block diagram of the modulator is depicted in Fig. 15a. In contrast to the SC oscillator proposed in [6], which uses an RC phase shift approach, the operation of the harmonic oscillator presented here is based on an analog computer implementation of a differential equation using stray-insensitive SC integrators. The circuit diagramm of the SC oscillator can be found in Fig. 15b. The programming of the two oscillation frequencies is accomplished by switching two capacitor pairs C_1, C_2 and C_4, C_5. In general, an oscillator is a nonlinear circuit which contains three basic functional blocks [7]: a gain stage, a frequency selective network, and an amplitude

400

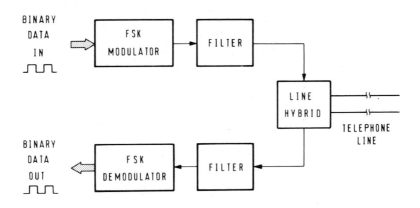

Fig. 14: Full-duplex FSK modem system

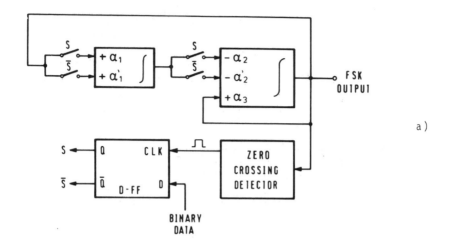

a)

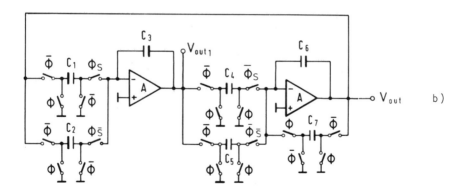

b)

Fig. 15: a) Block diagramm of the FSK modulator,
         b) circuit diagramm of the FSK modulator

limiter as a nonlinear element. As can be seen from Fig. 15b the gain stage and the frequency selective network have been built with a forward and a backward SC integrator [1]. Although there are many different possibilities to achieve a 360° phase shift or a multiple of this using SC integrators and SC amplifiers it turned out to be convenient to use only one forward and one backward integrator. In this case the oscillation frequency can be easily determined using the sinh-transformation known from "leapfrog" filters [1-3]. The new frequency variable Ψ is then defined as

$$\Psi = 2/T \sinh sT/2 \ , \tag{15}$$

with T = sampling period and s = common complex frequency variable. The circuit contains a positive feedback path (capacitor C_7) to guarantee oscillation. The maximum output voltage swing of the operational amplifiers serves as voltage limit.

Neglecting the positive feedback the critical loop gain in the case of ideal components (infinite voltage gain of the amplifiers, no parasitic capacitances, and zero offset voltage of the operational amplifiers) is defined under steady state conditions as

$$1 = \left(\frac{\alpha_1 z^{-1}}{1-z^{-1}} \right) \cdot \left(- \frac{\alpha_2}{1-z^{-1}} \right) \tag{16a}$$

$$= \frac{-(\alpha_1/2 \ \alpha_2/2)}{(\sinh sT/2)^2} = \frac{-\alpha_1 \alpha_2}{T^2 \psi^2} \tag{16b}$$

with $z = e^{sT}$, and $\alpha_1 = C_1/C_3$, $\alpha_2 = C_4/C_6$, and $\alpha_1´ = C_2/C_3$, $\alpha_2´ = C_5/C_6$.

In order to achieve a maximum dynamic range α_1 should be equal to α_2. Assuming real frequencies $s = j\omega$, $\Psi = j\phi$, and $\alpha_1 = \alpha_2 = \alpha$ the poles are purely imaginary ($\Psi_{1,2} = \pm j\alpha/T$), and the oscillation frequency f_{osc} is defined by

$$f_{osc} = f_c/\pi \ \arcsin (\alpha/2) \tag{17}$$

with $f_c = 1/T$ = sampling frequency. It was experimentally found that the ratio f_c/f_{osc} should be greater than 30 to avoid large phase jitter.

Due to the limited voltage gain of the operational amplifiers and the parasitic capacitances, though the effect of the latter is reduced by the open-loop gain of the op amps, the poles are slightly shifted into the left half-plane of the s-plane. This results in a damped oscillation because the overall system is stable. To ensure oscillation build-up the poles must be located in the right half-plane but near the imaginary axis in order to minimize the harmonic distortion. This can be easily accomplished by adding a regenerative feedback (capacitor C_7) as mentioned above. The inherent excitation by internal noise and clock feedthrough is always present. Due to the Barkhausen criterion the real part of the oscillator loop gain must be 1 while the imaginary part must be 0. Assuming linear finite open loop gain A_0 of the amplifiers and parasitic capacitances as indicated in the Appendix A and including the regeneration, the calculation for zero imaginary loop gain yields

$$\frac{C_1}{C_3} \frac{1}{A_0+1} + \frac{C_4-A_0C_7}{C_6(A_0+1)} = 0 \ . \tag{18a}$$

Thus the oscillation frequency f_{osc} is

$$f_{osc} \approx f_c/\pi \ \arcsin \left[(\alpha/2) \ A_0/(A_0 +1) \right] \ . \tag{18b}$$

As can be shown the influence of the parasitic capacitances can be neglected if the open loop gain of the amplifiers is sufficiently high. To ensure oscillation the capacitor C_7 has to be selected properly. The minimum value of C_7 which guarantees the oscillation depends on the minimum guaranteed open loop gain A_0 of the amplifiers [12]:

$$C_{7,min} = 2 \ C_1/A_{0,min} \ . \tag{19}$$

On the other hand C_7 should not be selected too large as this would move the poles further from the imaginary axis and therefore raise the total harmonic distortion. The relationship between distortion and oscillation frequency of an oscillator derived by Groszkowski states that the shift in oscillation frequency is proportional to the squared harmonic ratios [23].

The oscillation frequency is insensitive to variations of the power supply voltages, if the loop gain of the operational amplifiers is higher than the minimum loop gain required for a given C_7. Offset voltages of the op amps and clock feedthrough of the analog switches do not affect the oscillation frequency due to inherent large negative DC feedback. Inserting Eq. 18 in Eq. 19 gives the lower bound of the expected oscillation frequency as:

$$f_{osc} \simeq f_c/\pi \; arcsin \; (\frac{\alpha/2}{1+C_7/(2C_6)}) \; . \qquad (20)$$

It should be mentioned that Eqs. 18 and 20 do not hold if the positive feedback via C_7 is too large.

For oscillation frequencies far below the sampling frequency the following differential equation is approximately valid:

$$\ddot{v}_{out2}(t) + \xi \; \dot{v}_{out2}(t) + \omega_o^2 \; v_{out2}(t) = 0 \qquad (21)$$

with the solution

$$v_{out2}(t) = \hat{v}_{out} \; e^{\xi t/2} \; sin \; [\; \sqrt{\omega_o^2 + \xi^2/4} \; t \;]$$

$$\omega_o \simeq f_c \; \sqrt{\alpha_1 \alpha_2} \; , \qquad (23a)$$

$$\xi = \alpha_3 \; f_c \; , \qquad (23b)$$

and

$$\alpha_3 = \frac{C_7}{C_6} \; \frac{A_o}{A_o + 1 + (C_7 + C_{p6} + C_{p7})/C_6} \; . \qquad (23c)$$

Eq. 22 gives the upper bound of the expected oscillation frequency f_{osc}. Using Eqs. 23a and 23b, it can be readily shown that the sensitivity of the oscillation frequency to variations in α_3 (where $\alpha_3 \simeq C_7/C_6$) is $S_{\alpha_3}^{f_{osc}} \simeq (1/4)\alpha_3^2/(\alpha_1\alpha_2)$ for $\alpha_3 \ll \alpha_1$ and α_2. As stated above, the harmonic distortion is proportional to $\sqrt{\Delta f_{osc}/f_{osc}}$ according to Groszkowski and, hence, the distortion is proportional to $(1/2) \cdot (\alpha_3/\sqrt{\alpha_1\alpha_2}) \cdot \sqrt{\Delta\alpha_3/\alpha_3}$. This means, that the harmonic distortion of the oscillator is not very sensitive to variations in α_3 once the oscillation is ensured.

The FSK modulator has been integrated on a multiproject chip in the same CMOS silicon-gate technology as the interpolative A/D converter [20]. In order to save chip area only the oscillator was integrated while the crossover detector logic was breadboarded. The chip photomicrograph can be seen in Fig. 15c. The modulator occupies an area of 1.3 mm^2 without bonding pads. The circuit uses two dynamic op amps with 30 pF bias capacitances [21, 22]. The capacitors were formed between metal (Al) and n^+-diffusion with a deposited oxide layer as dielectric ($0.85 \times 10^{-3} \text{ pF}/\mu\text{m}^2$). The capacitor values were $C_1 = C_4 = 1.225$ pF, $C_2 = C_5 = 1.068$ pF, $C_3 = C_6 = 15$ pF, and $C_7 = 0.05$ pF for MARK and SPACE frequencies, respectively.

The circuit operated with power supply voltages in the range between ± 1.5 V and ± 6 V and at clock frequencies up to 1.9 MHz. The power consumption was 1.6 mW at 100 kHz clock rate and with ± 5 V power supply voltages. The measured data of the modulator are summarized in Table III. The total harmonic distortion (THD) was -37 dB and -38 dB for MARK and SPACE frequencies, respectively. This gives a deviation of the oscillation frequencies of less than 0.05% using Groszkowski's relationship. Thus the measured deviation of the oscillator frequencies was mainly due to capacitor mismatch. The capacitor mismatch could be substantially reduced if unit capacitors were used [24]. The variations in value of the capacitor C_7 (using an additional off-chip capacitor) by 250% caused a change of the SPACE frequency by 15 Hz and the THD increased by 3.1 dB. The MARK frequency changed by 14 Hz and THD increased by 1.8 dB. Figs. 16a and 16b show an example of an FSK waveshape.

4.3 FSK Demodulator

As a third example a design of an FSK demodulator fully integrated in CMOS silicon-gate technology will be presented that uses an SC phase-locked loop with an SC voltage-controlled oscillator.

Phase-locked loop (PLL) techniques are widely used for FSK demodulation [25-27]. The advantages of a PLL in comparison to a narrow bandpass filter are the capability of carrier tracking and excellent detection behavior for extremely noisy signals. A phase-locked loop is a feedback system with the property, that the frequency of an internal voltage-controlled oscillator (VCO) follows an incoming frequency within a fixed bandwidth. Generally, a PLL incorporates four functional blocks (see Fig. 17): a phase detector or phase comparator (PC), a lowpass loop filter (LPF), a loop amplifier, and a voltage controlled oscillator (VCO). When the PLL has locked on an incoming

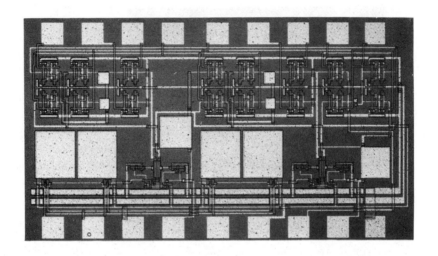

Fig. 15: c) Chip photomicrograph of the FSK modulator

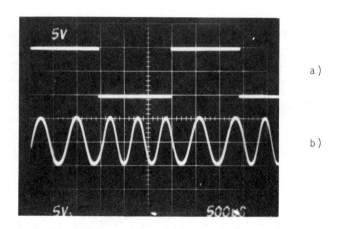

a)

b)

Fig. 16: a) Binary data input signal for
the FSK modulator (time-base 0.5 ms/ div.),
b) output signal of the FSK modulator modulated
with binary data

TABLE III

F S K   M O D U L A T O R

POWER SUPPLY VOLTAGES	±5 V
POWER SUPPLY VOLTAGE RANGE	±1.5 — ±6 V
CLOCK RATE	100 kHz
MAX. CLOCK RATE	1.9 MHz
MARK FREQUENCY (f_c = 100 kHz)	1323 Hz
SPACE FREQUENCY (f_c = 100 kHz)	1704 Hz
POWER SUPPLY SENSITIVITY	<2 Hz/V
THD:	
MARK FREQUENCY	−37 dB
SPACE FREQUENCY	−38 dB
POWER DISSIPATION (f_c = 100 kHz, V_{batt} = ±5 V)	1.6 mW
CHIP AREA	1.3 mm^2

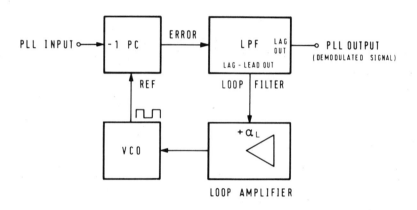

Fig. 17: Phase-locked loop

frequency, the phase comparator output signal is a function of the phase difference between the input and the VCO signal. After filtering and amplification, this signal serves as control voltage of the VCO.

A simple phase comparator has already been shown in Fig. 10. Depending on the square-wave VCO output, the incoming signal is directed to the output either inverted or non-inverted. The characteristic of a phase comparator usually depends on the input waveform. For a square-wave input signal this PC has a triangular characteristic and the DC component of the error signal V_{ERROR} depends linearly on the phase shift ϕ_{ERROR}:

$$V_{ERROR} = K_D \ (\phi_{ERROR} - \pi/2) \ , \tag{24}$$

where K_D is the phase comparator gain. Assuming a square-wave input signal with amplitude $\pm E$, the comparator gain is

$$K_D = \frac{dV_{ERROR}}{d\phi_{ERROR}} = 2E/\pi \ . \tag{25}$$

A voltage-controlled oscillator can be implemented using an SC integrator and an SC Schmitt-trigger (see Fig. 18). The input of the integrator is modified to incorporate a circuit called phase reverser [5,11,12]. This circuit changes the clock phasing of one input capacitor of the SC integrator to realize either an inverting or noninverting integrator with respect to that input. When the Schmitt-trigger output level is LOW, the clock phases are $A=\bar{\Phi}$ and $B=\Phi$; when it is HIGH, the clock phases are $A=\Phi$ and $B=\bar{\Phi}$ (see Fig. 18). The integrator integrates the output square-wave voltage of the Schmitt-trigger and, in addition, the VCO control voltage supplied through the phase reverser. Hence this voltage is added or subtracted, depending on the polarity of the square-wave voltage.

With no control voltage applied, the VCO operates at its free-running frequency f_0, which should be much smaller than f_c in order to avoid phase jitter:

$$f_0 = (\alpha_V/\alpha_A) \ f_c/4 \ . \tag{26}$$

Eq. 26 shows that the frequency is determined only by the two capacitor ratios α_V and α_A (see Fig. 18) and is not affected by the power supply voltage V_{BATT}.

411

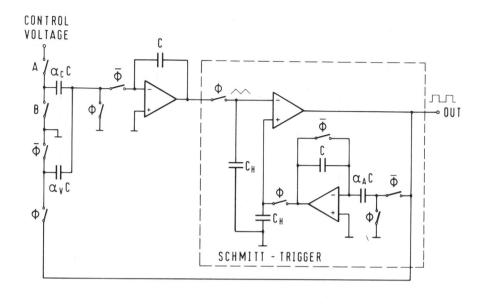

Fig. 18: Voltage-controlled oscillator

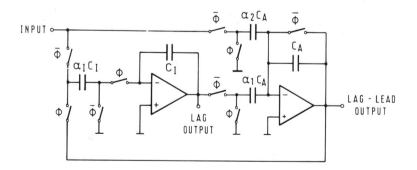

Fig. 19: Loop filter

Applying a non-zero control voltage, the oscillator frequency becomes:

$$f_{VCO} = f_0 \left[1 + \frac{\alpha_C}{\alpha_V} \frac{V_{CONTROL}}{V_{BATT}} \right] . \qquad (27)$$

The VCO gain is then defined as:

$$K_{VCO} = 2\pi \frac{df_{VCO}}{dV_{CONTROL}} = 2\pi \frac{\alpha_C}{\alpha_V} \frac{f_0}{V_{BATT}} . \qquad (28)$$

In order to guarantee PLL loop stability it is convenient to use an one-pole lowpass filter with an additional zero as loop filter [25,26]. This lead-lag filter has the following transfer characteristic:

$$F(s) = \frac{1 + s/\tau_2}{1 + s/\tau_1} , \qquad (29)$$

where τ_1 and τ_2 are the time constants associated with the pole and zero frequencies of the filter, respectively. A stray-insensitive SC realization of this filter is depicted in Fig. 19. It uses one SC amplifier and one SC integrator. The calculation of the characteristic frequencies yields for $1/\tau_1$ and $1/\tau_2$ much smaller than f_c:

$$1/\tau_1 = \alpha_1 \alpha_I f_c ,$$
$$1/\tau_2 = (\alpha_1/\alpha_2) \alpha_I f_c . \qquad (30)$$

In the PLL an additional loop amplifier (a stray-insensitive SC amplifier) is inserted in order to increase the PLL loop gain (see Fig. 17). This was preferred to making the DC gain of the filter greater than 1 which would have yielded too large capacitor ratios and signal clip-off at the lag output of the filter.

413

Assuming the DC gain of the loop filter to be unity, the PLL DC loop gain K_v can be written as:

$$K_v = K_D K_{VCO} \alpha_L , \qquad (31)$$

where α_L is the gain of the loop amplifier. In order to prevent signal clip-off in the loop filter we have to choose

$$\alpha_L \geqslant V_{BATT}/E . \qquad (32)$$

To eliminate the dependence of K_v on the signal amplitude E, the PLL has been preceded by a signal limiter which eliminates the amplitude modulation but retains the phase information. The output amplitude of the limiter is proportional to V_{BATT}. Eq. 31 yields with Eqs. 25, 28, and 32:

$$K_v = 4 f_0 \alpha_C/\alpha_V . \qquad (33)$$

The hold-in range of the PLL can be calculated using Eq. 33 [25]:

$$\omega_H = \pm K_v \pi/2$$

$$= \pm 2\pi f_0 \alpha_C/\alpha_V \qquad (34)$$

Assuming τ_1 much larger than τ_2 the PLL capture range can be estimated as:

$$\omega_c \simeq \pi \left[K_v(\zeta\omega_n + 1/(2\tau_1)) \right]^{1/2} , \qquad (35)$$

where

$$\omega_n = (K_v / \tau_1)^{1/2} \, , \tag{36}$$

$$\zeta = (1/2) \; \omega_n \; (\tau_2 + 1/K_v) \; . \tag{37}$$

ω_n is the natural loop frequency and ζ is the damping factor of the second order loop, respectively [25,26]. In this way Eq. 35 yields:

$$\omega_c \simeq \pi \; K_v \; \left[1/(K_v \tau_1) + \tau_2/(2\tau_1) \right]^{1/2} \; . \tag{38}$$

Fig. 20a shows a block diagram of the FSK demodulator. The demodulated output signal of the PLL is tapped at the lag output of the loop filter because the carrier is more attenuated here than at the lag-lead output, which supplies the VCO control via the loop amplifier (see Fig. 20a).

The PLL is followed by a 2nd-order SC lowpass data filter to suppress the carrier feedthrough. The last component is the comparator to restore the logic levels. To avoid drift effects, the input reference voltage of the comparator should track the DC level of the demodulated signal [27]. For this reason, the DC reference level has been extracted from the signal using a 10 Hz passive RC lowpass filter, which requires the only off-chip component in the demodulator: a large-value low-precision resistor. Note that the DC level extraction operates properly only if there are enough changes between zeroes and ones in the binary data stream within 100 ms.

The determination of PLL parameters must usually be done iteratively, since they influence each other. The following considerations therefore do not exactly reflect the design flow chart. Prior to the integration, the FSK demodulator was breadboarded in order to help determine design parameters [11]. It was experimentally found that the best FSK discrimination was obtained for the loop damping factor of $\zeta=0.28$. It was natural to choose the free-running frequency of the voltage-controlled oscillator at the channel center frequency, i.e. 1.5 kHz. In order to get a wide hold-in range (±1500 Hz), the PLL DC loop gain was set to $K_v = 6000 \; s^{-1}$ according to Eq. 34. With Eq. 33 and a clock rate of 100 kHz, this determined the VCO parameters as $\alpha_C = \alpha_V$. The VCO was designed with $\alpha_C = \alpha_V = 0.05$ and $\alpha_A = 0.833$. The gain of the loop amplifier was set to $\alpha_L = 15$. When fixing the PLL bandwidth ω_n, there is a contradiction between noise bandwidth and transient behavior requirements [25]. Selection of too narrow bandwidth ω_n improves noise performance but it can cause the PLL to loose the lock during the FSK transients. As a tradeoff we chose $\omega_n = 1000$ rad/s. Eqs. 36 and 37 then yield the pole and zero frequencies of the loop filter at 26.5 Hz and 400 Hz, respectively, and Eq. 38 yields $\omega_c/2\pi = \pm 700$ Hz for the PLL capture range. The data filter was designed for a selectivity of 1 and 300 Hz cut-off frequency.

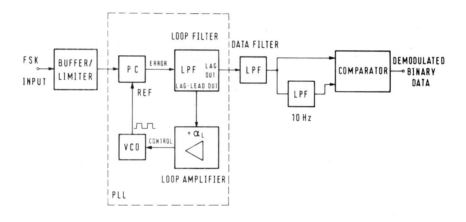

Fig. 20: a) FSK demodulator with the PLL

416

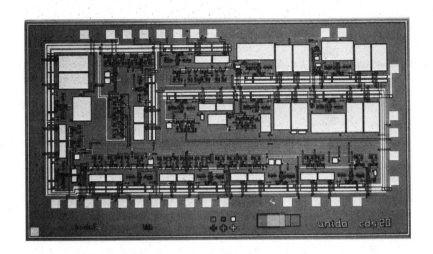

Fig. 20: b) Chip photomicrograph of the FSK demodulator

The FSK demodulator of Fig. 20a, except for the resistor of the 10 Hz lowpass filter, was integrated in the CMOS technology mentioned above [20]. A chip photomicrograph of the demodulator is shown in Fig. 20b. The smallest capacitor value was 2 pF. The chip contains a total of 12 dynamic operational amplifiers and the size is 6.6 mm^2 without bonding pads.

The actual measurement at the integrated prototype with a buffer at the input showed no appreciable differences for sine- and square-wave carriers. The data of the FSK demodulator are summarized in Table IV. Fig. 21 shows the measured VCO output spectrum as a function of the input frequency swept in both directions [28]. The input signal was corrupted by noise (S/N_{IN} = -6 dB), which had a flat characteristic and was bandlimited at 100 kHz. The verge of loss of lock occured at S/N_{IN} = -15 dB. Thus the loop is capable of recovering signals lower in level than the noise because the noise power in the loop is reduced by the ratio of loop noise bandwidth to input noise bandwidth (i.e. -22 dB in our case). This shows that the loop acquires lock for $(S/N)_{LOOP} \simeq 7$ dB. The calculation for voiceband 3.4 kHz yields minimum $(S/N)_{IN}$ = 0 dB necessary for lock acquisition. The expected error rate for $(S/N)_{IN}$ = 10 dB will be lower than $5x10^{-8}$ [29]. Reception in the 1200 baud modem channel was also possible. The demodulator dissipated 7.5 mW at 100 kHz clock rate and with power supply voltages ±5 V.

The demodulator adheres to the CCITT recommendation V.23 and is also compatible with the 202-type modem standard. Both devices can be used for full-duplex four-wire or half-duplex two-wire services. A complete modem would have to include an on-chip clock generator, a carrier detector, and a line termination. An example of data transmission using the integrated FSK modulator and demodulator at the 600 baud modem channel is shown in Fig. 22.

The PLL technique eliminates need for highly selective FSK filtering [30]. The PLL used in the demodulator achieves a sharp frequency separation by converting the signal frequencies into voltage levels which are much easier to evaluate. No functional trimming is required and only one off-chip low-precision resistor is necessary.

5 CONCLUSION

In this work, design principles of nonlinear MOS analog circuits have been presented. It has been demonstrated that such circuits can serve in number of applications and can be combined to form signal processing devices of higher complexity. Three examples, the interpolative A/D-converter, the FSK modulator, and the FSK demodulator have been successfully integrated in CMOS silicon-gate technology.

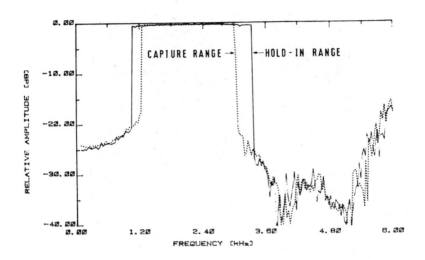

Fig. 21: PLL hold-in and capture range in presence of noise

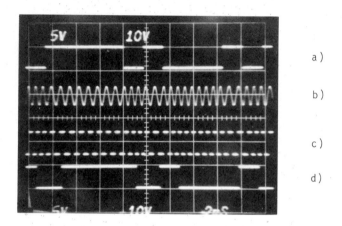

Fig. 22: a) Binary data input, 5 V/div.
(time-base 2 ms/div.),
b) FSK signal, 5 V/div.,
c) VCO output signal, 10 V/div.,
d) output signal of the FSK demodulator, 10 V/div.

420

TABLE IV

F S K   D E M O D U L A T O R

POWER SUPPLY VOLTAGES	± 5 V
CLOCK RATE	100 kHz
POWER DISSIPATION	7.5 mW
CHIP AREA	6.6 mm^2
MAX. TRANSMISSION RATE	1200 Baud
DEMOD. OUTPUT (± 10 % DEV. INPUT FREQUENCY)	0.7 V_{pp}
PHASE-LOCKED LOOP:	
PLL LOOP GAIN	6000 s^{-1}
DAMPING FACTOR	0.28
NATURAL FREQUENCY	160 Hz
NOISE BANDWIDTH	625 Hz

	CALCULATED	MEASURED
MIN. INPUT SNR		-15 dB
HOLD-IN RANGE (CARRIER 0.5 V_{rms})	0-3000 Hz	940-3470 Hz
CAPTURE RANGE (CARRIER 0.5 V_{rms})	800-2200 Hz	1200-2580 Hz
VCO: GAIN		300 Hz/V
FREE RUNNING FREQUENCY		1500 Hz
OVERALL LINEARITY (100-4000 Hz)		4 %
LOOP AMPLIFIER GAIN		15
LOOP FILTER: POLE		26.5 Hz
ZERO		400 Hz
DATA FILTER: CUT-OFF FREQUENCY		300 Hz
SELECTIVITY		1

6 ACKNOWLEDGEMENT

The author is greatly indebted to all colleagues at the Lehrstuhl Bauelemente der Elektrotechnik at the University of Dortmund for their support and encouragement. A careful perusal of this paper by W. Brockherde and K.-G. Dalßas is gratefully acknowledged.

7 APPENDIX A

For a closer analysis of the SC-amplifier and integrator, a model that includes parasitic effects, like linear parasitic capacitances C_{pi}, finite open loop gain A_0, offset voltage V_{off}, and finite bandwidth $1/\tau_1$ of the operational amplifier, has been chosen (in this paper a one pole model is used, which is a convenient assumption for op amps with internal frequency compensation). In this case, assuming parasitic effects as indicated in Figs. 2 and 3 and settled input voltages v_{in1} and v_{in2}, the output voltage of the SC-amplifier in the phase Φ is

$$v_{out}(t') = A_0/(A_0+1)V_{off} + \left[v_{out}((n-1)T)-A_0/(A_0+1)V_{off} \right] e^{-(1+A_0)t'/\tau_1}$$

$$(I)$$

where $t' = t - (n-1)T$, $A(s) = A_0/(1+s\tau_1)$, and

$$v_{out}(0') = v_{out}((n-1)T).$$

s is the common complex frequency variable. Similarly, the output voltage in the phase $\overline{\Phi}$ is given by

$$v_{out}(t') = C_A/C_F \frac{A_0 \left[v_{in1}(0') - v_{in2}(t') \right]}{A_0 + 1 + (C_A + C_p)/C_F}$$

$$+ \frac{A_o/C_F \left[(C_A + C_p)(V_{off} - v_{out}(0^-)) + C_F V_{off} \right]}{A_o + 1 + (C_A + C_p)/C_F}$$

$$- \left\{ C_A/C_F \frac{A_o \left[v_{in1}(0^-) - v_{in2}(t^-) \right]}{A_o + 1 + (C_A + C_p)/C_F} - v_{out}(0^-) \right.$$

$$\left. + \frac{A_o/C_F \left[(C_A + C_p)(V_{off} - v_{out}(0^-)) + C_F V_{off} \right]}{A_o + 1 + (C_A + C_p)/C_F} \right\}$$

$$\cdot e^{-(1 + C_F A_o/(C_F + C_A + C_p))t^-/\tau_1}$$

(II)

with $t^- = t - (n-1/2)T$ and initial conditions assumed as follows:

$$v_{in1}(0^-) = v_{in1}((n-1/2)T) ,$$

$$v_{in2}(t^-) = v_{in2}((n-1/2)T) ,$$

$$v_{out}(0^-) = v_{out}((n-1/2)T) .$$

The output voltage of the integrator can be calculated using the same model as above. To ease the calculation, we introduce a voltage v_s which appears at the inverting input node of the operational amplifier. If we assume $t^- = t - (n-1)T$ and the initial conditions

$$v_s(0^-) = v_s((n-1)T) ,$$

$$v_{out}(0^-) = v_{out}((n-1)T) ,$$

both voltages in the phase Φ are:

$$v_{out}(t') = \frac{A_o \{(C_{p2}+C_F)/C_F \ (V_{off}-v_s(0')) + v_{out}(0') \}}{1 + A_o + C_{p2}/C_F}$$

$$+ \{ v_{out}(0') - \frac{A_o \{(C_{p2}+C_F)/C_F \ (V_{off}-v_s(0')) + v_{out}(0') \}}{1 + A_o + C_{p2}/C_F} \}$$

$$\cdot \ e^{-(1+ C_F A_o/(C_{p2}+C_F)) \ t'/\tau_1} \ ,$$

$$v_s(t') = V_{off} - v_{out}(t')/A_o$$

$$- \left[\ 1/A_o \ (1 + C_F A_o/(C_{p2} + C_F)) \right.$$

$$\cdot \{ \frac{A_o \{(C_{p2}+C_F)/C_F \ (V_{off}-v_s(0')) + v_{out}(0') \}}{1 + A_o + C_{p2}/C_F} - v_{out}(0') \} \left. \right]$$

$$\cdot \ e^{-(1+C_F A_o/(C_{p2}+C_F)) \ t'/\tau_1} \ .$$

$$\text{(IIIa)}$$

The voltages in the phase $\overline{\Phi}$ can be determined in the same manner. Assuming $t' = t - (n-1/2)T$ and

$$v_s(0') \ \ = v_s((n-1/2)T) \ ,$$
$$v_{in1}(0') = v_{in1}((n-1/2)T) \ ,$$
$$v_{in2}(t') = v_{in2}((n-1/2)T) \ ,$$
$$v_{out}(0') = v_{out}((n-1/2)T) \ ,$$

the voltages are:

424

$$v_{out}(t') = \{\ C_A/C_F\ \frac{A_o\ [v_{in1}(0') - v_{in2}(t')]}{1 + A_o + (C_A+C_{p1}+C_{p2})/C_F}$$

$$+\ \frac{A_o/C_F\ (C_A+C_{p1}+C_{p2}+C_F)}{1 + A_o + (C_A+C_{p1}+C_{p2})/C_F}\ V_{off}$$

$$+\ \frac{A_o\ v_{out}(0')}{1 + A_o + (C_A+C_{p1}+C_{p2})/C_F}$$

$$-\ \frac{A_o/C_F\ (C_F+C_{p2})\ v_s(0')}{1 + A_o + (C_A+C_{p1}+C_{p2})/C_F}\ \}$$

$$\cdot\ \left[1 - e^{-(1+A_oC_F/(C_F+C_A+C_{p1}+C_{p2}))t'/\tau_1}\right]$$

$$+\ v_{out}(0')\ e^{-(1+A_oC_F/(C_F+C_A+C_{p1}+C_{p2}))t'/\tau_1}$$

and

$$v_s(t') = V_{off} - v_{out}(t')/A_o$$

$$-\ 1/A_o\ (1 + A_oC_F/(C_F+C_A+C_{p1}+C_{p2}))$$

$$\cdot\ [\ \{\ C_A/C_F\ \frac{A_o\ (v_{in1}(0') - v_{in2}(t'))}{1 + A_o + (C_A+C_{p1}+C_{p2})/C_F}$$

$$+\ \frac{A_o/C_F\ (C_A+C_{p1}+C_{p2}+C_F)}{1 + A_o + (C_A+C_{p1}+C_{p2})/C_F}\ V_{off}$$

$$+\ \frac{A_o\ v_{out}(0')}{1 + A_o + (C_A+C_{p1}+C_{p2})/C_F}$$

$$- \frac{A_o/C_F \ (C_F+C_{p2}) \ v_s(0^-)}{1 + A_o + (C_A+C_{p1}+C_{p2})/C_F} \} - v_{out}(0^-) \ ]$$

$$\cdot \ e^{-(1+A_oC_F/(C_F+C_A+C_{p1}+C_{p2}))t^-/\tau_1}.$$

(IIIb)

8 REFERENCES

[1] P.R. GRAY, D.A. HODGES, and R.W. BRODERSEN: Analog MOS Integrated Circuits, IEEE Press: New York, 1980.

[2] A. FETTWEIS, "Switched-capacitor filters: from early ideas to present possibilities," IEEE Proc. Int. Symp. Circuits and Systems (Chicago, Ill.), pp. 414-417, April 1981.

[3] G.C. TEMES, "MOS switched-capacitor filters - History and the state of the art," Proc. Europ. Conf. Circuit Theory & Design (The Hague), pp. 176-185, Aug. 1981.

[4] P.E. FLEISCHER, K.R. LAKER, D.G. MARSH, J.P. BALLANTYNE, A.A. YIANNOULOS, and D.L. FRASER, "An NMOS analog building block for telecommunication applications," IEEE Trans. Circuits Syst., vol. CAS-27, pp. 552-559, June 1980.

[5] K. MARTIN and A.S. SEDRA, "Switched-capacitor building blocks for adaptive systems," IEEE Trans. Circuits Syst., vol. CAS-28, pp. 576-584, June 1981.

[6] E.A. VITTOZ, "Micropower switched-capacitor oscillator," IEEE J. Solid-State Circuits, vol. SC-14, pp. 622-624, June 1979.

[7] J.G. GRAEME, G.E. TOBEY, and L.P. HUELSMAN: Operational Amplifier: Design and Applications, McGraw-Hill: New York, 1971.

[8] D.J. ALLSTOT and W.C. BLACK, Jr., "Technological design considerations for monolithic MOS switched-capacitor filtering systems," Proc. IEEE, vol. 71, pp. 967-986, Aug. 1983.

[9] M.V. HOOVER, "An introduction to the characteristics and applications of COS/MOS transistors in linear service," IEEE Circuits and Systems Magazine, vol. 10, pp. 2-11, Febr. 1976.

[10] K. MARTIN, "A voltage-controlled switched-capacitor relaxation oscillator," IEEE J. Solid-State Circuits, vol. SC-16, pp. 412-414,

Aug. 1981.

[11] B.J. HOSTICKA, W. BROCKHERDE, U. KLEINE, and R. SCHWEER, "Nonlinear analog switched-capacitor circuits," IEEE Proc. Int. Symp. Circuits and Systems (Rome), vol. 3, pp. 729-732, May 1982.

[12] B.J. HOSTICKA, W. BROCKHERDE, U. KLEINE, and R. SCHWEER, "Design of nonlinear analog switched-capacitor circuits using building blocks," IEEE Trans. Circuits Syst., vol. CAS-31, pp. 354-368, April 1984.

[13] C. HEWES, D. MEYER, R. HESTER, W. EVERSOLE, T. HIRI, and R. PETTENHILL,"A CCD/NMOS channel vocoder," Int. Conf. on the Application of CCD's (San Diego, Cal.), pp. 3A.17-3A.24, Oct. 1978.

[14] R. GREGORIAN, K.W. MARTIN, and G.C. TEMES,"Switched-capacitor circuit design," Proc. IEEE, vol. 71, pp. 941-966, Aug. 1983.

[15] D. BRODARAC, D. HERBST, B.J. HOSTICKA, and B. HOEFFLINGER, "Novel sampled-data MOS multiplier," Electronics Letters, vol. 18, pp. 229-230, March 1982.

[16] D. SOO and R.G. MEYER,"A 4-quadrant MOS analog multiplier," IEEE Int. Solid-State Circuits Conf. Dig. Tech. Papers, pp. 36-37, Febr. 1982.

[17] B.J. HOSTICKA and G.S. MOSCHYTZ,"Practical design of switched-capacitor networks for integrated circuit implementation," IEE J. Electronic Circuit and Systems, vol. 3, pp. 76-88, March 1979.

[18] J.C. CANDY,"A use of limit cycle oscillations to obtain robust analog-to-digital converters," IEEE Trans. Commun., vol. COM-22, pp. 298-305, March 1974.

[19] J.C. CANDY, W.H. NINKE, and B.A. WOOLEY,"A per-channel A/D converter having 15-segment μ-255 companding," IEEE Trans. Commun., vol. COM-24, pp. 33-42, Jan. 1976.

[20] B.J. HOSTICKA, W. BROCKHERDE, U. KLEINE, and G. ZIMMER, "Switched-capacitor FSK modulator and demodulator in CMOS technology," IEEE J. Solid-State Circuits, vol. SC-19, pp. 389-396, June 1984.

[21] B.J. HOSTICKA,"Dynamic CMOS amplifiers," IEEE J. Solid-State Circuits, vol. SC-15, pp. 877-894, Oct. 1980.

[22] B.J. HOSTICKA, D. HERBST, B. HOEFFLINGER, U. KLEINE, J. PANDEL, and R. Schweer,"Real-time programmable low power SC bandpass filter," IEEE J. Solid-State Circuits, vol. SC-17, pp. 499-506, June 1982.

[23] J. GROSZKOWSKI: Frequency of Self Oscillations, Macmillan: New York, 1964.

[24] J.L. McCREARY,"Matching properties, and voltage temperature dependence

of MOS capacitors," IEEE J. Solid-State Circuits, vol. SC-16, pp. 608-616, Dec. 1981.

[25] F.M. GARDNER: Phaselock Techniques, John Wiley & Sons, 1966.

[26] A. BLANCHARD: Phase-Locked Loops, John Wiley & Sons, 1976.

[27] E.N. MURTHI,"A monolithic phase-locked loop with post-detection processor," IEEE J. Solid-State Circuits, vol. SC-14, pp. 155-161, Febr. 1979. Aug. 1981.

[28] R.C. DEN DULK and M.P.I. HUISMAN,"Phase-lock loop operating ranges measurement," Electronics Letters, vol. 17, pp. 691-692, Sept. 1981.

[29] E. KETTEL,"Die Fehlerwahrscheinlichkeit bei binaerer Frequenzumtastung," Archiv der Elektrischen Uebertragung, vol. 22, no. 6, pp. 265-275, June 1968.

[30] R.K. MACDONALD, J.-L. SCHEVIN, and P. CARBON,"Single-chip 1,200-b/s modem melds U.S. and European standards," Electronics, vol. 56, no. 1, pp. 163-167, Jan. 1983.

PART III

REPRESENTATIVE SYSTEMS

INTRODUCTION

The last part of this book presents five integrated systems which use, among other circuits, many of the building blocks presented in Part II. These systems have been chosen both because each is important in its own right, and because taken together they represent a wide range of applications and points of view. Lothar Lerach overviews the field of modern telephony, in which heavy use is made of complex integrated circuits. The link between the telephone network and a computer, the modem, is then discussed in two chapters. Russell Apfel describes modems using digital signal processing; Kerry Hanson presents switched-capacitor modem techniques. Echo cancellers are presented by Eric Swanson as a prime example of adaptive filters. The last chapter of the book is representative of the system knowhow required in some of todays integrated circuit design: it deals with digital television, and is written by Rainer Schweer, Peott Baker, Thomas Fischer, and Heinrich Pfeifer.

13

LSI/VLSI FOR TELEPHONY

LOTHAR LERACH
SIEMENS AG
COMPONENTS GROUP
BALANSTRASSE 73
D 8000 München 80

1. INTRODUCTION

Although the principle - and advantages - of using pulse
code modulation (PCM) for digital voice transmission have long
been known, the breakthrough for fully digital, computer controlled telephone switching systems has come only with recent
advances in microelectronics. The structure of these mixed
analog/digital systems is highly influenced by the availability of standard VLSI devices, such as memories and microcomputers, and, with major impact on performance, flexibility
and operating software, by dedicated standard devices, e.g.
PCM codecs, as well as fully customized integrated circuits.

The main requirements for of the next generation of
systems will include the introduction of new voice/data services, increased modularity and flexibility of both hardware
and software, further decentralization of intelligence, improved reliability, simpler maintenance procedures and stream-
lining of costs.

In view of the large production quantities involved,
semiconductor manufacturers the world over are devoting their
attention to the subscriber telephone set, and the subscriber
line board at the exchange end of the subscriber loop. Nevertheless, LSI/VLSI devices will be used in almost every major
system application such as

- centralized or distributed concentration, switching and multiplexing
- a/d conversion, PCM coding, and analog or digital filtering
- link access protocol handling
- PCM regeneration and line synchronization
- tone detection and generation.

Due to the further evolution of automated office technology and the introduction of integrated services digital networks (ISDN), which support a wide range of voice and non
voice services in the same network, presently major concern
is given to full duplex data transmission methods, general
purpose data communication devices and local area network
controllers.

Present mixed analog/digital and future all digital
systems naturally offer a wide range of applications for digital signal processing (DSP). Moreover, further progress in
VLSI technology will also allow the replacement of many, traditionally used analog solutions by cost effective DSP based
integrated circuits.

For the successful development, production and marketing
of telecom ICs, a number of aspects have to be taken into
account carefully:

- recommendations set by international or European standard
  organizations (e.g. Consultative Committee for Telegraphy
  and Telephony, CCITT, and European Conference of Postal
  and Telecommunication, CEPT) and national postal administrations have major impact on device specifications and
  performance requirements
- close contact with system manufacturers is a must for the
  proper implementation of systems requirements
- severe environmental conditions, specific problem solutions
  and competition on the market make the use of highly optimized, advanced technologies imperative
- telecom ICs have to meet stringent quality levels
- in key system applications, second source or even multiple
  source of a product is important and may be a precondition
  for its acceptance and marketing success.

As a matter of fact, outstanding performance, excellent
quality and declining costs in microelectronics can be
achieved best with carefully optimized chips produced in high
volume. This correlation should have major impact on the IC
development and cooperation strategy of system houses.

It is the intention of this paper to give an introductory
overview of the function and application of the most important
dedicated ICs for telephony. After a brief introduction to
system structure, telephone set devices and specific digital
exchange components are discussed. Furthermore, the typical
use of DSP and devices necessary to establish digital voice/
data transmission in an ISDN will be emphasized.

2. SYSTEM STRUCTURE

Due to specific market requirements the telecomminications industry is offering a wide variety of telephone switching systems ranging from small Private Automatic Branch Exchange (PABX) systems to huge Public Network facilities serving tens of thousands of subscriber lines. While for PABX facilities shorter lengths of subscriber loops, lower system complexity, in house installation and private use of the equipment gives certain freedom to the system design, the characteristic of public switches principally differs in many aspects, strongly determined by national post office authorities.

The principal structure of a digital exchange system is shown in Fig. 1. The periphery consists of subscriber line boards (analog or digital), interoffice trunk modules and different special purpose units such as boards for code receiving,tone generation or conferencing. Subscriber line boards and trunk modules provide the link between the subscriber lines or public interoffice lines, and the switching network for troughswitching of the digitized voice channels based on time division multiplex (TDM) methods combined with space multiplex principles. Until now even for PABX systems the lines are almost exclusively analog in type. Besides line interfacing a basic function to be performed in the peripheral boards is a/d conversion and PCM coding. Data transfer to the switching network commonly accomplished using serial TDM highways with clock rates of 1.5036 or 1.544 MHz (24 time slots
per frame, US standard), 2.048 MHz (32 time slots per frame, European standard), or multiples of them.

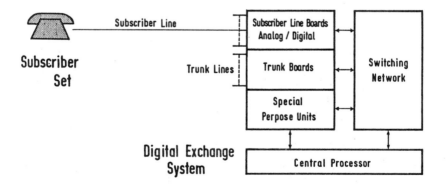

Fig. 1: Basic structure of digital exchange systems

Supervision and control of the entire system, including connection setup, maintenance and testing, dependent on the size of the system, is performed by a hierarchical structure of decentralized intelligence ranging from line board and group control level to system control level in a powerful central processing unit.

The ability to integrate in this type of modular digital exchange both data communications and telecommunications, driven by the efforts of national authorities, in the very near future even in the public domain makes feasible Integrated Services Digital Networks (ISDN) in which digitized voice, circuit switched data and digital signaling with packet switched data can be processed, based on existing wiring.

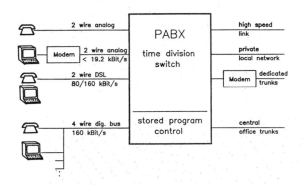

Fig. 2: Typical PABX termination environment

Adding digital end to end data transmission capabilities to their repertoires, PABX manufacturers presently are on the way to introduce third generation type of systems, which can handle not only conventional voice, but also synchronous as well as asynchronous data traffic at speeds as high as 160 kbit/s via ordinary twisted pair wiring. By further incorporation of gate ways to private local area networks, modem pools, decentralized cluster controllers, and high speed digital links to central office equipments, this type of system can economically serve a wide range of advanced business communication facilities in an automated office environment. The systems provide full availability of and access to all lines, high traffic capacity, and very low call blocking.

3. INTEGRATED CIRCUITS FOR ANALOG TELEPHONE SETS

The conventional analog voice transmission telephone set can simply be divided in the voice module (including 2/4-wire hybrid), dialling section and ringing module. Besides electronic replacement of these functions, the trend clearly ist towards high feature sets which incorporate extended intelligence and a host of new features such as number storing, automatic last number redialling, repertory dialling, handsfree speaking, use of optical display and cordless operation. Fundamental functions to be included for the electronic circuits are polarity guard, overvoltage protection and line voltage regulation. In general, low power consumption and low voltage operation are main requirements.

The first step in converting subscriber set functions to electronic operation will usually be the replacement of the rotary dial by a pushbutton keyboard operating in conjunction with either a pulse generator to originate the so called loop disconnect dialling, or a tone generator for dual tone multi-frequency DTMF) dialling.

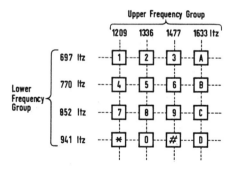

Fig. 3: Telephone pushbutton keyboard and associated DTMF dialling frequencies

In the DTMF technique each digit is represented by a different pair of frequencies within the voice band. The various frequencies are derived from a crystal controlled oscillator, followed by a sine wave synthesizer consisting of a programmble divider for the upper and lower frequency groups, and d/a-converter. Due to the combination of digital and linear circuitry as well as high driving capability on the same chip, the devices available on the market are primarily based on bipolar technology using integrated injection logic (I^2L) for the digital parts. Typical representatives are the TEA 1021 from Philips and the PSB 8591 from Siemens. Utilizing capability of modern technologies, low power CMOS integration with the addition of number memory and output low pass filters is in progress.

A consequent step even for simplistic telephones is the integration of the ringing module and the 2/4-wire hybrid in a tone ringer and a speech circuit, respectively. The tone ringer generates two periodic switchable tone frequencies and can drive a piezo ceramic converter, or loudspeaker.

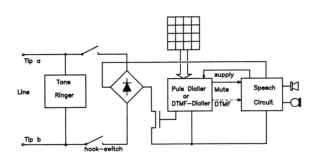

Fig. 4: Standard electronic telephone set

Linear speech circuits available or those under development are usually connected to the telephone line directly. Powered from the telephone line current the typical device contains receive and transmit amplifiers and means of switching inputs, voltage regulator, gain control facilities, exchange power supply voltage corrections, microphone input, loudspeaker output drive and power supply to external circuits. These components still require quite an amount of external circuitry for gain adjustment, stabilizing and side tone supression.

435

The handy feature of handsfree telephone operation permits the use of the telephone in the on-hook condition and, in its most advanced version, provides full two way loudspeaking capability. It requires a microphone to be connected, or mounted to the same enclosure as the loudspeaker. Possible acoustic feedback problems may be overcome by either voice switching facilities implemented in an extended speech circuit, or an anti-echo filter inserted in the speech path. The voice switching approach, in a cost effective way, simply compares the signal from the microphone with the received signal and allows the one with the highest level to be amplified, while the other is attenuated. A high quality anti-echo solution is expected for digital telephone sets, where it can be achieved by simple digital signal processing.

The functions discussed so far, to date are preferably integrated in linear bipolar and I^2L-techniques. On chip integration for a standard telephone module is feasible, and has been achieved e.g. by Motorola. Nevertheless, stringent power consumption limitations set by European postal administrations cause designers to pay more attention to integration in CMOS as a solution to this problem.

The step into the all electronic feature telephone is accomplished by adding to the subscriber set an optimized microprocessor, number memory and driving circuit for liquid crystal
display (LCD), preferably all in low power CMOS. Depending on the level of intelligence and the actual device features, a 4 or 8 bit microprocessor may be the optimum solution. Siemens Components Group has introduced a CMOS device family which includes an 8 bit microprocessor, a 2 kbit memory, and a 20 digit 7 segment LCD controller designed as an extremly flat "micropack" film carrier packaging. The whole family can be operated in a wide voltage range from 2.5 to 6 Volts. The microprocessor is a derivative of the popular 8048 and includes additional functions such as automatic key board scanning, idle mode and power on reset.

Obviously no single technology will statisfy all the criteria of high feature telephones, such as extremly low power consumption, low cost, optimization of analog and digital functions, and high driving capability. Consequently, also in a longer term, for feature sets we can expect a trend towards multitechnology concepts that satisfy a variety of system requirements in different types of telephones.

For the semiconductor industry, problems associated with the lack of international standardization coupled with national specifications and the various partitioning philosophies imposed by telephone manufacturers, continue to have a major influence on the design and market volume. Certainly this imimpedes progress towards large production volumes and low cost of electronic devices.

4. DEDICATED LSI/VLSI COMPONENTS FOR DIGITAL EXCHANGE SYSTEMS

4.1 Subscriber Line Board Integration

4.1.1 Line Board Functions - Each telephone station is connected to a set of line interface circuits, placed in a subscriber line board to interface the two wire analog subscriber loop with the PCM highway. A similar circuitry provides interfacing of analog interoffice trunk lines. The two wire connection can be considered as a "floating" high voltage loop consisting of a wire "a", the tip wire, and a wire "b", the ring wire. The subscriber interface functions include the battery feeding of the subscriber equipment, overvoltage protection against lightning, generation of the ringing voltage and feeding it to the subscriber, supervision of the subscriber loop e.g. "on-hook/off-hook"-detection, hybrid 2/4-wire conversion, and test access. High transmission quality is achieved by careful matching of the complex termination impedance, optimization of balance network and level control according to country specific line conditions. The cost effective and reliable replacement of these so called BORSHT-functions in a high voltage all silicon subscriber line interface circuit (SLIC) is one of the most difficult problems to be solved by the semiconductor industry. In the meantime first SLIC devices realized in bipolar technology are announced or made available by several semiconductor houses.

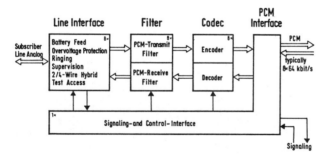

Fig. 5: Basic functions of a subscriber line board analog

Further fundamental tasks performed on the line board are a/d conversion of voice signals into time discrete digital equivalents in a companding encoder/decoder (codec), the accompanying limitation of the voice band in a receive/transmit filter, and PCM time slot assignment.

437

The principle of PCM coding is shown in Fig. 6. According to the sampling theorem, in any sampling procedure the input frequency spectrum is mirrored at the sampling frequency f$_s$. For PCM systems using a sampling sequence of 8 kHz, respresenting one voice sample at every 125 µs, a PCM filter therefore provides attenuation of frequencies above 3.4 kHz. Otherwise, when sampled at 8 kHz, these frequency components would be folded down to the passband and produce distortion. In addition to this anti-aliasing function, a 300 Hz highpass prevents low-frequency distortion. After passing the anti-aliasing filter in the encode direction, the analog signal is sampled in a input sample and hold amplifier, quantized and encoded. It is subsequently loaded in an output buffer and clocked to the PCM system. In the receive direction the PCM-filter acts as a reconstruction low pass filter. Since the decoder output sample and hold function has a sinx/x amplitude roll-off as frequency increases, this must be considered in the filter design.

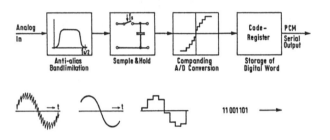

Fig. 6: Sequence of PCM coding

Traditionally, anti-aliasing and reconstruction filters have been implemented as analog circuits, and the codec samples the analog voice at the same 8 kHz rate used for digital data transmission in PCM systems. However, advances in large scale integration enable digital instead of analog filtering methods, resulting in a different sequence of PCM coding (see 4.1.4).

The codecs designed for telecom applications perform analog to digital conversion in nonuniform quantizing steps, using an increasing step size as the magnitude of the signal amplitude increases. This companding method provides a wide dynamic range over which the signal-to-quantizing noise ratio is approximately constant, and results in very small step sizes near the origin to improve low level resolution, suppression of crosstalk, signal enhancement and low idle channel noise. Theoretically the encoding scheme should follow a logarithmic transfer characteristic, which is approximated by the so called µ-225 law, standardized for North America and Japan, and the slightly different A-law, used in Europe. Significant for these approximations is a nonlinear binary weighted segmentation of 8 intervals per sign (called chords

438

or segments) and a linear quantization of each chord into 16 equal steps. This means that a chord and its companion step-sizes double in length as they move away from the origin. In total, the PCM word code format is described by 1 sign bit, 3 bits to specify one out of 8 chords, and 4 bits to specify one out of the 16 linear steps of the chord. The maximum resolution possible using an 8 bit PCM word is thus equivalent to a linear code of 12 bits and 13 bits for the A-law and μ-law, respectively.

The serial transmission of the 8 bit digital voice equivalents to the switching network via serial PCM-highways is performed by selecting one receive and transmit timeslot within the 125 μs frame for each subscriber. The number of channels per frame most commonly used is 24 in North America und Japan, and 32 in Europe; in accordance with this, the clock frequency used is 1.544 and 2.048 MHz. The time slot assignment (TSA) selection is done either by a fixed or a freely programmable strobe signal generated by on board TSA circuitry, an additional on chip TSA logic at each codec, which is programmable via a serial interface, or a special TSA device working as a multichannel PCM multiplexer for e.g. 8 codecs (see 4.1.2). Free selection of the time slot is necessary in system applications where the TSA is changed from one call to the next in order to upgrade the data transmission rate of the PCM highways. Control of the board is achieved via a signaling highway connected to a central control unit. Basic signaling information includes on hook/off hook and ground key in the transmit (subscriber to the exchange) direction, and power up/down, ringing and battery feed in the other. Extended line board control includes test routine initialization and supervision such as unbalanced line, specific status control, time slot assignment and others.

Optimized integration of the described functions in a SLIC, advanced codec filter and board controller is one of the great challenges for the semiconductor industry. The implementation of new features has a decisive influence on the performance capabilities of the system as a whole, as well as on its structure, operating software and cost.

4.1.2 <u>Advanced Line Board Architecture and Line Board Control</u> - The first generation of analog line boards being supplied in volume by system manufacturers is charcterized by the use of discrete or some sort of miniaturized hybrid SLIC which usually still includes a transformer, a single chip codec and filter and, in most cases, fixed time slot assignment. For signaling between line board and group control, sequential time division procedures with limited signaling capability are used as well.

In addition to declining cost, a increase in volume and enhancement of reliability, the technical goals for the next step of innovation are to introduce intricate switching functions, decentralized intelligence, software control of analog or digital functions, and extended testing capability into the subscriber line board with a high level of integration.

Depending on system structure, special applications in
digital traffic control may also require a signaling control
path realized by the use of a common channel signaling technique via PCM channels and access of the on board microprocessor to PCM voice/data channels. ISDN capability could be
provided by the addition of a powerful standard line board
interface to the rest of the system, which features improved
signaling capability between the line board and the control
host computer, and fits into a modular system architecture
fully compatible with voice/data communication.

To meet these goals, semiconductor companies including
Advanced Micro Devices, Intel, National Semiconductor and
Siemens are presently introducing chip sets which implement
extended capabilities for new system generations, following
different device solutions and line board architectures. The
comparison of the fundamental architectural solutions in
Fig. 7 indicates two mainstream directions: first, the implementation of programmable time slot assignment, PCM highway interfaces and a microprocessor interface into the codec
filter chip, and second, the concentration of line board specific standard functions, such as time slot assignment, message buffering, status signaling control, communication protocol handling as well as PCM and signaling interfacing, in a
single board controller for several subscribers. In detail,
of course, differences exist between the various solutions
concerning the actually realized functions, capabilities and
interfaces. Line board intelligence is in all cases accomplished by a microprocessor.

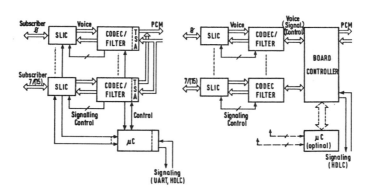

Fig. 7: Second generation subscriber line
board architectures

While e.g. AMD's Subscriber Line Audio Circuit
AM 7901 (SLAC) represents the first approach, the PEB 2050
Peripheral Board Controller (PBC) from Siemens and the

440

Digital Line Interface Controller (DLIC) from National Semiconductor are representativs of the board controller solution. Being the first device of a familiy of advanced telecommunication circuits, the Siemens Components Group supplied samples of the PEB 2050 during the first half of 1983 and started volume production in the early 1984.

Due to its exceptional problem solution including upward compatible functions towards voice/data communication, the PBC approach has been promoted by Intel from the very beginning, and meanwhile it is supported by a coalition of system houses and semiconductor companies. Moreover, the device will be produced by Intel as 2952 Line Card Controller; second source right also is granted to Philips.

The Peripheral Board Controller constitutes the interface between the subscriber line circuits (such as Siemens' PEB 2060 Signal-Processing Codec Filter, Intel's 29C51 Combo, or ISDN components being under development), fully redundant serial PCM lines, a high speed data link to a remote central control unit and, optionally, an on board microprocessor.
As a characteristic architectural feature the device permits efficient switching of data streams between all these interfaces and therefore ensures transparency between PCM channels and control/signaling data. This opens up attractive technical possibilities such as common channel signaling via PCM time slots and microprocessor access to PCM data. In addition, the line board partitioning chosen reduces chip area and pinning for the subscriber line devices, allows modular pc board design and features simple backplane wiring.

For the connection of sophisticated subscriber line circuits, the PEB 2050 defines a ping pong type of subscriber line data link protocol (SLD-bus), which allows full duplex transfer of two 64 kbit/s data/voice channels as well as one byte for signaling and one byte for feature control per 125 µs frame at any of the 8 serial links.

Besides the realization of the described physical interfaces, essential functions of the PBC are freely programmable time slot assignment for up to 16 PCM channels, processing of feature-control data, indication and signaling, support of microprocesor access to the different types of data, and message handling between the line board and a remote control unit via the serial signaling highway or PCM time slots. Hard-wired control of these functions permits simultaneous processing of different complex realtime procedures.

For the exchange of information between a remote control unit and the PBC working as a "slave" in a point-multipoint configuration, the device supports a subset of CCITT's High Level Data Link Control (HDLC) communications protocol. Upon initiation via special user commands within the information field of the HDLC frame, the PBC independently compiles or distributes requested or received data packets.

It is able to autonomously evaluate received HDLC commands
and to generate reactions depending on the internal status. A
CPU interrupt request only occurs if there are data available
for the microprocessor. The device is produced in an advanced
n-channel silicon gate technology with 3.5 μm channel length.
It consists of 22,000 transistors on a chip area of 37.9 mm².

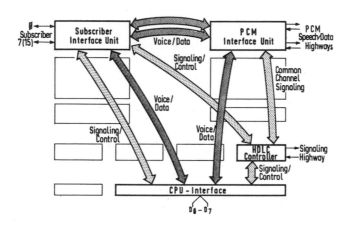

Fig. 8: Possible data streams within the PEB 2050
Peripheral Board Controller (Siemens)

Further significant impact on line board structure
results from the integration of a SLIC, the realization
of additional functions in an extended codec filter device
and the actual realization of interfaces.

The described SLD-bus approach supports SLIC control via
interface pins at the codec filter, or direct coupling to the
bus without cross wiring. Nevertheless, the use of μP-ports,
e.g. recommended by National Semiconductor, or some sort of
mixed approach may also be appropriate.

4.1.3 First/Second-Generation Codec and Filter - Essential
factors which describe the performance of a monolithic filter combo-device are

- stop band characteristic

- passband ripple

- inband signal to noise ratio

- harmonic out of band distortion

- idle channel noise contribution

- gain tracking

- dynamic range

- power supply rejection and crosstalk coupling

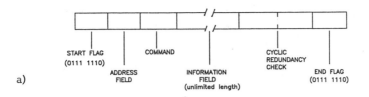

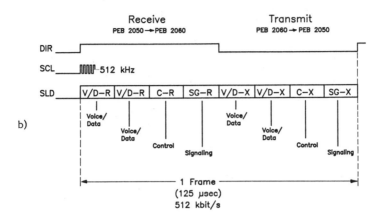

Fig. 9: HDLC-protocol frame structure (a)
and SLD-Bus configuration (b)

Boundary figures are specified by international system
recommendations (CCITT standards and D3/4 channel bank speci-
specifications). Improvements concerning fast and precise
testing of these parameters is still one of the challenges
in device production and incoming quality control.

For codecs, fundamental architectural differences arise,
first, from whether it is shared by many lines using an analog multiplexer or to be used in a single line, and second,
from the filter to be chosen. Due to the fact that shared co-
codecs have drawbacks concerning reliability, speed requirements and channel crosstalk, an increase in cost effectiveness over the past years gave per-channel codecs major attention.

Traditionally, anti-aliasing and reconstruction filters
have been implemented as analog circuits, and the codec samples the analog voice at the same 8 kHz rate used for the PCM
system. In this solution encoding is performed in an successive-approximation sequence for the definition of sign
bit, chord- and segment bits. In a fixed algorithm of setting
bits in a succesive approximation register, companding d/a
conversion of its digital information and comparison of its

analog equivalent with the input signal in a high resolution
comparator, the register content is successively matched to
represent a digital equivalent of the input voltage.

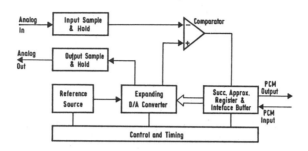

Fig. 10: Simplified block diagramm of a
successive approximation codec

One of the key problems which had to be overcome for
codec integration was to realize complex digital logic and
high precision analog circuitry, such as operational input/
output amplifiers, comparators, voltage reference and precision matched components, on one chip. While first types
with analog capability were preferably realized in bipolar
I^2L or mixed technology, advances in MOS analog capability
also made monolithic codecs in NMOS and CMOS technology possible. Regarding analog circuitry, MOS is highly suited to
voltage- and charge switching, allows simple sample/ hold realization and, furthermore, inherently produces capacitors of
high linearity and precision. A key feature is its compatibility with switched capacitor filter technology. Therefore second generation combo devices are exclusively realized in
MOS. Due to its low power consumption and better analog
performance, the trend clearly is towards CMOS technology.

State of the art fully integrated PCM filters delivered
to system houses provide switched capacitor ladder or CCD
transversal filters. Basically, both techniques are similar
in some respects since they are both analog sampled data filters, which make use of charge storage capacitors. However,
while CCD transversal filters use only a great number of
transfer delay zeros to shape the frequency response,
switched capacitor ladder filters are recursive, using poles
and zeros. The delay elements in CCD filters consist of
chargecoupled-devices which are capable of either storing or
shifting analog information.

The basic idea of the switched capacitor approach was to meet the high precision requirements for the integration in an active filter design by replacing resistors with switched MOS capacitors of high switching rate, where the switches are realized by MOS-transistors. Following this principle the band edge of a simple RC network is determined basically by the matching accuracy of capacitors in the order of 0.1 %; the range of capacitor values typically employed is from 1 to 100 pF.

A comparsion of performance characteristics shows that the switched capacitor approach has advantages in response accuracy, noise, dynamic range and process sensitivity. Due to this, switched capacitor filters are world wide the most commonly used filters.

Typically, the transmit filter section consists of a third order RC-active anti-alias filter with a nominal 3 dB frequency of 30 kHz to prevent distortion from high frequencies, a second or third-order monotonic high-pass switched capacitor filter suppressing frequencies below 300 Hz, and a fifth-order elliptic low-pass switched capacitor ladder filter.This is followed by a simple second-order smoothing filter. The receive reconstruction filter also consists of a fifth-order low-pass elliptic switched-capacitor filter which is similar to that in the transmit filter, but provides sinx/x correction by optimizing pole and zero locations.

4.1.4 High feature third generation codec-filter - The combo-chips discussed so far provide the standard functions of coding and filtering, and on chip voltage reference. Presently, the semiconductor industry is introducing third generation telephone line integrated circuits which implement further BORSHT and service functions in a highly programmably manner. This results in reduced discrete on board circuitry and wiring, enhanced transmission quality as well as improved reliability and testability.

As an advanced step to its 2914 combo-chip, Intel's 29C51 feature control combo, based on switched capacitor technique in addition to the codec and transmit/receive filter, supports programmable gain adjust, selectable μ/A law coding, power down, and selction of one of three on-chip or two external balance networks for 2/4 wire conversion. Moreover, it adds programmable gain, loop back mode, secondary analog input and conferencing capability. Entire control of signaling to the SLIC is handled via the device by selectable interface leads.

A codec filter approach that is quite different from conventional solutions results from a digital filtering concept and digital signal processing techniques implemented in AMD's AM 7901 Subscriber Line Audio Processing Circuit (SLAC) and Siemens' PEB 2060 Digital Signal-Processing Codec Filter (SICOFI). The basic idea of this DSP approach is to provide simple anti-aliasing filters and a/d conversion circuitry for speech digitizing by using some sort of oversampling noise shaping coder, followed by digital signal processing to produce PCM and band limitation. Fine resolution is achieved by averaging over a great number of coarse quantizations utilizing the fact that there is a correlation between sampling rate and necessary accuracy of the a/d-conversion. The noise shaping coder procuces noise in a wide frequency range, which afterwards is limited to its voice band part by digital filters. In this way, the a/d converter can provide an excellent dynamic range and the resulting signal to noise ratio exceeds CCITT requirements considerably.

The DSP approach potentially enables the control of the device's analog behaviour by digital signal processing in a wide range of programmability. It takes advantage of typical DSP-features such as highly predictable performance, tolerance to parameter fluctuations, low crosstalk and flexibility. Furthermore, the new approach also has significant impact on device-testing philosophy. By testing the signal processor, all digital filters and the programmable functions are confirmed, provided the a/d converter reaches its full performance. Therefore, testing in a very fast method can primarily be reduced to the seperate test of a/d-, d/a converters, voltage reference and the digital part. Moreover, system test under operation can be supported by different loop back functions implemented in the microprogramm.

The functional concept of both devices, the SLAC and SICOFI, is quite similar. The encoder basically consists of a low order anti-aliasing filter, an interpolative oversampling a/d converter, and a digital filter section performing downsampling to the 8 kHz PCM sampling rate and PCM band limitation. For decimation to a lower sampling rate, possible foldover distortion has to be taken into account. The result is an equivalent linear digital sample code of the analog signal at each 125 µs frame, which is then compressed in an additional step. The encoding direction includes linearization, band limitation and interpolation, d/a conversion at a high sample rate, and reconstruction post filtering.

Incorporation of programmable filters enable software controlled adjustment of the effective termination impedance and hybrid balancing for different applications, which traditionally have been achieved individually using discrete networks. Moreover, level control and fequency response correction can be easily implementd.

The impedance matching (z) filter implements a feedback loop, which modifies the effective termination impedance of either the SLIC or transformer hybrid towards optimized return loss. The hybrid balance (b) filter is a filter path from the receive to the transmit side, which, properly programmed, cancels the receive echo signals and thus provides a high degree of balancing. In fact, the b-filter matches the full echo loop and generates a compensation signal, which is then subtracted in the transmit path.

The digital filtering is performed by a microprogrammed digital signal processor having a highly optimized architecture with CSD (canonic sign digit) code arithmetic. This method provides multiplication with simple shift and add operations.

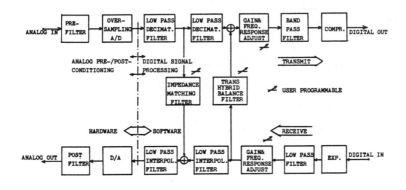

Fig. 11: SICOFI signal flow (Siemens)

In spite of the similar functional structure used internally, the SLAC and SICOFI are completely different concerning device architecture and implementation. While the SLAC implements a 512 kHz delta modulator and separate signal processors for receive and transmit in a low voltage n-channel MOS technology, the SICOFI design is based on a 3 μm double poly CMOS technology, highly optimized for telecom applications. This results in dense digital parts in spite of 10 V analog capability. The noise shaping, predictive a/d converter used provides high linearity at a sampling rate of 128 kHz resulting in a signal to noise ratio which does not suffer from intermodulation signals. Using an internal 4 MHz clock generated by an on chip phase locked loop from an external 512 kHz system clock, a single DSP performes all digital filtering including decimation and interpolation. For the recursive type of band limitation filters, the device implements so called Wave Digital Filters developed in close cooperation with the University of Bochum, Germany. This filters are proved to be stable under all conditions. All filter algorithms are contained in a 10 kbit microprogram.

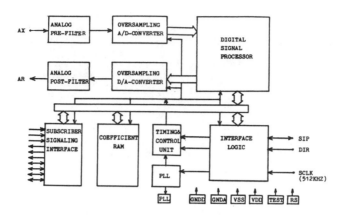

Fig. 12: SICOFI blockdiagramm (Siemens)

Feature control and read back of coefficients are handled by
the Peripheral Board Controller via the SLD bus. Since the
device uses no external discrete components, a small 22 pin
package has been achieved, in spite of supporting 10 pins for
SLIC control and device test in a test mode. Further device
features worth of mention are on-chip reference, selectable
A/μ law coding, three-party conference support, hard and
software reset, supply voltage supervision, and power down
mode.

        The fourth generation subscriber line codec filters
will primarily be aimed at the integration of further BORSHT
functions onto silicon. An intermediate step is the optimization of the electronic SLICs towards present solutions.

4.1.5 Subscriber Line Interface Circuit (SLIC) - The BORSHT
functions described in chapter 4.1.1 are traditionally realized with discrete components by the use of a transformer, relays, passive components and operational amplifiers. Depending on subscriber line termination resistance, battery feeding of the subscriber telephone requires voltage levels of
up to and above 60 V. In the case of short lines, a high voltage must be dropped accross a series on board resistor. To
avoid high power consumption, electronic current limitation
or constant current feeding is to be used. Overvoltage protection has the objective to prevent damage to equipment by
overvoltages, e.g. lightning, which are reduced by protective
devices to about 1000 V for 700 μs. Furthermore, longitudinal
common mode line voltages such as 50 Hz-signals induced from
sources adjacent to the line have to be suppressed. Longitudinal signal rejection is currently performed by the transformer, which simultaneously provides isolation from

foreign potentials. Ringing with an ac-voltage of up to 70 Vrms and test access to the subscriber line is generally performed by interrupting the line with a relay; supervision of loop state is done via the subscriber loop current flow to detect the "off hook" state or periodic dialling interrupts. Since the subscriber is connected to a 2-wire line, 2-wire/ 4-wire conversion is achieved by a hybrid, which also provides high attenuation between the transmit and receive channels of the 4-wire port. The solution is basically a Wheatstone bridge, which is balanced by matching a balancing network to the line impedance. Traditionally, the hybrid is implemented by a transformer with multiple windings.

While an all silicon replacement of the interface functions discussed so far has in principle been shown to be possible, the implementation for public exchange systems suffers from the combination of extremly stringent requirements, including:

- capability of handling voltages in the range of 150 V

- high driving capability

- large on chip power dissipation of up to 4 W

- high longitutinal balance exceeding 52 dB

- excellent reliability.

In any case, protection of low level electronics against high voltage transients (e.g. 1000 V, 700 µs) must be provided externally, e.g. by series elements and clamp diodes placed at the tip and ring leads.

Potentially, the most suitable technology for monolithic SLIC integration is a high voltage bipolar process. To meet the requirements concerning common mode rejection, methods to implement high precision resistors with an accuracy below 0.1 % are of utmost importance.

To date several integrated versions of a SLIC have been introduced that perform some or all of the BORSHT functions for central office or PABX applications in conjunction with some external elements such as relays, power devices or transformers. In close cooperation with system houses extensive work on fully electronic approaches is done by a variety of semiconductor makers such as AMD, Harris Corp., Motorola, Texas Instruments, RIFA (Sweden) and SGS-ATES (Italy). It was shown, that the optimized implementation of central switch demands can best by achieved by a two chip approach using a high voltage bipolar technology for the realization of the high voltage functions, and standard bipolar or MOS technique to implement the low voltage parts, in conjunction with on chip or external realization of the precision resistors. Based on its dielectrically isolated, complementary high voltage bipolar technology and its conventional bipolar junction isolated process, e.g. Harris has come up with an experimental prototype two chip solution which also integrates the ringing function on the chip. In production for the past two

years, the HC 5502 monolithic SLIC from the same company provides -48 V battery feed, switch hook and ground key detection, on chip ring relay drivers, 2/4 wire conversion, and
overvoltage protection with some external devices for PABX
applications. This type competes with products from Motorola
and others. Quite similar to Harris, Italy's SGS-ATES presented its own relayless two chip solution for public exchange
systems which uses the company's multiwatt package.

The SLIC solutions introduced so far are primarily designed with standard codec filters in mind. Optimization with
respect to the extended functions of the described third generation devices will reduce complexity. Moreover, integra-
gration of all low voltage functions together with the codec
filter as a next step, eventually also for public exchange
applications, may result in a two chip subscriber line
integrated solution.

4.2 PCM Switching Network Components

Representative of further problem oriented dedicated
VLSI's used in PCM systems, integration of the switching
network will be discussed briefly.

The objective of the switching network in a communications system is to establish connections between two or
more subscribers. In fully analog configurations, this is
performed by electromechanical or electronic physical links
in a space division switch, which for n subsribers in its
straight forward form requires n^2 crosspoints of excellent
analog transmission capability. Highly suited for memory
oriented implementation, the basic principle of PCM time division multiplex (TDM) circuit switches is to transfer digital voice/data samples from the PCM-highway time slot attached to the "transmit" subscriber to that of the "receive" subscriber. This merely requires an 8 bit memory location for each time slot of a PCM-highway, the content
of which is clocked to the outgoing highway with relation
to the new timing and, if required, new clock frequency. The
ability of connecting incoming time slots to several outgoing PCM highways results in a mixture of TDM and space division switches. Traditionally this type of switching network has been realized using standard LSI/VLSI circuits including shift registers, counters and memories. Dedicated
VLSI devices implementing an 8x8 (which means 8 incoming and
8 outgoing serial PCM lines) and a 16x8 switching matrix have
been introduced by Mitel and Siemens, respectively. A block
diagram of the Siemens PEB 2030 Memory Time Switch is shown
in Fig. 12. 16 incoming PCM lines are converted from serial
to parallel and stored in the 4 kbit speech memory with
16x32=512 words. For this write procedure, the address is
generated by an input counter. Interleaved with the writing
function, a speech memory location is addressed by one of 256
words of the connection memory (just in time with the

outgoing time slot) and the information is then clocked to
one of the 8 output lines. This is due to the fact that the
location of each speech memory address stored in the connection memory defines the output time slot, while during read
cycles the connection memory is addressed by a synchronously
running counter. The connection memory is controlled by a
asynchronously working microprocessor. Two of the devices
simply allow the realization of a nonblocking switch for 512
subscribers, replacing quite a number of hitherto separate
components. Fabricated in NMOS-technology, the PEB 2040 implements 40.000 transistors on a 34 mm^2 die; power consumption is 350 mW.

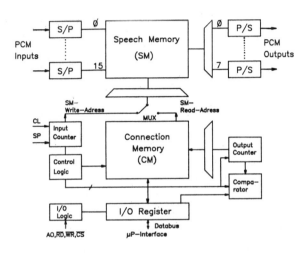

Fig. 12: Block diagramm of the PEB 2040
         Memory Time Switch (Siemens)

5. DIGITAL SIGNAL PROCESSING IN PCM SYSTEMS

With the advances in integrated circuit technology,
digital signal-processing (DSP) methods became more and more
attractive. Moreover, the introduction of end to end voice
and data capability is creating a nutural variety of tasks
wich can be efficiently solved by using VLSI-oriented DSP

The objectives of DSP techniques include

- signal generation and modulation
- digital filtering to eliminate unwanted signal components
- digital encoding for bit rate reduction and speech synthesis
- sample rate conversion
- adaptive signal processing

In the following, several major applications using this capabilities are emphasized.

Typical examples of the digital representation of signals are the generation of audible tones (e.g. dial tones), the generation and detection of specific test signals for analog channels, and the generation of compressed speech.

An attractive usage of digital filtering in PCM-systems is the detection of DTMF dialling tones (see chapter 3.1) to define the dialled number. In this application a signal processor connected to the speech channel performs linearization, seperation into upper and lower frequency groups and definition of the frequencies by passing the signals through an ensemble of band passes related to the dial frequencies. A comparison of the individually filtered signal power permits the received digit to be identified. Cost effective implementation in a DSP-VLSI allows decentralization of this function resulting in increased system reliability.

The impact of DSP on subscriber line circuits implementing highly flexible codec filters which include further parts of the BORSHT functions is discussed in chapter 4.1.4. This trend is considered to have a major impact on the transmission quality in telephone systems.

Modulator/Demodulators (MODEMS) are at present widely used to support transmission of digital data via analog lines of a public network using inband modulation techniques as a first step towards network integration (e.g. telex/teletex). Typical modulation schemes are frequency shift keying (FSK), phase shift keying (PSK) and quadrature amplitude modulation (QAM). The baud rates most commonly used in a variety of speeds are 300, 600, 1200, 4800 and 9600 bit/s. Depending on modulation scheme and baudrate, full douplex and half douplex operation is possible. While at present most monolithic approaches use switched capacitor techniques, DSP performing the tasks of digital scrambling, filtering, modulation, signal generation and channel equalization becomes more and more attractive.

A typical example of adaptive signal processing is the adaptive echo cancellation used in full duplex two wire digital subscriber loops. This is covered in more detail in the next chapter.

6. VLSI COMPONENTS FOR ISDN

The integration of voice data communication in an integrated services digital network which allows simultanous transmission of digitized voice, circuit switched data and digital signaling based on existing wiring, is presently one of the major development projects being tackled by the telecommunication industry. Moreover, the subscriber access is extensively discussed in international standard organizations and national administrations. According to CCITT recommendation 1.310, the architecture for the subscriber access to an ISDN consists of an exchange and line termination (ET, LT), a local two wire loop (U-interface) to connect the network termination (NT), and a four wire S-interface bus to the subscriber terminal. The terminals will provide a variety of voice and a data processing facilities. Various study group recommendations to date define performance objectives, realization issues, network structure, interfaces, protocols, and further details of the network.

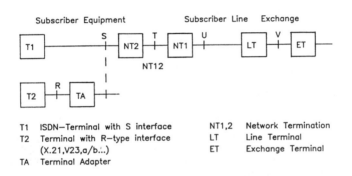

T1 ISDN—Terminal with S interface
T2 Terminal with R—type interface
    (X.21,V23,a/b...)
TA Terminal Adapter

NT1,2 Network Termination
LT   Line Terminal
ET   Exchange Terminal

Fig. 14: ISDN architecture according to CCITT

Due to this recommendations, the channel structure is defined with 64+64+16 kbit/s (B+B+D), where the D channel is attached to signaling, end to end packet switched low rate data and telemetry. The D channel protocol architecture is in agreement with the 7-layer model proposed by the International Standard Organization (ISO). In the lower levels this includes the physical layer of data transfer (1), the link layer (2) for protocol classification, and the network layer (3) e.g. for user part identification. The type of link access protocol used is similar to CCITT's X.25 high level data link control (HDLC) communication protocol, which meanwhile is well established in data communications.

The 4 wire S bus provides separate receive and transmit loops working at 192 kbit/s. Supporting a bus configuration in the range of 150 m, up to 8 terminals can be connected which in the network will be reached under the same dial-number.

In point to point applications, the transmission range is above 1 km.

Since specific figures such as baud rate, transmission code, pulse shape, frame synchronization, activation and deactivation, timing recovery and contention resolution are well defined, this bus can be considered as an important tool towards standardization of data comminication equipments.

To keep customer access to ISDN to an affordable level, a whole set of VLSI devices for the terminal, the network termination and the digital exchange is required based on advanced technologies. The basic functions to be implemented include the physical level of data transmission at the U and S interface, the data link access protocol, as well as switching and multiplex functions. As administrations usually demand emergency remote power feeding there is a need to keep power consumption as low as possible.

One of the greatest problems to be solved is to provide two wire bidirectional transmission at the effective data rate of 144 kbit/s over existing wire pairs without extra investment in subscriber loop plant. In principle, several methods can be applied: frequency separation, hybrid separation, time division (ping-pong) and echo cancellation. Concerning performance and VLSI realization, the latter techniques are most promising. Moreover, there are a number of contsraints which give major attention to echo cancellation principles. Major reasons are:

- with respect to time division and propagation delay, ping pong methods require a high baud rate causing severe linear distortion and high attenuation which limits line length significantly; typical baud rates necessary are in the range of 400 to 500 kbaud/s

- echo cancellation offers continuous data transmission with low bandwith. Therefore digital subscriber loops of up to 8 km can be achieved based on bauderates in the range of 120 to 160 kbaud/s.

Nevertheless, integration of an echo cancellation receiver/transmitter represents a twofold challenge to the semiconductor industry, first, complexity and speed of digital parts and, second, anlog performance requirements with respect to linearity, resolution, signal to noise ratio and speed of the a/d conversion as well as linearity of d/a. The second may be well understood taking into account that line attenuations in the order of 45 dB may exist in combination with nearly directly reflected echos. Since the useful incoming line signal represents the difference between large signals, the precision for reconstruction of the echo needs to be in the order of 0.1 %.

The principle of an echo canecllation based data-transmission circuit is shown in Fig. 15. Using an adaptive type of transversal filter the echo signal is precisely reconstructed from the transmit signal by means of DSP and then subtracted from the receive signal on a digital or analog basis. Basic differences in the architectural approach for these circuits arise from whether signal processing functions such as equalizing are performed in analog or all digital form, due to the type of adaption agorithm, timing recovery, pulse shaping, coding principle, a/d conversion, and the type of hybrid to provide directional isolation.

The typical ISDN voice/data terminal connected to the S bus in its simplistic version will include a/d conversion and PCM coding of the voice in a codec filter, level 2 and 3 of protocol handling, and the physical link to the 4 wires. While a more chip approach may be most effective in the first step, monolithic integration of these functions in a microprocessor compatible device can be expected in the longer term. To meet specific customer services, a spectrum of terminals having powerful voice, data and image processing capabilities can be expected.

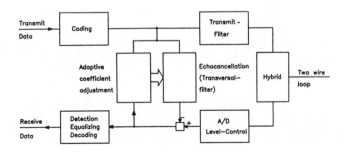

Fig. 15: Principal signal flow for an
echo cancellation data transceiver

7. CONSLUSION

Advanced state of the art telephone systems are potentially powerful communication and data processing tools. Introduction of ISDN will extend the capabilities significantly and also will open up the door for a generation of network oriented equipments taking advantage of integrated voice/data transport technology. The speed in further evolution concerning system performance, services, reliability and cost similar to the rapid change during the last decade will be determined to a great extent by the advances made in microelectronics. However, proceeding in system integration necessitates a high degree of cooperation between system houses, semiconconductor companies, national authorities and international standard organizations.

8. REFEENENCES

/ 1/   K.Goser, "The challenge of the VLSI technique to telecommunication systems", IEEE J. Solid-State Circuits, vol. SC-15, pp. 406-610, Aug. 1980

/ 2/   P.R.Gray, and Messerschmitt, "Integrated circuits for telephony", Proc. IEEE, vol. 68, pp. 991-1009, Aug. 1980

/ 3/   D.K.Melvin, "LSI helps telephones go digital", IEEE sepctrum, pp. 30-33, June 1980

/ 4/   L.Lerach, "Integrated circuits for telephony", ESSCIRC Dig. Tech. Papers, Freiburg, West Germany, Sept. 1981, p. 109

/ 5/   W.D.Pace, "A one-chip telephone", Telecommunications, p. 43, Apr. 1984

/ 6/   L.Lerach, "A novel device concept for subscriber line boards in digital exchange systems", NTG-Fachtagung Dig. Techn. Papers, Baden-Baden, West Germany, Mar. 1983

/ 7/   L.Lerach, G.Geiger and M.Strafner, "Pheripheral Board Controller for digital exchange systems", ISSCC Dig. Techn. Papers, New York, Febr. 1983, p. 76; to be publ. in IEEE J. Solid-State Circuits, vol. 18, Aug. 1984

/ 8/   W.Twadell, "Telecommunications ICs", EDN, pp. 139-152, Mar. 1984

/ 9/   R.Apfel, H.Ibrahim and R.Ruebush "Signalprocessing chips enrich telephone line card architecture", Electronics, vol. 5, pp. 113-118, May 1982

/10/   B.K.Ahuja, W.M.Baxter and P.R.Gray, "A programmable CMOS dual channel interface processor for telecommunication applications" ISSCC Dig. Techn. Papers, New York, Febr.84

/11/   D.Vogel and W.Pribyl, "CMOS digital signalprocessing codec-filter with high performance and flexibility", submitted to ESSCIRC, Edinbourgh, Sept. 84

/12/   E.Schmid and J.Reisinger, "High performance oversampling a/d-converter for digital signal processing applications" submitted to ESSCIRC, Edinbourgh, Sept. 1984

/13/   L.Gaszi, "Hardware implementation of Wave Digital Filters using programmable digital signal processors", ECCTD Dig. Techn. Papers, The Hagne 1981, pp. 1052-1057

/14/   D.P.Laude, "A monolithic subscriber line interface circuit", IEEE J. Solid-State Circuits, vol SC-16, pp. 266-270, Aug. 1981

/15/   R.M.Sirsi, P.J.Mezza, D.P.Laude and R.C.Strawbrich, "Integrated electronic subscriber line interface with ringig on chip" IEEE J. Solid-State Circuits, vol. SC-18, pp. 665, Dec. 1983

/16/   S.Horvath, "Digital signal processing in communiation and transmission" EUSIPCO Dig. Techn. Papiers, Erlangen, 1983, pp. 515-522

/17/   R.Lechner, "Voice signal processing in digital local exchange", Globecom Conf. Rec. Vol. 3, pp. 1566-1571, 1983

14

DIGITAL SIGNAL PROCESSING MODEMS

RUSSELL J. APFEL
ADVANCED MICRO DEVICES
SUNNYVALE, CA 94088   USA

1.   INTRODUCTION

Modern day communications systems require the ability to transmit digital information between remote locations. This is normally accomplished using a voice band MODEM (modulator/demodulator) that works over the public switched telephone network (PSTN). These voice band modems employ several different techniques for taking digital data and modulating it on a band limited analog carrier which can be successfully transmitted over the PSTN and then demodulating the data at the far end of the system to restore the digital information. Modems of various speeds ranging from 300 bits per second to 9600 bits per second with various performance features (such as full-duplex and half-duplex modems) have been successfully developed for use on the PSTN. A variety of design techniques are available for designing voice band modems. These modems combine both analog and digital techniques in order to achieve high performance in a difficult environment. This paper will describe some of the aspects of modems and how digital signal processing techniques can be employed to develop high performance cost effective modems.

2.   MODEM SYSTEMS

2.1   Why Use Modems

Modems are required because the telephone system has a very narrow bandwidth and is not an efficient means of transmitting digital information. Digital information, in the form of square waves, has a very wide harmonic content and is not the ideal type of information to send over the public telephone network. The spread of frequencies and the attenuation of signal on the telephone lines makes it very difficult to insure error free transmission of data over the systems. Modems have been developed to alleviate some of these problems. Modems encode the digital data and modulate this data using analog carriers in order to transmit the data.

The telephone system for all practical purposes has about 3 kHz of usable bandwidth, roughly from 300 Hz to about 3.3 kHz. All information should be encoded within this frequency band. The telephone system has significant attenuation of signals in excess of 3 kHz. Signals below 300 Hz are greatly interfered with by coupling from power lines at 60 Hz. Another practical limitation is that PSTN transmission systems multiplex multiple conversations onto a single trunk carrier using either Frequency Division Multiplex (FDM) of Time Division Multiplex (TDM) techniques. Frequency division multiplexing takes the analog signal, places it upon a carrier and transmits it. Signals that exceed the 4 kHz bandwidth may fall into the adjacent channel's carrier and will create noise and difficulty in obtaining good transmission quality. Time division multiplexing samples the signal, digitizes it and transmits the digitized data using high speed digital bursts on a digital carrier. Since the sampling rate is 8 kHz, any signal in excess of 4 kHz would be folded back on top of other signals creating distortion, noise, and errors in transmission. Therefore, it is important to remember that all signals transmitted by modems should be band limited and not have any components in excess of 4 kHz.

Modems, using a variety of techniques, allow digital information to be transmitted on the public network. The various techniques allowing coding of one or more bits of information to enable highly efficient transmission rates up to 9.6 kbits per second can be successfully transmitted within the 3 kHz band width of the PSDN.

2.2 Modem Types

Modem systems (Fig. 1) employ a modulator (transmitter) that interfaces on one side to a computer system normally through a standard interface such as an RS232C or CCITT V.22 interface and on the other side to a public switched telephone network through some telephone line access module. The modulator takes the incoming digital data and converts it to the appropriate analog carrier system for transmission. The data may be synchronous or asynchronous data depending on the modem system. On the far end of the system a demodulator will interface with the telephone line access module, process the analog signals and demodulate them to restore the original digital information and then interface to the computer system again through some type of standard interface such as V.24 or RS232C. There are several possible types of modem systems. The most common ones are:
   a) Simplex transmission systems where there is only one way communications from a source which has a transmitter to a source which has a receiver.

   b) Half-duplex systems where each end of the system has a transmitter and a receiver but only one end transmits while the other end receives during a period of time. After transmission is completed there is a procedure called "line turnaround" whereby the direction of the transmission is reversed and the other end of the system can transmit information.
   c) Full-duplex transmission where both parties are transmitting and receiving simultaneously. Since the same 2-wire environment is being used, there must be some sort

459

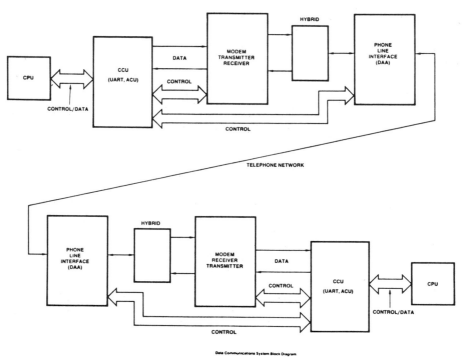

Data Communications System Block Diagram

Figure 1

460

of separation of data between the transmitter and the receiver.
These separations are usually achieved by FDM techniques where one party transmits within one frequency band and the other party transmits in a different non-overlapping frequency band. Other techniques that provide full-duplex operation would include echo cancellation and time compression multiplexing.

d) Half-duplex with back channel systems are one way in nature, but allow a low speed channel for the reverse direction for sending control information. This may be viewed as a full-duplex system where one party gets a high data rate and the other party gets a much lower data rate. It can be considered half-duplex or a subset of full-duplex transmission and will be discussed later in terms of standard modem types.

There are three popular techniques for encoding data in voice band modems. These are Frequency Shift Key (FSK) modems, Phase Shift Key (PSK) modems and Quadature Amplitude Modulated (QAM) modems.[1] Frequency Shift Key modems allow data to be transmitted as discrete frequencies and a 1 and a 0 is differentiated by the frequency of transmission. The frequency of the modulator is controlled by the incoming digital data. FSK modems are the simplest type of modem and operate at lowest speed. Phase Shift Key modems use a single carrier and encode information by changing the phase of the carrier. The information can be encoded one bit at a time or multiple bits at a time by breaking up the Phase Shifts in increments of less than 180 degrees. Because multiple bits can be encoded within a single transition, PSK modems offer the possibility of higher data rates on the telephone network. Quadature Amplitude Modulation modems combine phase shift key techniques with amplitude modulation to allow even a wider number of encoding possibilities within each transition of the modem and offer the highest possible data rates for modem systems.

2.3 Telephone System Limitations on Modem Performance

The PSTN system places a number of constraints and requirements on modem performance. As previously mentioned, signals in excess of 4 kHz should not be transmitted on the telephone systems. Telephone systems have a normal amplitude loss curve as shown in Fig. 2. The telephone system has relatively flat performance for signals up to 1.5 kHz, but there is significant loss for signals in the 3 kHz region. The telephone lines also vary greatly in their performance and care must be taken in designing modems to make sure that they can operate over a wide range of telephone-like characteristics. The output of a modem transmitter is normally fixed at a level of -9dB, which nominally insures that at the telephone office the signal level is less than -12 dBm. Losses in the telephone system will create a typical receive input signal that is about 20dB less than the transmit signal. Modems must work over a wide range of incoming amplitude signals. Normally there is a carrier detection requirement that says the modem must detect an incoming signal as

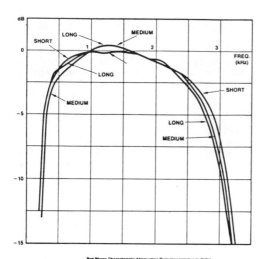

Bell Phone Characteristic Attenuation Distortion (relative to 1kHz)

Figure 2a

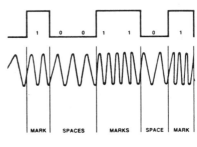

1 : 1 Modulated FSK Signal

Figure 3

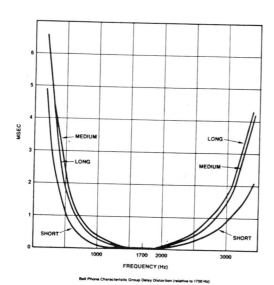

Bell Phone Characteristic Group Delay Distortion (relative to 1700 Hz)

Figure 2b

a valid carrier if the amplitude is greater than -43dBm and must say that the carrier is not valid if the amplitude is less than -48dBm. The carrier detection circuit also has a 2dB histeresis requirement to guarantee that small amounts of noise do not cause flickering in the carrier detect indication. Therefore, we can see that the telephone administrations assume that the modem signal can be attenuated from its -9dBm transmit level by as much as 37dB. The modem receivers are therefore required to operate with good performance over a 37dB dynamic range. The telephone system not only attenuates the amplitude of the incoming signal, but also provides phase distortion. The phase distortion creates uncertainties in frequency shifts and frequency measurements as we modulate signals. The normal telephone line systems can put up to 500 microseconds of delay uncertainty onto the transmitted signal before it gets to the modem receiver. Some modems specify a compromise equalizer that compensates for the telephone line's nominal amplitude and delay characteristics. This equalizer is usually split so that half of the equalizer is done in the transmitter and half in the receiver. These modems will have the best performance on nominal, not ideal telephone lines. In some systems, such as FDM systems, where the signals are modulated onto transmission carrier and then demodulated back down to their normal signal, frequency offsets occur that can be as large as 7 Hz and must be accounted for in the system. Telephone systems also add numerous impairments to the systems such as creating phase jitter, impulse noise, amplitude hits, and line dropouts which all must be handled by a modem to ensure good data integrity.

The normal measure of the integrity of the data coming in and the performance of the modem is bit error rate. Bit error rate as the name implies, is a measure of the number of errors per number of bits transmitted. Bit error rate must be measured under specific operating conditions. The major conditions to specify are the signal to noise that is generated on the line, an adjacent channel level, a receive carrier level and other impairments on the telephone line. The bit error rate versus signal to noise measurement determines the range of conditions that the modem can successfully operate under. Normally bit error rates less than 10^{-5} are desirable for high quality transmission. Modems are typically designed to provide this level of performance. FSK modems should be able to provide that level of performance with signal to noise ratios of 10dB. As the speed of the modem increases, the signal to noise ratio at which it can achieve the error rate of 10^{-5} normally increases.

2.4 FSK Modems

Frequency Shift Key (FSK) modems are the simplest form of modems. In the modem system (Fig. 3) the carrier is modulated between a mark and a space frequency. The most popular modems of this type are the Bell 103 modem,[2] a 300-bit per second full-duplex modem, Bell 202 modem,[3] a 1200-bit per second half-duplex modem, the CCITT V.21 modem,[4] a 300-bit per second full-duplex modem and the CCITT V.23 modem,[5] a 1200-bit per second half-duplex modem with a 75-bit per second back channel. Fig. 4

shows the frequency bands chosen for transmission in these modems. The full-duplex modems use an FDM technique to allow two frequency bands, an originating band and an answer band, within the 3 kHz bandwidth of the telephone system. 300 bits per second modems require that each band be somewhat wider than 300 Hz in bandwidth. The 103 modem uses an originate band which has frequencies at 1070 Hz for space and 1270 Hz for mark. The answer band, has 2025 Hz for space and 2225 Hz for mark. The modulation between space and marks creates a spreading of the frequency spectrum beyond that of the space and mark frequencies. In fact, the major amount of energy is at the center frequency between the mark and the space frequencies. A filter is required after the modulator for two major purposes. First, the filter guarantees that energy does not spill over into the other band i.e., energy from the originating modem that would fall within the answer band, is attenuated so as not to distort information being sent in the other direction. Secondly, the filter is required to guarantee that out-of-band energy, energy above 4 kHz, does not exceed the specifications of the PSTN. Public administrations have different requirements on

how much out-of-band energy is allowed to be transmitted on the telephone. Typically however, requirements are such that energy should be decreasing with frequency and no energy in excess of -25dB should be transmitted above 4 kHz with a roll-off versus frequency so that above 16 kHz there is no energy greater than -55 dB on the telephone line. This ensures that the modem does not interfere with other telephone systems.

Twelve hundred-bit per second modems require a much wider bandwidth than 300-bit per second modems and a full-duplex 1200 bit per second FSK modem is not possible. The Bell 202 and CCITT V.23 modems use a center frequency of 1700 Hz with mark and space frequencies of 1200 and 2200 for the 202 modem and 1300 and 2100 Hz for the V.23 modem. Because the 1200-bit per second modem cannot support a second high speed channel and can only be operated in the half-duplex mode, a lower speed channel is made available. This lower speed channel occupies a frequency band around 400 Hz and has a modulation rate of 5 bits per second for 202 and 75 bits per second for the CCITT V.23. The V.23 uses a space frequency of 450 Hz and a mark frequency of 390 Hz while the Bell 202 modem uses a frequency of 387 Hz format and the absence of any carrier to denote a space. The back-channel in the V.23 modem can be used for low speed data such as in the case of an interactive terminal talking to a computer. The interactive terminal whose inputs are generally a human typing and cannot generate data at a rate faster than about 75 bits per second, so all the user needs is a 75 bits per second channel going from a terminal to a computer and the higher speed 1200 bit per second channel returning from the computer to the terminal. The 5-bit per second back-channel of the Bell 202 modem is not really useful for data channel and is only useable to let the person who is transmitting at 1200-bits per second and receiving at 5-bits per second to know that the other party is still connected and to begin and initiate line turnaround procedures.

Fig. 5 shows a simplified block diagram of an FSK transmitter. Digital data is input from a digital source to the modem where it is fed into a sine synthesizor. The sine synthesizor generates either the mark or the space frequencies

464

dependent on the incoming data. It is important that the transition between frequencies be smooth and that the no phase non-linearities occur as the frequency changes. This type of transition is known as phase continuous FSK transmission. The output of the sine synthesizor is fed to a digital bandpass filter which as earlier mentioned, band limits the signal so that it does not interfere with adjacent channels and does not produce noise which exceeds telephone line system requirements. The output of the digital bandpass filters can then be converted to an analog signal using a digital to analog converter and fed to the telephone lines. Normally, some simple analog post filter may be required to guarantee that resampled noise at the sampling rate of the D to A converter does not interfere with meeting the telephone system requirements for out-of-band energy. The design of this filter is a function of the sampling rate of the data fed to the digital analog converter.

A simplified block diagram of an FSK receiver is shown in Fig. 6. The FSK receiver must provide complementary functions to the FSK transmitter. An analog pre-filter is required as an anti-aliasing filter prior to the A to D conversion which takes the analog signal and converts it into a digital signal to be processed by the digital signal processor. The digital bandpass filters are used to remove energy that is not in the band of interest. This energy is noise that might be generated or picked up from the telephone line as well as the signal that could be generated and fed through from a local transmitter which is in a different frequency band from the signal that is being received. The digital bandpass filter only passes signals that are in the frequency band centered around the center frequency of the modem band that is being received. The output of the digital bandpass filter goes in two blocks. The first block is a carrier detection block which measures energy in this frequency band and determines whether there is a carrier present, i.e., whether there is data being transferred from another modem to this modem for reception. This output is provided to the system as one of the handshaking signals to indicate that the system is functioning properly. The data is then fed into a automatic gain control (AGC) block which adjust the final amplitude to a constant predetermined level. Data is also fed into a digital demodulator which restores the digital information that was transmitted from the analog carrier. Different types of demodulation schemes can be used for an FSK modem.

There are two basic classes of demodulation techniques known as coherent and noncoherent demodulation. A coherent demodulator recovers the center frequency or carrier of the transmission, that is the center frequency between the mark and space from the incoming data and uses the recovered carrier as part of the demodulation process. A noncoherent demodulator does not recover the carrier frequency but rather processes only the incoming signal. Coherent demodulators offer a somewhat higher degree of performance when properly implemented but require a carrier recovery system which could be implemented by use of a phase locked loop. The output of the phase locked loop is a DC signal which is proportional to the incoming signal and therefore is easily useable to recover whether the incoming signal is greater or less than the center frequency between the mark and space,

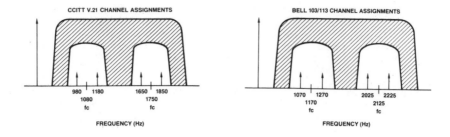

Figure 4   Figure 5

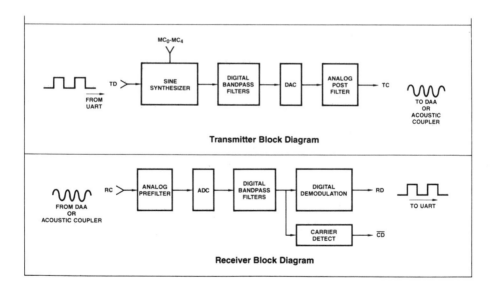

Figure 6

466

depending on whether it is a mark or a space frequency. Noncoherent techniques include FM demodulation techniques and frequency counting techniques.

2.5 PSK Modems

Phase Shift Key modulation modems involve an analog carrier of a constant frequency whose phase is modulated to provide signal information. The instantaneous phase of the constant carrier is changed at the data boundary while leaving the frequency constant. The digital data inputted to the modem will control the phase shift which the modem performs on this carrier in such a manner that the receiver can recover the digital data by measuring the phase shift. Typically, PSK modems use a differential phase shift encoding scheme where rather than looking at the absolute phase of a signal to determine the information being transmitted, one must only look at the change of phase from one time period to another time period to enable the information to be demodulated. Fig. 7 shows a wave form of a differential PSK (DPSK) signal. In this particular signal the phase shifts that are performed are 0°, 90°, 180° or 270°. Since there are four possible phase shifts in this scheme, it is known as differential quaternary PSK (DQPSK) modulation. The four possible phase shifts mean that at each data boundary, 2 bits of information can be transmiss. FSK systems are able to run synchronously or asynchronously, but DQPSK systems must be operated in a synchronous manner because information must be available to know when to look for phase shifts relative to previous signals. The most popular DQPSK systems are the 1200 bit per second modems in the US and Europe under the specifications of the Bell 212A in the United States, and the CCITT V.22 modem in Europe. These modems use 1200 Hz as one carrier and 2400 Hz as another carrier. The data is encoded two bits per baud, so for a 1200 bit per second modem, the effective modulation rate is 600 baud, and the frequency spreading is approximately ± 300 Hz around the center frequency. This allows enough of a band separation between the high end of the low frequency band, (approximately 1600) and the low end of the high frequency band, (approximately 2000 Hz). To provide good separation, very steep roll-off filters are required in the DQPSK system.

A DQPSK transmitter is shown in Fig. 8. The modem can operate only in a synchronous mode, so an asynchronous to synchronous converter is required as part of the modem for use in asynchronous systems. Normally these modems can accept asynchronous data at rates that are within 1 percent of the 1200 bit per second rate, and the asynchronous to synchronous converter inserts or deletes start and stop bits at an appropriate rate so that the data transmitted on the line is 1200 bits per second. The 1200 bit per second data must be encoded with character field lengths of 7, 8, 9, 10 or 11 bits. The output of the asynchronous to synchronous converter is fed into a data scrambler. The data scrambler is used to randomize the incoming data in a manner such as to keep the frequency of information transmitted, spread evenly over the available bandwidth. Scrambling techniques, known as self synchronizing scrambling can be descrambled reliabilty with a

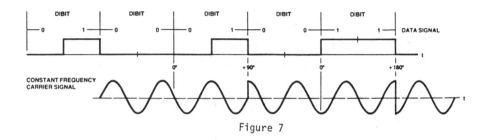

Figure 7

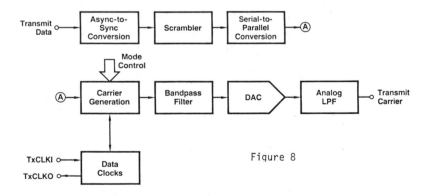

Figure 8

synchronized descrambler at the other end of the system. The data is then taken two bits at a time and encoded into what is known as di bits which are the actual inputs determining the phase shifts for the modulator. Data is then fed into the modulator which generates the DQPSK signal. The modulation implementation for DQPSK falls into two classes; the first class generates a 1200 or 2400 Hz carrier and actually has non-continuous phase shift at the baud interval boundaries. The second scheme is to generate two carriers, a cosine term and a sine term which are 90° out of phase and which can be multiplied by plus one, minus one or zero. The four phases can then be generated as a plus sine, plus cosine, minus sine and minus cosine term by appropriately manipulating the coefficients by which the sine and cosine terms are multiplied. The modulated carrier is then fed into a digital bandpass filter which is required to bandlimit the signals power to transmission. The digital bandpass filters need to have very good phase characteristics, that is very low group delay distortion. If large delay distortion exists, it will cause great difficulty in the demodulation process and the error rate of the system will increase. The digital bandpass filters also normally have part of a compromise equalizer built into them. A compromise equalizer is included within the modem to compensate for the nominal amplitude and phase distortion of the PSTN (Fig. 2). The line characteristics are fairly well known and rather than putting all the line equalization in the receiver it is split so that some of the equalization is put into the transmitter and some in the receiver. This equalization normally causes some gain to be present in the high band, to compensate for amplitude loses in the 2400 Hz and above range of the telephone lines. The DQPSK modem also requires a shaping characteristic. The shaping characteristic is normally a 22% raised cosine characteristics at the baud rate. This shaping characteristic allows a smoother transition by attenuating the signal at the phase transition boundary to reduce the amount of energy being transmitted and restoring the information to full amplitude in the center region between the data boundaries. This shaping signal is used to help in the demodulation process by allowing the timing to easily be recovered from the shaping signal. The shaping can be done in the pass band in the same block as the digital bandpass filters or to be done in the baseband. Baseband shaping can be accomplished in the quaduarte modulation scheme, where we use cosine and sine characteristics, by multiplying the sine and cosine terms with a time varying characteristic rather than the constants, 0 and ±1.
The output of the bandpass filters is then fed into a digital analog converter which converts the digital information into analog format and then into a simple analog post filter to remove any of the resampled information at a sampling rate to prevent high frequency energy from being put on the telephone line. The data is then transmitted over the telephone system. The transmitter clock can be locally generated, provided from a terminal or obtained from the receiver clock.

A DQPSK receiver is shown in Fig. 9. This receiver is more complex than an FSK receiver. The incoming data is fed through the analog prefilter and then to the analog to digital converter to create digital information. The digital information goes into a digital bandpass filter, which removes the local transmitter carrier information, which is not desired and any noise picked up on the telephone wire. The output of the digital bandpass filter

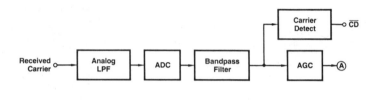

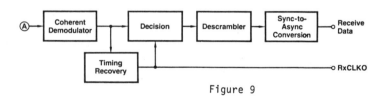

Figure 9

470

is a signal within the band of information that is desired for demodulation. The digital bandpass filter, additionally, has one half of the compromise equalizer for the telephone line system built into it. In the case of DQPSK modems, performance can be improved by including an adaptive equalizer as opposed to a fixed equalizer. A fixed equalizer will equalize towards the compromise telephone line and its output still have group delay distortion and amplitude distortion due to telephone line characteristics which are different from the nominal line. An adaptive equalizer is a digital filter which has adjustable coefficients and adjustable characteristics so as to adapt to the actual line characteristics and equalize for the actual line to remove much more of the group delay and amplitude distortion than is possible to be removed by compromise equalizer. The reduction of distortion can lead to significant improvement in bit error rates out of the modem, especially on severely distorted telephone lines. The output of the bandpass filter again goes two blocks, a carrier detect block and demodulation block. The carrier detection block is similar to the FSK modem in that it measures the energy of the received carrier and determines whether there is a carrier on the line. The receive carrier goes through an AGC block, which adjusts the signal to a constant amplitude, independent of the loss on the telephone line. The signal then goes into a demodulator block. The demodulation for DPSK modems normally includes splitting the signal into an in phase (I) and quadature (Q) phase signal. This is done by multiplying the incoming signal by a two carriers that are phase shifted 90° relative to each other. The signal that is used to do this multiplication can come from a carrier recovery block, if the system is a coherent modulator, or can be a locally generated carrier if it is a noncoherent system. The I and Q terms that are generated through these multiplications are low-pass filtered to obtain the baseband information and then fed into a decision device which decides what the differential phase shifting was. Very often, the adaptive equalization is done in the baseband in

the demodulator as opposed to the digital bandpass filter block which precedes demodulation. Carrier recovery is accomplished, for coherent demodulators, within the demodulation block. The decision device also requires a timing recovery block which can recover the 600 baud timing information in order to determine when is the proper time to make a decision on the phase shift between baud intervals. The receiver always recovers the clock of the received signal to do demodulation and the receive clock is also provided as an output to the modem. The receive data in dibit form is then decoded into a serial stream, descrambled, and is available for the outputs to the system.

DQPSK modems use scrambling to spread the spectrum of the system and therefore require a training sequence to allow the scramblers to synchronize to each other. During the training sequence, it is also possible for adaptive equalizers to adapt to the telephone line characteristics. The Bell 212A modem in the United States, also includes a fall-back mode to the Bell 103 modem, which is a 300-bit per second FSK modem. During the training sequence the modem can be used to determine whether the modem at the other end of the line is a 1200-bit per second or a 300-bit per second modem, an automatic adjustment of speed can be done during this training period.

471

2.6   QAM Modem

Quadature Amplitude Modulated modems are the most complex modems used on the PSTN. QAM modems are used for 2400 to 9600 bits per second transmission. The most popular modems of this type are the 2400-bit per second full-duplex modems (V.22 bis and V.26 ter), the half-duplex modem running at 2400, 4800 and 9600-bits per second (CCITT V.26, V.27, and V.29 and Bell 208 and 209 modems) and the new 9600-bit per second full duplex modem, CCITT V.3. QAM techniques involve both phase shifting the signal such as is done in PSK modems, as well as modulating the amplitude of the modems. This creates a constellation of possible phase and amplitude conditions as shown in Fig. 10. The constellation in Fig. 10 is that of a V.22 bis modem. This modem is an extention of the V.22 constellation used for 1200-bit per second DQPSK modem. In the V.22 bis implementation, each quadrant has four possible points designated by the amplitude (1/3, 1/3), (1/3, 1), (1, 1/3) and (1,1). This gives three possible amplitude values and three different phase values within each quadrant. The V.22 bis modem uses the same baud rate, 600 baud as the 1200-bit per second modem, but encodes 4 bits per baud rather than 2 bits per baud giving us a 16 point bit constellation, as opposed to the four possible values in the PSK modem. The higher speed modems that use QAM techniques, use a combination of more bits per baud and higher baud rates to generate the higher bit per second rates. For example, 9600 bits per second modems use 1600 baud modulation with 6 bits encoded per baud or a 64 point constellation. This has 16 points in each quadrant. The requirements of both the modulator and the demodulator increase significantly for QAM systems. The system becomes more sensitive to amplitude and phase distortion on the telephone line and therefore many adaptive characteristics are required within the modem to compensate for the telephone line characteristics.

The QAM modems require very strong digital signal processing to enable good performance to be achieved using adaptive filter techniques. Some QAM modems, such as V.22 bis, use FDM techniques for achieving full-duplex transmission on a telephone line. That is, there are separate bands for the transmit and receive frequencies of the modem. Other high-speed modems are beginning to use echo cancellation techniques to allow the transmit and receive modems to occupy the same band. The advantage of this technique is that a wider bandwidth is now available for transmission. For instance, 9600-bit per second modems require a 1600 Hz baud rate and therefore take up most of the bandwidth available on the telephone line. With echo cancellation, the signal of the transmitter which is well known by the modems is cancelled out from the signal on the telephone line leaving only the received signal. Echo cancellation requires a high precision adaptive filter to adapt to the telephone line characteristics to be able to cancel the echo of the transmitted signal which is returned into the receiver. This paper will concentrate on the V.22 bis modem as a QAM modem, to look at the basic techniques required for building a QAM modem, and will not go into the issue of echo cancellation.

472

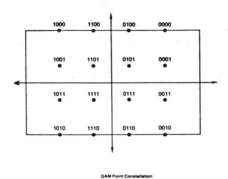

QAM Point Constellation

Figure 10

QAM modems require a more complex training and learning sequence than PSK modems because more features must adapt to the telephone line and the timing recovery and carrier recovery are significantly more complex. FSK and PSK modems contain significant information only in the frequency and the phase of the incoming signal and therefore, a simple AGC technique can be used in the receiver to adjust the signal level prior to demodulation. QAM modems require a great deal more care because information is held in the amplitude of the incoming signal. Therefore, amplitude distortion as well as phase distortion characteristics are much more critical. Because phase shift of much less than 90° must be detected along with amplitude changes, the QAM modem requires much more care in its design.

Fig. 11 shows a QAM modem transmitter. The modem transmitter is similar in nature to the DQPSK transmitter. The incoming data goes to a asynchronous to synchronous converter if it is asynchronous data and is fed into the transmitter. The transmitter clock rate, like the Bell 212 and V.22 modems, can be provided internally from the modem, can be provided externally, (for synchronous transmission), or the receiver clock can be used for generating the clock for transmitting data. The data is scrambled using similar scrambling techniques as DQPSK modem, and then is encoded into four bits per baud interval. The modem transmitter uses a single carrier and can be implemented a number of ways. The transmitter requires shaping similar to the DQPSK modem so that at the baud interval boundaries, we attenuate the phase and amplitude shift and then restore the data in the middle of the baud interval. Shaping can be done in the baseband by modulating the I and Q components or can be done in the passband. The modem transmitter can be implemented with a modified form of the DQPSK modem. That is, to generate sine and cosine terms and have the transmitted carrier, (A cosine WT + B sine WT), with the values for A and B being other than ±0. In the case of the V.22 bis modem, the values can be zero, $\pm1/3$ or ±1 for A and B, giving us all possible combinations for transmission. Baseband shaping would be accomplished by modulating the values of A and B with the baseband shaping characteristic. The transmitter can also be implemented by generating the carrier frequency and adding in phase shift and amplitude modulation and then providing passband filtering and passband shaping of the signal. The digital bandpass filters require more rolloff and less distortion than required by the lower speed modems. The transmitter also includes one half of the compromise equalizer as is required in the transmitter for the Bell 212A and V.22 modems. The output of the digital bandpass filter is fed into the D to A converter to a post filter and then sent out on the line in the system.

The receiver for the QAM modem is shown in Fig. 12. This receiver is significantly more complex than the receiver for the DQPSK modem. The front-end requires an analog to digital converter that must have a wider dynamic range than the converter that is used in a DQPSK modem because there is a wider amplitude range of incoming signals due to the amplitude modulation in this modem. The QAM modem also requires a higher signal to noise ratio in the A to D converter in order to guarantee low bit error rate than is required by the other modem type. The A to D converter requirements imply three to four bits more A to D resolution and

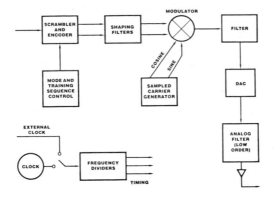

Figure 11

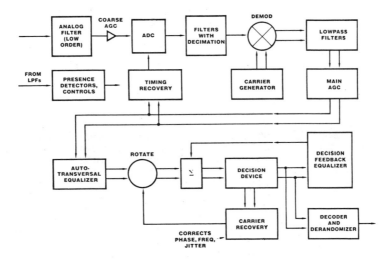

Figure 12

475

accuracy than is required by the other modems. The output of the A to D converter again goes into bandpass filters which remove the adjacent channel and any noise or energy that is picked up on the system. The output of the bandpass filters goes into the same traditional blocks - the carrier detector and the demodulator. The carrier detection for QAM modems is complicated again by the carrier whose amplitude is modulated. This places more filtering requirements on the carrier detection circuit to ensure that we have made an accurate measurement of the energy of the signal.

The demodulator for the QAM modem is always built as a coherent demodulator. The incoming carrier is recovered through a phase lock loop or a very sharp filtering of the signal. The carrier that is recovered is used to multiply the incoming signal and generate the in phase and quadature terms. These terms are then applied to low-pass filters and equalizers which correct for the telephone line filters. The timing recovery can be accomplished on the output of the demodulated signal or by processing the incoming signal in a separate path. There is no true information at the timing level so special processing techniques are required to extract the timing information. In addition to doing equalization for amplitude, phase adjustment must also be made. Corrections for frequency offsets and phase jitter in the telephone systems must be done in QAM modems to get high performance. The output of the timing recovery block is again used in the decision device to determine the optimum point in making a decision. QAM modems use differential decision devices similar to DQPSK modems. The output of the decision device is then decoded, descrambled and provided to the system as the most output.

3.   DIGITAL SIGNAL PROCESSING

3.1 Sample Date Systems

Digital signal processing systems are sampled data systems. That is, the analog signals are represented as time samples in the digital domain. Sampled data systems must meet Nyquist requirements for good performance. The Nyquist requirement states that a sampled data system can only represent signals that are less than 1/2 of the sample rate of the system. Signals that are greater than 1/2 of the sampling rate of the system will be folded back into the frequency band less than 1/2 the sampling rate of the system because they are indistinguishable from the lower frequency signals. Sampled data systems are normally analyzed using z transforms which is the sampled data equivalent to Fourier transforms for continuous time systems. Sampled data systems, using digital signal processing allow high performance signal processing to be performed using digital rather than analog circuitry. Digital sampled data systems require a certain amount of overhead that is not required in analog systems because since the normal input to the system and output to the system are in analog form, analog-to-digital and digital-to-analog converters are required to make the transitions between the analog and digital world. A digital signal processor will normally consist of an arithmetic processing unit, random access memory to store data values, a memory to store coefficient values and a sequencer that allows the processor to go through the signal processing procedure. Digital sampled data systems only become cost effective when significant signal processing is required to justify the initial overhead that is required to set up the digital sampled data system.

3.2 Digital Filters

Filtering digital signals is a combination of mathematical processing of the signals. Digital filters fall into two basic classes, finite impulse response (FIR) filters and infinite impulse response (IIR) filters.[6] Finite impulse response filters are filters where the output value is a function of a series of input samples and there is no feedback in the filter. The output of the filter is the summation of N input samples times an approriate coefficients for each input signal and it has a finite impulse response in the sense that when the input values all go to zero, the output goes to zero and the stability is guaranteed. Infinite impulse response filters are filters that have feedback. The output value is a function of a number of input signals plus a number of previous output samples multiplied by appropriate coefficients to provide a new output sample. Infinite impulse response filters do not necessarily have responses which go to zero when the input samples go to zero. IIR filters typically can be designed with much steeper roll-off characteristics than FIR filters because of their feedback, but care must be taken to guarantee stability of these filters because of the feedback. Digital filters have no variation with time or temperature and no unit to unit variations. Exact pole zero cancellation can be achieved and the same structure can be used to generate (low pass, high pass or bandpass) any filter characteristic. The hardware can be multiplexed so that many filters can share common hardware.

Finite impulse response filters offer many advantages which are useful in modem systems. FIR filters can be designed to have linear phase characteristics, that is, the phase varies linearily with frequency and the delay is constant independent of frequency. Because the phase group delay distortion is zero in FIR filters they are very useful in modems where group delay distortion is very critical to the design. FIR filters are also very useful in modifying the sampling rate of a system. Many modem configurations require processing to be done at different sampling rates. The analog-to-digital and digital-to-analog conversion are usually done at high sampling rates to minimize the requirements on the analog circuitry while the complex digital signal processing functions are normally done at lower sampling rates to enable the signal processor to do more functions. Translation between sampling rates requires the use of decimation filters to reduce the sampling rates and interpolation filters to increase the sampling rates. Decimation filters are used to decrease the sampling rates[7] and are low-pass filters which insure that the signal is filtered and no components will fold back to a lower frequency band. Decimation can be accomplished by eliminating samples, or setting the sample values to zero. For instance, a sampling rate reduction for two can be accomplished by dropping every other output signal of a filter assuming that no energy exists in that filter at frequencies less than 1/4 of the sampling rate. FIR filters are useful in decimation because the output of the filter (since it is only dependent on input samples) only needs to be calculated at the lower sampling rate and therefore signals processing requirements are reduced. An IIR filter, since it is dependent on feedback, (and previous output values) would have to be calculated at the input or higher sampling rate output would just not be used. Using FIR filters for decimation, the unused outputs do not have to be calculated. Interpolation filters are similar to decimation filters except that now the output sampling rate is higher than the input sampling rate. Interpolation filters are also low-pass filters that filter the resampled components that are inherent in sampled data systems while increasing the sampling rate. Interpolation filters use zero insertion techniques where the new values are set to zero and a filter is used to interpolate between the incoming signals. FIR filters are useful as interpolators because many of the sample values would be zero and do not have to be calculated when doing the interpolation. IIR filters, again because of the feedback effect have more non-zero terms and are more complex to calculate than the FIR filters.

IIR filters provide high Q filters with very steep roll-off characteristics with much less requirements in memory and computation than FIR filters. The IIR filter is the most efficient filter implementation in digital domain and can significantly reduce the amount of memory and signal processing required within the system. IIR filters require careful design to guarantee stability under all conditions. IIR filters also are not linear phase filters and therefore create group delay distortions. However, the group delay distortion can usually be corrected with an equalizer built out of an all pass section which does not affect the amplitude characteristic and corrects for the delay characteristic. Normally complex filters built out of IIR filters have a number of filter sections which provide the amplitude characteristic desired, and then additional sections which correct for the group delay distortion created by the filter that was required to generate the amplitude characteristics. IIR filters are almost always designed as cascaded second order sections. A second order section involves storing two previous sums and having two input samples available to calculate the new output value. Second order sections are easy to design, can be designed stably under all conditions and are easily cascaded to create any characteristic. IIR filters are similar in nature to traditional analog filters and standard filter types such as elliptic filters can be designed using IIR structures.

The high speed digital modems require a number of adaptive features that are implemented using adaptive filters. Adaptive filters vary from the normal filters in that the coefficients and the characteristics of the filter change to adapt to the telephone line characteristics to give the best possible output from the demodulator. The output of the demodulator is used to help calculate the characteristics required for the filter to improve the performance of the demodulator. Most adaptive filters are implemented as FIR filters because as the coefficients change, it is difficult to guarantee that IIR structures will be stable, while FIR filters are inherently stable. Adaptive filters require constant recalculation of the coefficient values used in the filter adding additional burden to the signal processor.

3.3 Digital Signal Processing Architecture

Digital signal processors are specialized arithmetic computers optimized to do multiplications and additions, as well as performing logical functions and comparisons. The major processing requirement in most signal processors is the ability to multiply a data value by a coefficient and add that to a sum. This implies some type of special structure to do multiplication.

Multiplication is achieved in one of three ways; a parallel multiplier structure, a shift and add structure or look-up table techniques. Parallel multipliers offer the highest level of performance (as far as speed is concerned) but require a large amount of silicon complexity. A parallel multiplier is an array of adders used to create the product of two input signals, in a single cycle. Most DSP processors for modems require 16-bit arithmetic and a 16 x 16 multiplier requires 128 adders and is a large complex structure. However, from the performance and programming point of view, multipliers are very desirable structures to include within the digital signal processing architecture. Shift and add multiplication is done by taking the multiplier one bit at a time and either adding the multiplicand to a product sum or not adding it and then shifting the multiplicand and examining the next bit of multiplier. Shift and add multipliers require n cycles to do an n bit multiplication and therefore are much slower than parallel multipliers but require much less silicon hardware. A compromise structure which does not shift one bit at a time, but only does additions where there is a one in the multiplier can be used to significantly reduce the computation time required for shift and add multiplier. Shift and add structures are slower, more difficult to program and less versatile than parallel multipliers but are less expensive to implement. Look-up table techniques can also be used to do multiplication. In a look-up table, a ROM is used and the inputs are the multiplier and the multiplican. Typically large ROMs are required to generate the multiplier values and therefore this tends to be an impractical solution for complex signal processing functions.

Fig. 13 shows a block diagram of a digital signal processor using a multiply structure. The multiplier is directly in the data path. Data values are stored in a data RAM, coefficients are stored in a coefficient memory and a sequencer is provided to control the operation of the multiplier, accumulator and memory. The data memory and coefficient memory can exist as separate memories or can be combined into a single memory. If a single memory structure is used, care has to be taken to enable information to be read in and out of memory in an efficient manner, or else the expensive multiplier will sit idle while data is being accessed in and out of memory. The bus architecture of the signal processor must allow flexibility so that data values can be input into the multiplier, into the accumulator or the adder and the output of the accumulator should be routable back to the memory, to the multiplier or directly back into the adder structure. Temporary buffer registers are often useful to signal processing architectures to store data from the accumulator to be used to input to the ALU ports, multiplier or memory so as not to limit the throughput to the system. The ALU must be able to perform certain logical functions such as comparisons (greater than zero, less than zero, equal to maximum values, or minimum values). These comparisons will be used in a number of signal processing blocks such as the automatic gain control, carrier detection and adaptive filters of the modem. The architecture is designed for high throughput rate and pipelining of data is important as the data memory is always random access memory, its values are constantly being updated and never have to be stored for a long period of time. The coefficient memory can be ROM, RAM or combination of both. Normally fixed coefficient filters have their coefficients stored in ROM and programmable filters have their coefficients stored in RAM. The programmable coefficients may be variable coefficients such as in the case of adaptive filters or non-variable coefficients that are loaded from an external source. If adaptive features are required, and the coefficients are recalculated and updated in the ALU structure, the data path from the accumulator to the coefficient memory must be provided to allow this updating to take place. Normally, a shifter is provided for scaling of the output of the ALU or accumulator before storage or reuse of the values. This scaling is required to optimally use the dynamic range of the signal processor and to provide the best performance with the highest signal to noise ratio possible.

A simplified architecture is shown in Fig. 14 that uses a shift and add structure instead of a multiplier. Although this structure is simpler in silicon implementation, it is more complex in concept. The multiplier is replaced with a programmable shift register. The coefficients are encoded as shift positions. The coefficients can be encoded canonic sine digit arithmetic and special techniques of minimizing the number of ones in the coefficient. Canonic sine digit arithmetic is a representation of a value using not only the values of zero and one but the values of zero, and ± 1 and minus one. Canonic sign digit representation can be used to minimize the number of ones used to represent a digital value. For instance, the value of 15/16 can be represented digitally as .1111 in binary representation, or in canonic sine digit arithmetic as $1.000\bar{1}$, where the $\bar{1}$ represents the value of minus 1/16 and the canonic sine digit number is equal to 1 -1/16 rather than the binary representation (1/2 plus 1/4 plus 1/8 plus 1/16). The shift and add multiplier structure is encoded so that the multiplier is shifted n number of positions representing the distance between successive ones in the coefficient and then the value is added or subtracted to the sum depending on whether the one is a plus one or a minus one. The efficiency generated by canonic sign digit arithmetic guarantees that the number of ones in an n bit representation is less than n divided by two. Special techniques have been developed so that most filters can be designed with fixed coefficients that have either one, two, or three ones in the canonic sine digit

480

High Speed Multiply Filter

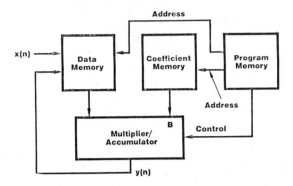

Figure 13

Shift and Add Filter

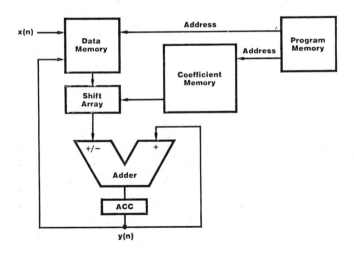

Figure 14

481

representation. This means that most multiplication can be done in less than three addition cycles and therefore can be implemented in an efficient manner using this type of architecture. The canonic sign digit shift and add structure works quite well when fixed coefficients are multiplied by data values, however, all modems require some demodulation with multiplication of two data values or a data value times some adjustable coefficient. If these are done infrequently, a traditional shift and add multiplication can be done using the same basic hardware structure. The shift and add multiplier requires that coefficients be encoded in a special technique and have a data path that goes into the shifting portion of the structure and not into the normal ALU data path. The shift and add structure is very efficient for fixed filters, but when adaptive filters are required, it is very difficult to calculate out new values, and create data paths that get the values in and out of memory. The shift and add structure requires similar logical operations as the multiplier structure for the digital signal processing system.

4. MODEM FUNCTIONS USING DIGITAL SIGNAL PROCESSORS

4.1 Types of Functions

Modems require a number of different signal processing functions to be performed. Filtering is a major portion of the modem signal processor. Modems, depending on the type, require fixed coefficient filters, programmable coefficient filters and adaptive filters. In the case of adaptive filters, not only must the filtering function be performed, but the calculation of the coefficient and the updating of the coefficient must be performed by the signal processor. Modem transmitters require a sine synthesizer within the modem as well as some kind of shaping characteristic implemented either in the baseband or passband. Modem receivers require signal processing functions such as automatic gain control, carrier detection, carrier recovery, timing recovery, demodulation and slicing. In this section, we will examine the various different signal processing requirements of a modem and how they might be implemented with a digital signal processor.

4.2 Modem Filters

Modems require a number of different filter types. Fixed filters are required throughout the modem to provide bandpass filtering of both the transmitter and the receiver as well as filter blocks for doing equalization of the nominal line characteristics, filters in the demodulator to remove unwanted information generated in the demodulation process, filters in feedback loops for phase lock loops used to do carrier recovery

482

and timing recovery and filters used as part of the automatic gain
control and carrier detection to smooth out and convert the modem
signal to a DC level which can be used in determining the value of
the incoming signal level. Fixed filters in modems are
combinations of finite impulse response and infinite impulse
response characteristics to achieve optimum performance. Many
functions within the signal processor will operate at different
sampling rates and decimation and interpolation filters can be
used to modify different sampling rates. The very high
performance filters are normally implemented using IIR techniques.
However, IIR filters, even with group delay equalizers in some
cases still have too much delay distortion at the band edges and
FIR filters are used in those places where group delay distortion,
is absolutely critical. Programmable filters differ from fixed
filters in that the coefficient filters are provided by the user
and stored in RAM and can be updated by the user. The computation
of the use of these filters is identical to that of fixed filters
and will not be considered further.

Adaptive filters require extra signal processing capabilities
beyond fixed filters. Adaptive filters require that the
coefficients be calculated during the training sequence and often
updated during the signal processing. Adaptive filters require
operation upon the coefficients and require that the coefficients
be able to be read out of the coefficient memory, processed and
rewritten into the coefficient memory. This puts requirements on
the data path relative to the coefficient memory. Adaptive
coefficients are recalculated by looking at some of the signal
outputs of the demodulator and comparing how close these outputs
are to the expected or ideal outputs and then using the difference
between the ideal and the actual outputs. The coefficients are
adjusted in a manner that try to move the results closer to the
ideal results. The coefficients are normally calculated in one of
two manners. The first manner is using LMS type algorithms and
multiplication of error signals values to create the coefficient
values. This type of processing requires complex multiplications
and complete multiplications without the use of short
coefficients. It is very difficult to implement in any structure
other than the parallel multiplier structure. Another method for
adaptive coefficients involves determining the direction the
coefficient needs to change (increase or decrease) and adding
small incremental deltas to the coefficient until it converges to
the proper value. This type of technique, since it only requires
the addition (or subtraction) of a fixed incremental change can be
implemented within an adder structure. In fact, because of the
encoding in the shift and add structure a special counter can be
constructed where the coefficient is incremented or decremented in
the counter rather than using an adder and reduces the complexity
of the signal processing requirements. In this type of scheme,
there is no need to take the encoded coefficient, convert it to a
normal data value, add the constant and then reconvert it back to
an encoded format.

4.3   Sine Wave Synthesis

Sine waves can be generated in a number of ways in a signal
processor. The most common way of digitally generating a sine

wave is to divide down a clock and use the basic frequency information of the signal processing clock to create the frequency information of the sine wave. A sine wave can be constructed from either a square wave or a triangle wave or any wave form whose period is equal to the period of the sine wave. The fundamental of that wave form will be the desired sine wave and the wave form contains additional harmonics of the fundamental which can be

filtered out to provide a pure tone if that is desired or necessary. The frequency division techniques have certain limitations because normal frequency division techniques only can generate frequencies which are integer dividends of the clock frequency. This requires a clock frequency that is very high to be used to generate the desired signal. Special circuitry in the form of programmable counters are used to generate tones in this manner. A signal processor can be used to generate a triangle wave by generating a signal that goes between two limits that are detected by the ALU structure. The period is set by adding a constant to a sum where the constant is the difference between the two limits divided by the ratio of the sampling rate to the desired frequency. When the limit is reached and detected, the sign of the constant is inverted and it is subtracted from the sum until the lower limit is generated. A linear triangle wave is therefore generated which has a period dependent on the sampling rate and the constant added and different frequencies can be generated by adding different constants to this generator. A triangle wave can then be filtered by a digital filter to provide a sine wave.

A third technique for generating a sine wave in a digital signal processor uses a phase accumulator and a look-up table ROM.[8] Sine waves have linear phase relationship so an accumulator stores the phase angle of the sine wave and a phase increment is added at the given sampling rate that is equal to the change in phase angle during that period. When the phase angle reaches 360° the sum flips back to zero. The phase angle is then used as an address to a ROM which stores the value of the sine wave. The phase angle could also be used as the input to a Taylor-series expansion to calculate out the value of the sine wave at that phase angle. The look-up table ROM is extremely efficient because since sine waves are periodic and symetrical there is only only need to store look up values for the first quadrant of the sine wave and then by manipulating the two most significant bits (to determine which of the four quadrants of the sine wave is represented by the phase angle) and using simple logic, the other three quadrants can be generated from the first quadrant input signal. This technique gives extremely high frequency accuracy and very low harmonic distortion sine waves.

4.4 Demodulation

Modem demodulators are the most complex section of the modems. All demodulators require AGC, carrier detect and some sort of multiplication. Most modems require some fixed phase shift characteristic to be generated, carrier recovery, timing recovery and equalization.

Automatic gain control is required in modems to remove the amplitude variations from the telephone lines. FSK and DQPSK modems do not have any information conveyed in the amplitude and in order to keep the performance of the modem optimal (the signal to noise as low as possible), the output of the bandpass filters is scaled, using automatic gain control, before being input to the demodulator. The AGC block detects the energy level of the signal (the output of the bandpass filters) and increases the gains to a desired level. Simple AGC can be done just by shifting the signal in 6dB increments. This is acceptable for FSK demodulations. DQPSK demodulation requires a more continuous AGC with smaller step sizes in order to ensure good performance. Energy detection is normally done by either squaring the signal or full wave rectifying the signal and then low pass filtering the results to measure the DC component of the signal. Full wave rectification is accomplished in the signal processor by inverting the signal when negative samples are sensed. Squaring a signal requires a full multiplication of the input signal times itself. In both of these processes, the frequency spectrum is doubled and care must be taken in the signal processor to ensure that the sampling rate is more than four times the maximum frequency outputted by the digital bandpass filter, or else signals will fold back into other bands during the squaring/rectification process. The low pass filter on the rectified/square signal then measures the DC energy and based on the energy, the AGC coefficient can be determined. The AGC function can be performed directly on the output of the bandpass filter and the energy measurement can be used as the input calculating the AGC coefficient. The AGC coefficient can be derived from a look-up table, or can be calculated using the energy value. Another possible strategy is to provide AGC and measure the energy on the output of the energy circuit and then adjust the coefficient to keep the AGC output constant. In this case, the output energy is used as an update to modify and correct the coefficient in the multiplication for AGC.

Carrier detection requires similar circuitry as automatic gain control. The energy of the system must be measured and accurately compared against limits to determine carrier detect. The carrier detect can use the same rectification/squaring circuit as the automatic gain control but requires different circuitry to make decisions. The automatic gain control coefficient, if it is continuous, can be used and compared to carrier detect limits to determine if the carrier is detected. Carrier detection is normally accomplished by measuring the energy level and subtracting the threshold level from that level and comparing the results to zero. If the result is greater than zero, then the energy excedes the threshold level, if the result is less than zero, the energy is less than the threhold level. Threshold levels must be selected depending on whether the carrier has already been senseded or not due to the hysteresis requirement of

the system. A high threshold level is initially chosen until carrier detection has been accomplished. A bit is set within the logic of the system which then selects a different threshold level for future comparisons so that carrier detection will not go away until the signal drops to a lower level than that used for detecting the carrier initially. FSK and DQPSK systems have relatively simple carrier detection schemes. QAM systems require much more complex carrier detection because the amplitude varies.

This puts requirements on the low pass filter used on the output of the rectified/squared signal to make sure that the energy is accurately measured. Whereas rectification are squaring of both acceptable techniques in FSK and DQPSK carrier detection, QAM carrier detection normally requires that the signal be squared and not rectified to ensure better accuracy.

Coherent modems use a recovered carrier as part of the demodulation process. The recovered carrier provides a higher performance of demodulation because it is a clean signal which is used to multiply the incoming carrier and the resultant demodulated signal has less noise than if the incoming carrier where a delayed version of the incoming carrier since the incoming carrier is noisy. Carrier recovery gives approximately a 3dB improvement in signal to noise in the demodulation scheme and is essential for higher speed modems. Normal carrier recovery techniques use phase lock loops with a voltage control oscillator (VCO) to recover the carrier. The high speed modems have very accurately specified carrier frequency but due to the FDM nature of the public telephone system, and varying clocks in the system, the frequency of the carrier can be changed or modified during transmission of the system. Therefore, carrier recovery systems need to have a range over which they can recover the carrier. A typical phase lock loop system is shown in Fig. 15. This is a simple demodulator for the DQPSK modem system and the carrier recovery is part of a feedback loop where the output of the VCO (which is the recovered carrier) is generated along with quadature term. These signals are used to multiply the incoming carrier to generate quadature demodulated signals. These demodulated signals should have no components of the carrier remaining on them and only baseband information. If the output of the VCO does not match the incoming carrier there will be a time varying output of the I and Q low pass filters which is then fed into the loop filter which controls the input to the voltage control oscillator. A VCO is the heart of the phase lock loop and can be implemented as a digital voltage control oscillator. This implementation is done using some of the techniques described in the sine systhesis section for transmitter. The VCO can use a phase accumulation scheme where the phase increment is adjusted by the output of the loop filter to increase or decrease the output of the VCO or a triangle wave increment can be modified to also change the output.

The triangle wave output of the VCO would go through a low pass filter before being used as a demodulator input. Other techniques for doing carrier recovery include squaring of signals, low pass filtering and then recovering the carrier through complex signal process techniques.

Demodulation in modems is also shown in Fig. 15. Demodulation techniques differ depending on the modem. FSK modems commonly use either coherent and non-coherent demodulation schemes. The non-coherent demodulation schemes used in FSK modems would include limiting the signal and counting the time differential between zero crossings. This is be accomplished in the signal processor by examining the flag bits which are set by whether the sample is greater than or less than zero and triggering a counter when that bit makes a transition between greater than zero and less than zero. The counter would count the zero crossings and determine the frequency time between zero

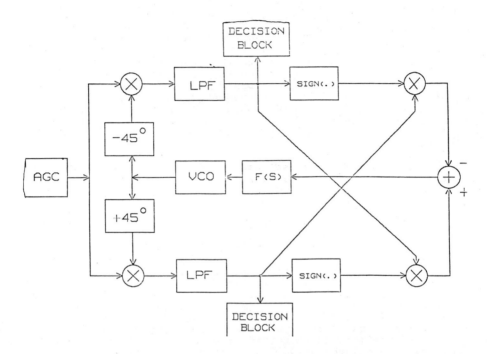

Figure 15

crossings. This scheme is subject to noise and is not the most accurate scheme for FSK demodulation. A more accurate non-coherent demodulation scheme uses an FM demodulation scheme where the incoming carrier is phase shifted or time delayed by a time which is equivalent to a 90° phase shift at the center frequency between the mark and the space frequency. If the phase shifted signal is multiplied by the incoming signal and low pass filtered, the remaining signal has a DC component which is proportional to the incoming signal and is positive that the incoming signal has a frequency less than the center frequency and is negative if the incoming signal has a frequency greater than the center frequency. This technique will be discussed in more detail in Section 6. A coherent FSK demodulator would use a phased lock loop scheme to recover the carrier and the input to the VCO would determine what the incoming frequency is. If the VCO is set up so that a zero output is equal to the center frequency between the mark and the space, then an input which drives the frequency less than a center frequency would imply that the incoming frequency is less than the center frequency and likewise a signal that makes the frequency greater would imply an incoming frequency that is greater than this signal is a demodulated output of the modem. The coherent demodulator is the most accurate FSK demodulator.

High speed modems all use coherent demodulation and carrier recovery. The demodulation is always done by separating the in phase and quadature terms and processing them separately. The generation of in phase and quadature terms can be done using Hilbert transform filters or by generating two outputs from the VCO that are 90° apart in phase. Since a digital VCO is available, it is very easy to accurately generate the 90° phase shifted signal. If a phase accumulator is used, then two values are read out of the look-up table whereas if a triangle wave system is used, a phase shifting filter with high accuracy can be used to provide the 90° phase shift. When the incoming carrier is multiplied by the output of the VCO at the carrier frequency, the products have baseband signal information and information at twice the carrier frequency. The low pass filters after the multiplication remove the unwanted information leaving only baseband information relative to the in phase and quadature components. This information is then provided to the decision devices. Often the information is first processed through an adaptive equalizer to equalize in the baseband for distortion created by the telephone line and other parts of the system.

Adaptive equalizers used for baseband equalization in the modem demodulator are traditionally transversal equalizers that are very good at removing intersymbol interference. The equalizer coefficients are calculated by examining the inputs of the decision devices and the output decisions that are made. The inputs of the decision devices compared to what ideally would be expected for such a decision and new coefficients are calculated using LMS type algorithms to reduce the error in the signal prior to the decision device. Adaptive equalizers are complex structures that must adapt fairly quickly on initialization, but must slowly adapt during normal operation. Telephone lines are often hit with transient conditions such as phase hits, amplitude hits and line drop outs. The adaptive equalizer should not respond too quickly to these impairments and should be able to restore itself to its original value quickly after the removal of the impairment.

The output of the demodulator goes into decision devices that decide on the digital value of the signal. In the case of the asynchronous FSK modems, the decision device must constantly be looking for the frequency change because there is no knowledge of when the change is going to occur. Many modem systems operate at low sampling rates in order to allow the complex signal processing to take place. However, examining low sample rate signals creates uncertainty in the instant of when the frequency changes. Therefore, FSK modems require some interpolation of the output of the demodulator before a decision is made. In a FSK modem, the decision is made by looking at the sign of a DC value by (either the input to the VCO in the phase lock loop or the output of the product demodulator in the FM demodulation scheme). Since only the sign bit is examined, the decision circuitry is relatively simple. In a PSK modem, the decision is made by looking at the sign of the information from the I and Q paths to make a decision on the phase of the signal and the phase of the previous signal must be stored within the system. Because the higher speed modems are synchronous, the exact instant of decision making is known. In order to determine this instant, timing information must be recovered from the signal. Timing information is inherent in the signals and is embodied in the shaping that is done on the modem signal. Timing recovery can be done using phase lock loop techniques to phase lock onto the timing signal and recover the timing clock. The baseband signal does not really have a component at the baud rate in it and therefore, some non-linearity needs to be introduced into the signal prior to the phase lock loop. This is normally done by introducing a delay equal to one half of the baud interval and multiplying the delay signal by the undelayed signal to generate a signal which does have components at the baud interval. This is used as the input to the phase locked loop. The recovered timing signal is used to make the decision at the maximum or optimum point in the baseband signal spectrum. The decision is made at the optimum point which is at the peak of the shaping characteristic. The decision should not be made near the points in the shaping characteristic where the amplitude is very low.

5.   A TO D AND D TO A CONVERSION FOR DIGITAL SIGNAL PROCESSING

5.1   Oversampled A/D Converters

The A to D converter used as a front end of the digital signal processing system must meet several stringent requirements. The sampling rate of the A to D converter should be as high as possible to simplify the antialiasing filter requirements that precede it. The sampling rate must exceed 8 kHz, because the information coming in to the modem can be as high as 3 kHz. However, an 8 kHz sampling A/D converter would require a very complex antialiasing filter and would defeat some of the purposes of having a digital signal processor. To use simple antialiasing

489

filters, the sampling rate must be in excess of 64 kHz. At 64 kHz, a three pole antialiasing filter is required, and this is still more complexity than normally would be desirable in a digital technology. Sampling rates above a few hundred kilohertz are required in order to simplify the antialiasing filter to be a simple non-critical RC low pass filter. Over sampled A/D converters can benefit from the fact that there are many more samples generated than required by the bandwidth constraint of the incoming signals. Several over sampled techniques are available which reduce the requirements and complexity of the analog circuitry.

Traditional A/D converter designs, using successive approximation techniques are really not viable for digital signal processing modems without complex antialiasing filters. Successive approximation A/D converters built onto complex chips can digitize signals in the range of 8 khz to 64 kHz, but are not useful for signals greater than 64 kHz. Successive approximation converters require very accurate digital analog converters to get high performance. These digital analog converters are very difficult to build in LSI technologies.

The A/D converter in the modem system must operate on analog signals that can vary over 40 dB range and must provide signal to noise ratios of 30 to 40 dB after the conversion (QAM modems, where the amplitude of the signal varies an additional 12 dB). This requires the A/D converter to have between 70 and 90 dB dynamic range depending on the type of modem. The A/D requirements can be relaxed if the user has an analog filter in front of the A/D converter to remove the adjacent channel signal that is present on the line and an analog AGC circuit to then take the received carrier signal and adjust it to a constant level. This is a very complex analog signal processor and one of the goals of designing DSP modems is to eliminate the complex analog circuitry. In order to do this, high performance, simple A/D converters need to be developed for a modem system. Two such converters are delta modulators and interpolative converters.

Delta modulators are a traditional class of over sampled A/D converters. A delta modulator is a very simple feed back loop to generate quantized approximation of the input signal and compare it to the analog input signal. The difference signal is fed to an analog comparator and the output of the comparator causes the quantized signal to increase or decrease in magnitude to track incoming signals. The output of the delta modulator is a one bit code at a high sampling rate. Delta modulators have been used in signal processing systems for many years and many improvements and different types of delta modulators are available including sigma delta modulators, continuously variable slope delta (CVSD) modulators and adaptive step size delta modulators. Delta modulator devices require very high sampling rates relative to the signal rates. Simple delta modulators for handling voice band signals run as high as several megahertz in sampling rates while CVSD systems running at 64 to 128 kHz provide reasonably good performance on voiceband signals. New techniques in delta modulation include modifying the feedback path filters for generating the quantized approximation of the input signal and have very high levels of performance. An improved class of converters used on single-chip DSP modems and based on delta modulation systems are called interpolative converters.

The interpolative converter[9,10,11] is a class of sigma delta modulators where the quantized approximation of the input signal is subtracted from the input signal and the different signal is integrated. The output of the integration is then compared to make a decision on whether increasing or decreasing the quantized values is required. The interpolative coder is shown in Fig. 17. Unlike traditional delta modulators where the quantized approximation to the input signal is generated in analog means, the interpolative coder takes the output of the comparator and digitally processes it to generate a digital word which is used as the input to a D/A converter that generates the quantized output. The interpolative encoder uses an n bit digital analog converter in the feedback loop but only provides plus or minus n levels out of that converter. The filtering algorithm used in the feedback loop causes the output of the D/A converter to change at every sample instance, and restricts the changes to three possible changes. The output can double in value, be halved in value or change sign. These are the only possible transitions and therefore can be implemented with a simple shift structure that shifts the input to the D/A converter one position left, one position right or changes the sign bit of the D/A converter. The algorithm used for determining the next state of the quantized output has to make decisions on when to change the sign bit. The A/D converter only uses six of the levels out of a D/A converter or any one cycle. A counter counts the maximum level used, and when the signal is six levels below the maximum level, instead of decreasing the amplitude any further, the sine bit is inverted. A signal at a low clock rate is sent to count down the peak value at a random rate to ensure that amplitude changes in the incoming signal can be tracked. This scheme allows a wide dynamic range withing running into traditional slope overload problems. Large amplitude signals only use the largest six levels of the D/A converter and small step sizes around Ø which do not contain significant information are skipped over allowing fast transitions through Ø. As the signal amplitude decreases, smaller step sizes are used. The A/D converter runs at a highly over sampled rate and the integrator forces the long term average value of the integral of the difference between the incoming signal and the quantized signal to be Ø. The quantized signal is available as a parallel digital word and with low pass filtering and averaging techniques, the high frequency noise generated can be removed and the fundamental signal restored. This type of A/D converter has been built in silicon and commercially used on several products[1,2,3] and has achieved performance of 9Ø dB dynamic range and a signal to noise ratio of 66 dB for large amplitude signals. This performance is more than adequate for FSK and DQPSK modems but is slightly lower signal to noise than is required for high performance QAM modems. Therefore, QAM modems need an improved version of the A/D converter or some analog processing in advance of the A/D converter.

The interpolative A/D converter has a relatively simple analog structure which can be implemented in digital MOS technology without additional processoring. The comparator is a sampling comparator and can be easily implemented using standard MOS comparator designs. The op-amp (used in the integrator) is not a very critical op-amp, but should have relatively good slew rate (2 to 3 volts per microsecond) and should have symmetrical

Interpolative A/D Converter

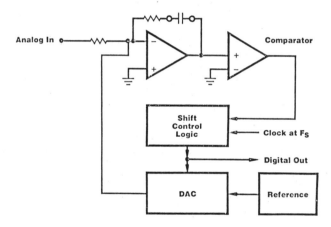

Figure 17

slew rates for positive and negative transitions. With the feedback loop tracking the input signal, the changes in the output of the integrator amplifier are not very large and therefore the amplifier does not have to slew at a very fast rate. The digital to analog converter used in this design is an R2R ladder built out of MOS transistors using the on resistance of the MOS transistor as a resistance. No critical resistors or capacitors are used in the entire A/D converter design. The shift register input into the digital-to-analog converter is a shift register where a series of ones is shifted in from right to left. When a new one is shifted in, the value of the register is doubled. When a zero is shifted in, from left to right, the value of the register is half. With this technique, the D/A converter magnitude changes are accomplished by turning on one bit switch or turning off one bit switch and there is no glitch content in the transition between adjacent levels. The D/A converter accuracy requirements are also reduced because each level and the adjacent level which would be twice as high, have all the same current switches on except for one. The only critical analog matching is that adjacent should have a ratio of two to one and by having the lower level current included within the higher level output current the matching of the analog components is reduced by a factor of two. The A/D converter works well with current outputs that match to 1%. The over sampled nature of the A/D converter spreads the spectrum of randomly spaced noise throughout the entire sampling rate of the A/D converter. Low pass filtering and decimation to reduce the sampling rate after A/D conversion down to a lower sampling rate prior to complex digital signal processing removes most of the noise generated in the conversion process. Every doubling of the sampling rate reduces the noise in-band by 3 dB. This reduction in noise greatly relaxes the requirements of the analog components in the A/D converter.

5.2 Over Sampled Digital to Analog Converters

Digital to Analog converters used as outputs for modem transmitters can be implemented with traditional D/A converters with parallel inputs and voltage outputs. The D/A converter could be similar to the D/A converter described in the feedback path of the interpolative A/D converter or current source output D/A converters can be used. Some digital signal processing systems convert digital words into delta modulation codes and use a simple delta modulation feedback loop (receiver) to generate analog output signals. This feedback loop is a very simple system which is the analog to the feedback path in the delta modulator encoder. It achieves similar signal to noise performance as delta modulator. This requires high speed signal processing to generate the delta modulated symbol. Most DSP modems have simple enough requirements that a standard DAC can be used. Care must be taken to make sure that the DAC has low second harmonic distortion and does not add noise to the system. Second harmonic distortion is the most important parameter because many low band transmitters have receiver bands that include the second harmonic of the transmitter.

493

6. AN FSK MODEM IMPLEMENTATION

6.1 Modem Description

An FSK modem that uses DSP techniques is the Am7910[12,13,14] produced by Advanced Micro Devices, Inc. This single chip modem uses digital signal processing to achieve a device that meets four different modem standards. In developing the Am7910, some key goals and objectives were established. The overriding goal was to produce a single-chip modem which contains a high level of performance and a low price. This involves including on-chip as many as possible of the functions and features required to integrate the modem into the final system. In addition, it was important to build a modem which could be used in as many applications world-wide as possible.

Among these items are the inclusion of:

1) pin programmable selection of all of the widely used FSK modem specifications (CCITT V.21, CCITT V.23, Bell 103, and Bell 202). These four modem specifications may be operated in 9 different data mode configurations,

2) all traditional modem functions such as modulation, demodulation, and filtering. It was important that the modem be a true single-chip solution requiring that no critical external components be supplied by the user - particularly if they were required for implementation of the traditional modem functions,

3) all analog-to-digital and digital-to-analog conversion on chip,

4) the traditional RS-232-C/V.24 functional interface for modem control; this allows the user a simple interface to the DSP modem which is compatible with other modems with which he may already be familiar,

5) the ability to perform loopback testing of the modem for all of the modes in 1) above; loopback mode provides ten additional operational configurations for testing the nine data modes for a total of 19 different operational modes,

6) the ability to assist the user in an auto-answer mode for automatic answering applications; this relieves the user of the need to take an active part in the automatic answering process. It was also important that the end user for the modem not be required to possess knowledge about either modems in general or the PSTN. The digital logic designer should be able to design the traditionally analog modem function into his system without additional knowledge. Continuing along this line, the modem should require as few external parts as possible and no active external parts. A crystal and a single resistor and capacitor meet this criterion.

6.2 Modem Implementation

Fig. 18 shows a functional block block diagram of the Am7910 FSK modem. Notice that the transmitter and receiver are shown as separate blocks in the diagram. This is how they are implemented in silicon. The transmitter and receiver each contain their own instruction ROM, RAM, and ALU for performing the signal processing. There is no interaction between the two processors. The interface block controls the operation of both the transmitter and receiver as a function of the programming of the mode control pins provided by the user, and the states of the handshake (RS-232-C or V.24) pins. The transmit data provided by the user is a direct input to the transmitter, and the demodulated data is provided as an output directly from the receiver.

Fig. 19 shows a block diagram of the transmitter signal processing blocks. The frequency of the carrier to be generated is a function of the modem selected on the mode control pins and the state of the transmit data (TD) pin. The phase constant selection, serial adder, and phase-to-amplitude conversion blocks comprise the frequency-shift sinewave generation block.

The phase accumulator technique of sinewave generation is used in the Am7910. Hence, in generating a sinewave, a phase constant is added to the last phase value stored in the phase accumulator (implemented using a serial adder). This highly sampled signal must have its sampling rate reduced (decimated) down to a reasonable sampling rate (7.68kHz) prior to bandpass filtering. This is accomplished with a series of low pass FIR filters. If filtering was performed at 122.8 kHz, the speed requirements of the ALU would be too great to allow any significant signal processing to be performed. The bandpass filter is an 8th order IIR eliptic filter.

After bandpass filtering, the FSK waveform is interpolated up to a high sampling rate again (122.8 kHz) for output to the DAC. The interpolation is performed to allow a simple analog low pass filter to be used as the reconstruction filter on the analog output. If the output to the DAC was provided at the low sampling rate of 7.68 kHz at which the ALU operates, the analog filtering requirements of the reconstruction filter would dictate a high order filter in order to eliminate harmonics which are generated at multiples of the sampling rate. Additionally, the higher sampling rate eliminates the need for any sin x/xcorrection.

Fig. 20 shows a block diagram of the DSP modem receiver. The analog low pass filter eliminates energy which may alias back into the passband after sampling by the analog-to-digital converter. Because of the all digital nature of the Am7910, the received signal is converted to a digital format prior to entering the remainder of the signal processing chain. Similar to the operation of the transmitter DAC, the receiver ADC operates at a very high sampling rate (496 kHz). This lessens the filtering requirements of the analog pre-filter. In fact, the pre-filter is a single pole (-6dB/octave) RC lowpass filter.

495

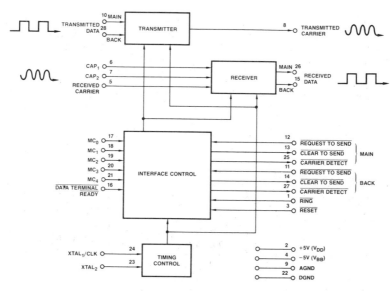

Am7910 Block Diagram

Figure 18

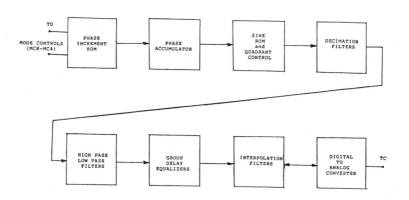

Am7910/11 MODEM TRANSMITTER BLOCK DIAGRAM

Figure 19

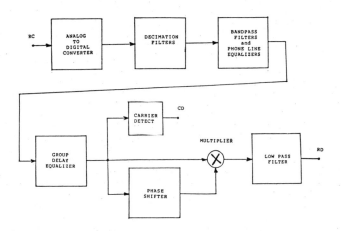

Am7910/11 MODEM RECEIVER BLOCK DIAGRAM

Figure 20

The decimators (low pass FIR filters) reduce the sampling rate from 496 kHz to the nominal 7.68 kHz signal processing rate used by the receiver signal processor. Decimation needs to be performed within the receiver for precisely the same reason that it was performed within the transmitter - to reduce the sampling rate in order to allow more signal processing to be performed. Note that the decimation required to reduce the sampling rate from 496 kHz to 7.68 kHz is 64. This decimation by 64 is performed half in hardware and half in microcode within the DSP modem. The first decimation filter (decimator for 496 kHz to 62 kHz) is performed in hardware and the second decimation (62 kHz to 7.68 kHz) is performed within the microcode.

After decimation, the received signal is filtered to remove the out-of-band energy which may pollute the demodulation process. The bandpass filtering is performed entirely within the receiver ALU. The filters are cascaded second-order IIR stages which are equalized after filtering. After filtering, the automatic gain control (AGC) block ensures that the received signal is of the proper signal strength to perform the demodulation. A low level received signal is amplified to a higher level to maintain the largest dynamic range possible within the ALU.

The Am7910 demodulator known as a product demodulator. The major advantage of this type of demodulator is that it is less susceptible to phase (group delay) distortions present on the telephone network relative to other types of demodulation schemes. It consists of phase shifting the incoming signal by 90° at the center frequency between the two frequency shifts and multiplying this phase shifted signal with the incoming signal. The result is a baseband term containing the desired digital data information, and a double frequency term containing no useful information. This double frequency term is filtered out by the low pass filter following the demodulator.

The interpolator has the job of accurately determining where the zero crossing transitions of the demodulated and lowpass filtered signal occur. These zero crossings determine changes in the ste of the data. The zero crossings should be accurately defined to reduce the amount of jitter observed in the received data waveform. The signal processor provides only 7680/300 (sampling rate divided by bit interval) = 25.6 samples per bit interval out of the low pass filter. If a zero crossing occurs between two of these samples, the jitter out of the modem contributed by the decision process alone would be about 4% (unacceptable) without interpolation. (Jitter is defined as the deviation from the ideal change of state instant by the received data out of the modem.) Jitter may be contributed by many sources in the system, but jitter should not be introduced by a source such as the interpolation/decision process which should inherently provide zero jitter.

The interpolator divides the interval between signal processing samples into eight equal smaller samples. If a zero crossing occurs between signal processing samples, the interpolator chooses which of the eight smaller intervals lies closest to the zero crossing. Hence, the interpolation by eight reduces the percentage of jitter contributed by the decision process by a factor of eight to about 0.5%. The decision process

498

then is simply a comparison process to change the state of the received data when a sign change is detected from the interpolation block.

The interface control is a state machine which provides the control to the transmitter and receiver as a function of the programming by the user. The five mode control pins select the operational mode of the modem, while the handshake pins control the dynamic modem operation. The handshake pins provided are: $\overline{DTR}$, main and back channel $\overline{RTS}$, main and back channel $\overline{CTS}$, main and back channel $\overline{CD}$, main and back channel TD, and main and back channel RD.

The function of the decision block is simply to restore the sharp edges to the demodulated waveform. The decision block samples the output of the interpolator. If a zero crossing is detected, the state of the received data pin is changed to reflect the new data. The carrier detection block within the DSP modem determines if the received signal contains sufficient energy to indicate a valid carrier. If a valid carrier is detected, the $\overline{CD}$ pin will turn ON, otherwise it will be turned OFF. If a valid carrier is not being received, the remainder of the interface pins are either ignored if they are input pins or set to an appropriate initialized logic state if they are output pins.

6.3 FSK Modem Circuit Design

The Am7910 has a dual processor architecture; both the receiver CPU and the transmitter CPU process data independently at 2.4576 MHz. The two central processors have an identical architecture which is shown in Fig. 21. The architecture consists of a two-port ALU, three data buffers, an ALU result bus, and a bidirectional bus for access to data RAM. All data operations are performed in one's-complement arithmetic. After data is read from RAM onto the bidirectional bus, it can be read into the B port of the ALU or into the data buffer. The data buffer has three output paths: one to the write buffer, and one each to the A and B ports of the ALU. The A port has a shift register which is controlled by three bits of the microcode word. Data at the input ports of the ALU is added or subtracted on every instruction cycle. The result of this arithmetic operation is latched onto the ALU result bus. The ALU result bus has return paths to the A port of the ALU and to the bidirectional bus. In addition, the ALU result can be transferred into the temporary buffer. The timing of the ALU bus and the temporary buffer is such that results can be passed through the buffer to the ALU B port on a single microcode cycle. Data can also be read onto the bidirectional bus from the write buffer.

By using the same filter and demodulator structures for every modem type on the Am7910, it is possible to use the same microcode instruction set in every case. Only the filter coefficients are changed from modem type to modem type. Two microcode ROMs are employed on each CPU. The first ROM contains the microcode for the filter and demodulator structures, while the second ROM contains the filter coefficients for each modem type. By

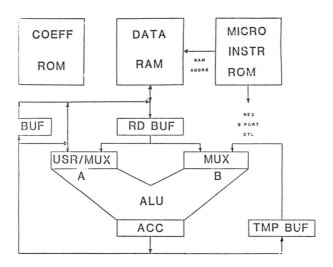

Figure 21

separating the microcode into two ROMs, the Am7910 is able to operate in nine different modem configurations with only 10% of the die dedicated to the modem type-dependent coefficient ROM. Some microcode blocks such as the decimators, interpolators, and multiplier are common to all modem types. When the address of these common blocks is decoded, coefficients for the filters are not chosen from the coefficient ROM; they come from an extra set of coefficients stored in the instruction ROM. This technique of coefficient selection reduced the size of the coefficient ROM by approximately 100 words per modem type. Because of the limited number of microcode instructions, it was necessary to place two of the receiver decimators in a separate ALU from the main CPU. The hardware decimators could be implemented because of the simplicity of the coefficients for the two functions. Samples from the decimator 1-2 ALU are strobed onto the bidirectional bus by the main CPU at a 61.44 kHz rate.

A single multiply was necessary to demodulate the input receive carrier. The basic DSP CPU does not have the capability to do a multiplication since its filter coefficients are stored as Shift-and-add codes in ROM. For this reason, some hardware modifications were made to the receiver CPU to allow multiplication. By using a modified Booth multiplication algorithm and controlling the multiplier with strobes derived from unused RAM locations, an eighteen-bit product was accumulated from two nine-bit operands.

Another interesting hardware feature of the Am7910 receiver CPU is the output interpolator. If the output of the signal processor was sampled at the 7.68 kHz rate of the CPU, then the receiver output would have 15.5% jitter error at a 1200 bps transmission rate. To reduce this error, it is necessary either to increase the sample rate by interpolation, or to simulate the interpolation process in hardware. On the Am7910, the latter option was chosen. By linearly interpolating between the amplitude of 7.68 kHz samples, it is possible to reduce the theoretical jitter error to less than 2%. When a sign change is sensed in the output of the receiver low-pass filter, the difference of the two amplitudes is taken. This difference is divided by 8. The fractional difference is added to the last LPF output sample up to eight times, and an interpolator strobe clocks the sign of the resulting addition into an output shift register. The interpolator strobe signal is created by addressing another non-existant RAM location.

The method of sine wave generation on the Am7910 employs two ROMs, a serial adder, and an accumulator external to the transmitter CPU. The mode control pins and the transmit data pin select a value in a 28 word phase ROM. The 11-bit output of the phase ROM is serialized and fed into the serial adder. The other input to the serial adder comes from the phase accumulator. The sum of the phase constant and the phase accumulator is stored in the phase accumulator. The accumulated phase is used to address a sine ROM which converts the phase to an amplitude. Due to the symmetry of the sine function, it is necessary to store amplitudes for only one quadrant of a sine wave. The two most significant bits of the phase accumulator are used to control the quadrant function of the sine wave. The output of the sine wave generator is strobed into the transmitter CPU by setting a microcode bit.

501

Approximately 15% of the die area was dedicated to test logic. The RAMs, ROMs, ALUs and analog sections are separated into components for test procedures. The test logic allowed each component of the Am7910 to be tested individually without resorting to time consuming system tests.

All of the circuit functions developed external to the main CPU structure would have added a large amount of unnecessary circuit complexity. By tailoring each function to work with the CPU, it was possible to implement the system without an unwieldy circuit.

6.4 FSK Modem Algorithms

The Am7910 system design called for both IIR and FIR filters. The IIR filters were to be implemented as standard direct form II second-order sections, while the FIR filters were simply linear phase filters developed using the McClellan-Parks algorithm.[15] Each IIR second-order section requires two memory locations and has five coefficients. One memory location and one coefficient are required per filter order for the standard ladder implementation of the FIR filters. The memory locations are allocated from data RAM on the CPU, but the filter coefficients are stored in the instruction and coefficient ROMs of the CPU. The multiplication of data can simulated by adds and shifts in the ALU, but the Am7910 also has the ability to represent coefficients by subtraction. A canonic sign-digit representation was chosen for the coefficients in ROM. Canonic sign-digit allows grouping of bits in a binary representation to be replaced by one addition and one subtraction. An example of an eight-bit coefficient is given below in both binary and canonic sign-digit.

$-0.11100111_2 = -0.90234375_{10}$

in canonic sign-digit :

$-1.00\overline{1}01001 = -0.90234375_{10}$ where $\overline{1}$ in canonic sign-digit means subtract

on the Am7910 :

$-2^0(1-2^{-3}(1-2^{-2}(1-2^{-3}))) = -0.90234375_{10}$

From the example above, the exponents are the shift codes for the ALU. Note that the binary representation would have required six add-and-shift operations, while the canonic sign-digit representation takes only four operations.

Another important algorithmic consideration was the product demodulator. The concept behind the product demodulator is fairly straightforward. The signal into the demodulator is phase shifted by some amount and then multiplied by itself. The phase shift is determined by the frequency centered between the two modulating frequencies. At the center frequency, the phase shift is exactly $90°$, yielding a zero product out of the demodulator. At frequencies below and above the center frequency the phase shift is less than and greater than 90° respectively. The equations below illustrate the principle using a simple sine wave input. Assume q is the phase shift angle provided in the demodulator.

$$[A \sin(wt + p)] \; [A \sin(wt + p + q)] =$$

$$A^2/2 \; [\cos q - \cos q \cos 2(wt + p) + \sin q \sin 2(wt + p)]$$

The higher order frequency terms are filtered out by the receiver low-pass filter leaving only the $A^2/2 \cos q$ term as a digital output.

6.5 FSK Modem Performance

Several performance axes are important to a modem designer, but two of the best performance indicators for FSK modems are bit error rate (BER) and isochronous distortion. Bit error rate is a measure of the number of bit errors divided by the total number of bits sent. By varying the power level of the signal in the adjacent frequency band and the power level of the noise in the channel, the BER can be determined over a wide variety of simulated telephone conditions. Amplitude and phase hits as well as shifts can occur in the network environment. These conditions are also simulated on the bench for BER performance measurements. Typical BER curves (shown in Fig. 22) are plots of probability of error versus signal-to-noise power in the channel. The graphs show the BER performance of the CCITT V.21 (300 bps) modem types at an input signal level of -24 dBm and an adjacent channel level of -9 dBm.

Isochronous distortion is a measure of the difference in transition edges between the transmitted digital data and the received digital data. This distortion is unique to asynchronous transmission. Isochronous distortion is given as a percentage deviation of the received data period compared to an entire baud interval. Two components make up isochonous distortion: jitter distortion and bias distortion. Jitter is a measure of the absolute latest and earliest transitions in a receive data stream when compared with the transmit data. A skew towards one logic level or another in a data stream is bias. In the lab, isochronous distortion is determined by sending synchronous data through the modem and then measuring the actual transitions with a clock that is running much faster than the transmit data. Isochronous distortion numbers for the Bell 103 modem types are tabulated below. This performance was measured with a -24dBm input signal.

modem type	adjacent channel signal	
	none	-9dBm
Bell 103 ANSWER	11%	16%
Bell 103 ORIGINATE	10%	20%

503

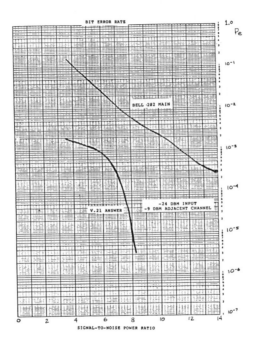

BIT ERROR RATE PERFORMANCE

Figure 22

One of the most important features of DSP and digital
components in general is their high tolerance of environmental
conditions.

The Am7910 distortion performance varies by a very small
percentage over the entire operating temperature and power supply
range. The transmit carrier level is held to within 0.5 dB of the
nominal -3 dBm level over all temperatures and power supply levels
on any given chip. The chip-to-chip variation is no more than
0.7 dB from nominal, which allows most customers to set their
transmit carrier level with a few fixed-value resistors. The
frequency deviation of any tone generated on the Am7910 is less
than 0.5 Hz.

7. DPSK MODEM IMPLEMENTATIONS

7.1 System Design

There are presently no DSP modem chips available for DPSK
type modems. DPSK modems have been implemeted using general
purpose DSP chips and/or single chip microcomputers. Several
companies have single chip DPSK modems under development using DSP
techniques.

A block diagram of a DSP implementation of a Bell 212A (or
CCITT V.22) modem is shown in Fig. 8. This modem can be built
using a dual processor architecture such as the Am7910 or a single
processor architecture with a higher speed processor. A parallel
multiplier structure can be used if a single processor
architecture is desired. A shift and add structure can also be
used, if care is taken in the design because the number of
multiplication (other than by fixed, optimized coefficients) is
low. A number of different techniques could be used to design
this type of modem and one of those approaches will be discussed
in detail.

7.2 Transmitter Design

The Bell 212A modem includes a 300 baud FSK modem as a
fallback mode so it is desirable to design the DQPSK modem with
the fallback mode in mind. The transmitter of the modem uses this
philosophy.

The DQPSK modem transmitter is synchronized to a 1200-bit per
second ($\pm$.01%) clock after the asynchronous to synchronous
converter and does not need a very high sampling rate. However,
the FSK modem, which operates asynchronously requires a high
sampling rate transmitter to minimize the isochronous distortion.
The transmitter uses a high sampling rate sine synthesizer,
similar to the Am7910 and generates the phase shifts by doing an
extra addition at the baud rate dependent on the incoming dibit.
This addition adds a $0°$, $90°$, $180°$ or $270°$ phase increment to

505

phase accumulator to generate the DQPSK signal. The sine
constants for the FSK frequencies and DQPSK carriers (1200Hz and
2400Hz) are stored in the phase constant ROM and are selected by
the mode chosen and the input data (FSK only). The high sampling
rate signal is decimated to a lower rate (around 8kHz) where band
pass filtering and shaping are done. The baud rate shaping
characteristics is introduced in the passband to allow the same
transmitter circuitry to be used for the FSK and DPSK modems.
Baseband shaping of the signal will be discussed in section 8.
The bandpass filter uses 7 second order sections to achieve
shaping, attenuation of out of band signals and group delay
equalization. An additional section is added for a compromise
equalzer to compensate for telephone line characteritics. The
DQPSK modem requires less than 100us group delay distribution, so
care must be taken to minimize this distortion. The FSK
transmitter does not need as complex a filter but for simplicity
the entire filter is used and unneeded sections are set to unity
by using a coefficient of 1 for the input signal and zero for the
feedback coefficients. The filter output is interpolated back up
to a high sampling and outputted to the DAC where it is converted
to analog and is sent to the telephone line interface.

7.3   Receiver Design

        The DQPSK receiver (Fig. 9) uses a coherent demodulator
designed so that the FSK demodulator can be obtained as a subset.
The interpolative analog-to-digital converter provides more than
adequate performance as a front end of the receiver. Low pass
decimation filters, similar to the Am7910 are used to reduce the
sampling rate to the 8kHz range. The bandpass filter that follow
are similar to the transmitter filters and must provide at least
60 dB rejection of adjacent channel and unwanted signals. These
filters also include one half of the compromise equalizer, like
the transmitter. These filters have critical group delay
distortion specification and require precision signal processing.
The FSK filters are simplier and can be implemented by choosing
different coefficients. Analog implementations of Bell 212A
modems attempt to use the same filters for both the modem types
since the frequency bands of the FSK modem (1070Hz to 1270Hz and
2025Hz to 225Hz as 2400Hz) and DQPSK modems are similar. This
usually causes compromising the FSK performance. The DSP approach
allows optimized filters to be used for each modem type.

        The output of the bandpass filter goes into the AGC and
carrier detection blocks. The DQPSK has a more stringent
requirement on the AGC than an FSK modem so some form of
continuous AGC is normally used. This AGC can be accomplished by
measuring the output of the bandpass filter by squaring the
signal, low pass filtering the result (one second order IIR
section) and using the output to determine an AGC constant (from a
look up table in ROM). The AGC constant can be pre-endcoded with
a minimum number of ones if a shift and add structure is used.
The AGC constant can also be for comparison with stored constants
to determine carrier detect level.

The DQPSK demodulator is a coherent demodulator with a phased locked loop for a carrier recovery with the phase constant modified to adjust the frequency. The phase increment is the sum of a phase constant (at the center frequency) and a modifier. The range of modifier values determines the capture and lock range of the VCO. The VCO generates two outputs which are 90° out of phase by addressing the sine ROM twice using the phase accumulator output. The incoming carrier is multiplied by the two VCO outputs to generate the I and Q components of the baseband signals. These are processed summed and filtered to generate the VCO input. When the VCO tracks the incoming signal the sum of the terms should cancel. The FSK demodulator is a subset of this loop where only one phase from the VCO is required and only one multiplaction is needed. Care must be taken in selecting the sampling rate of the demodulator to guarantee that the double frequency terms generated by the multiplication do not fold back with the bandwidth of the low pass filters. The low pass filters remove the double frequency terms, leaving the baseband information.

The decision device of the DQPSK modem is very different from the FSK modem. The FSK modem examines the demodulated output (which is the input to the VCO loop) and looks for zero crossings of the signal. The asynchronous nature of the FSK modem requires that the signal be interpolated up to a higher sampling rate in order to minimize isochronous distortion. The DQPSK modem must recover a timing signal (600 baud) and use this to determine the decision timing. Additionally since the modem uses differential encoding it must compare the phase value to the prior baud period rather than to zero. The timing information is not directly obtainable from the baseband data but can be restored by processing the data. The baseband signal has information at one half the baud rate so that by squaring the signal the baud rate timing can be regenerated. This signal can be inputted to a simple PLL (similar to the carrier recovery loop) and the timing signal restored. The decision device then uses the timing signal to sample the demodulator output and compare to the last sample. A comparison of the sign bits of the I and Q outputs will result in the output dibit. The dibit can be descrambled and outputted to the system directly or through a synchronous to asynchronous convertor depending on the application.

7.4 DQPSK Implementation Issues

A DQPSK modem requires more DSP complexity than a FSK modem but it is pratical to implement such a modem on a single chip. The signal processing requirements for the transmitter are only slightly (about 10%) more complex while the receiver requires about twice as much signal processing. The receiver can be implemented with five low sample rate multiplications (two for demodulation, one for AGC and two for timing recovery) and therefore it is possible to use a shift and add architecture to complement a DQPSK modem.

507

8. QAM MODEM IMPLEMENTATIONS

8.1 System Design

QAM modems are much more complex than FSK or DQPSK modems and require very complex, adaptive signal processing. These modems are very sensitive to telephone line impairments and make extensive use of adaptive equalizers to correct for signal degradation due to the PSTN. Single chip implementation of a QAM modem has not been practical until recent advances in technology which now allow adequate processing power and memory to be placed on one chip. Whereas a DQPSK is one and a half to two times the complexity of an FSK modem, a QAM modem requires four times the signal processing power and three times the memory of a DQPSK modem. The adaptive filters cannot be practically implemented with any structure other than a parallel multiplier. The analog requirments of the QAM modem are also much more demanding than for lower speed modems and require higher performance than is achieved by the components developed for use by lower speed modems. Lower speed modems have been developed using almost all digital or all analog approaches but QAM modems probably require a true marriage of analog and digital technologies. Since there is no existing IC modem for QAM applications, the V.22 bis will be examined for possible implementation.

8.2 V.22 bis Transmitter

The V.22 bis modem is 16 point QAM modem that has a 2400-bit per second data rate and a 600 baud modulation rate. It is a superset of the V.22 modem which is a 1200-bit per second DQPSK modem. There is no 300 baud FSK mode required and this significantly supplies the mdoem design. The QAM modem transmitter is shown in Fig. 11. A quaduature type transmitter is used where the sine and cosine carriers are multipled by incoming data after encoding and shaping. The V.22 bis signal is encoded as 4 bits per baud while the V.22 signal has 2 bits per baud. The shaping filter generates separate (but related) baseband shaping signals for the sine and cosine terms and includes the raised cosine shaping filter. If these filters are sampled at a sample rate that is an integer multiple of the baud rate, then the filter shapes can easily be stored in a small ROM. The carrier signals (sine and cosine at 1200 Hz and 2400 Hz) can be generated with a phase accumulator/look-up table or if the sample rate is also an integer multiple of 2400 then they could also be stored in a small look-up table. In fact, the whole front part of the transmitter could be stored in ROM.

The output of the modulator is bandpass filtered and also filtered with one half of the compromise line equalizer, then converted to analog and fed to the telephone line. The QAM has a sophisticated traning sequence that requires four phases in order to allow the scramblers to synchronize, the equalizers to adapt, the timing and carrier recovery to achieved and modem type to be identified.

The training sequence is stored in a state machine that interacts between the transmitter and receiver. This machine provides input sequences to the transmitter and monitors the receiver outputs in order to decide how to procede. Training of QAM modems takes about 1 sec.

8.3 V.22 bis Modem Receiver

A simplified drawing of the V.22 bis modem receiver is shown in Fig. 12. This receiver is too complex to discuss in detail in this paper, however, a few highlights will be reviewed.

The analog front end requires more performance than the present modems achieve. This major problem is noise (in the convertors) generated by a large adjacent channel that falls into the receiver bandwidth. This is commonly seen when the transmitter uses the 1200 Hz carrier and second harmonics are generated in the 2400 Hz band. Three alternatives are being explored - improved A/D convertors, simple analog filtering, and echo cancellation. The A/D convertor can be improved by analyzing the causes of second harmonic distortion and correcting them. It appears that most of the second harmonic distortion in the interpolative A/D is caused to be three factors - asymetrical op amp slew rate, asymetrical levels for positive and negative inputs and offset voltage induced non-linearities. Each of these effects can be significantly redcued by new design techniques and auto calibration methods and an improved A/D convertor is possible. A second alternative is to reduce the adjacent channel signal seen at the A/D input by using a simple analog filter. The filter needs to reduce the signal by 10-20 dB and does not have critical passband requirements because the adaptive equalizer will correct for any passband distortion on the input signal. This filter, however, does have low frequency poles and requires at least a third order filter. This is only practically implemented with a switched capacitor technology. This requires extra processing of non optimized technology for the DSP engine. Once a switched capacitor filter is used, one must question if it is not more efficient to do more complex analog filtering and simplifying the A/D and DSP requirements further. A third alternative is to use an echo cancellor in the analog domain. The transmitter output will be fed into an adaptive digital filter that tries to match the line feed through of the adjacent channel. This signal is converted to analog and used to cancel out the adjacent channel energy of the incoming signal. A digital filter in the receiver can measure the energy (at the receiver) in the tranmsitter band and this output is used to update the coefficients of the echo cancellation filter. This filter, again, only requires about 10-20 dB cancellation and is not very complex.

The QAM receiver is similar to the DQPSK with the additional requirements of adaptive equalization and amplitude detection in the demodulator. Often several types of equalizers are combined to achieve optimum performance. Rotation of the signal constellation to correct for phase and frequency jitter is required in QAM modems because their multiple point constellation are much more sensitive then the 90° phase shifts of a DQPSK

modem.   Carrier detection   and   AGC   need   more   sophisticated
techniques because of the amplitude modulation. Carrier detection
is   extremely   difficult   because   the   complex   filter   used must
respond to quickly enough to loss or gain of carrier to meet   the
CCITT   specification for   carrier detect   delay times.   The entire
modem   needs   a   signal   processor   with   approximately   a   10MHz
induction   cycle, several   hundred words   of RAM   and several kilo
words   of progress ROM.   The adaptive   filters require that a high
bandwidth memory structure be devised in order to access data more
rapidly.   In   lower   speed   modems,   almost   all the computations
involve   data   values   fines   fixed   coefficients and these can be
accessed simultaneously from separate memories (RAM and ROM).   QAM
modems require many multiplication of two values stored in RAM and
cannot tolerate two cycle memory access.

9.   CONCLUSION

     Digital signal   processing techniques have been proven viable
for   implementing voice   band modems   for use   on the PSTN.   These
modems cover   the range of   300 bits per   second to 9600   bits per
second.   As VLSI design techniques and   technology improve it will
be possible to build   single chip modems at ever   increasing speed
and performance.

References

[1] J.G. PROAKIS, Digital Communications, McGraw-Hill, 1983, pp. 116-117, 162-171, 183-190.

[2] Data Sets 103J, 113C, 113D, Interface Specification, Bell Systems Technical Reference, Pub. 41106, April 1977.

[3] Data Sets 202S and 202T, Interface Specification, Bell Systems Technical Reference, Pub. 41212, July 1976.

[4] "200-Baud Modem Standardized For Use in the General Switched Telephone Network", CCITT Recommendation V.21, Volume VIII.1, 1976.

[5] "600/1200-Baud Modem Standardized For Use In The General Switched Telephone Network," CCITT Recommendation V.23, Volume VIII.1, 1976.

[6] A.V. OPPENHEIN, R.W. SCHAFER, Digital Signal Processing, Prentice-Hall, 1975, Ch. 5.

[7] R.E. CROCHIERE, L.R. RABINER, "Interpolation and Decimation of Digital Signals - A Tutorial Review", Proceedings, Of The IEEE Vol. 69, No.3, pp. 302-331, March 1981.

[8] H.W. COOPER, "Why Complicate Frequency Synthesis?", Electronic Design, July 19, 1974.

[9] J.C. CANDY, H.W. NINKE, B.A. WOOLEY, "A Per Channel A/D Converter Having 15-Segment u-255 Companding", IEEE Transactions on Communications," Vol. COM-24, No.1, pp. 33-42, January 1976.

[10] G. ERICSON, "An Interpolative A/D Converter with Adaptive Quantizing Levels", National Telecommunications Conference 1980, Conference Record, pp. 56.6.1-56.6.6.

[11] R. APFEL, H. IBRAHIM, R. RUEBUSCH, "Signal-Processing Chips Enrich Telephone Line-Card Architectures", Electronics, McGraw-Hill, May 5, 1982.

[12] D.M.TAYLOR, "Single Chip FSK Modem Streamlines Modem Design", Digital Design, October 1982.

[13] D.M.TAYLOR, "Inside A Single Chip Modem," Data Communications, November 1982.

[14] M.K. STAUFFER "Single Chip FSK Modem Expands Your Design Choices", EDN, May 1982.

[15] J.H.McCLELLAN, T.W. PARKS, L.R. RABINER, "A Computer Program for Designing Optimum FIR Linear Phase Digital Filters," IEEE Transactions on Audio and Electroacoustics, Vol. AU-21, pp. 506-526, December 1973.

15

SWITCHED CAPACITOR MODEMS

KERRY HANSON
TEXAS INSTRUMENTS, INC.
P.O. BOX 1444
HOUSTON, TEXAS 77251
USA

1   INTRODUCTION

The increasing need for low cost data communications has forced a continuing evolution toward more highly integrated modems. Recently technology has reached the point where complete data communication systems can be implemented in monolithic form. This chapter will describe some of the techniques that are used to design these systems. Since the technology is developing so rapidly this description is not intended to be complete but rather is an indication of the current state of the art.

The basics of data communication over voice grade networks will be described first. Modems that are commonly in use today will be described along with a brief review of the most popular modulation techniques. In addition the most common network characteristics that affect modem performance will be mentioned. This will be followed by a detailed analysis of FSK modem design. Included are system level block design and various methods of implementing these blocks with switched capacitor circuits. The relative merits of each design are included. This is followed by a similar analysis of PSK and QAM modulation techniques. A section will then be included on advanced modulation techniques including adaptive equalization and adaptive echo canceling. It will be shown that equalization and echo canceling are two different versions of the same problem. Finally, some suggested future technology directions will be described including the tradeoff between analog and digital design as well as some fundamental limits to analog integration. The chapter ends with a description of a next generation voice band modem.

2 BASICS OF TELECOMMUNICATIONS

2.1.Voice Band Modems

The word modem is an acronym for modulator-demodulator. A voice band
modem is a device that transmits and receives a serial digital data stream
over a telephone network.   The device works by modulating a carrier
frequency in the voice band by the digital data at a rate anywhere from
zero to 9600 bits per second (BPS). Specially conditioned phone lines can
send data at up to sixteen thousand BPS.   The modulation can be any
combination of amplitude, frequency and phase modulation.   The only
requirement is that the total frequency content of the signal be within the
voice band range of approximately 300 Hz to 3400 Hz.

Table 1 shows examples of several typical modems currently available
on the market in both the United States and Europe. The Bell modems are
the standard used in the United States and the recommendations by the
International Telegraph And Telephone Consultative Committee (CCITT) are
the standard used by the European countries.

Model	Data Rate (BPS)	Duplex	Modulation
Bell 103	0-300	Full	FSK
CCITT V.21	0-300	Full	FSK
Bell 202	0-1200	Half	FSK
CCITT V.23	0-1200	Half	FSK
Bell 212	1200	Full	4-PSK
CCITT V.22	1200	Full	4-PSK
Bell 201	2400	Half	4-PSK
CCITT V.26	2400	Half	4-PSK
CCITT V.22 bis	2400	Full	QAM
Bell 208	4800	Half	8-PSK
CCITT V.27	4800	Half	8-PSK
Bell 209	9600	Half	QAM
CCITT V.29	9600	Half	QAM

Table 1 - Typical Modems Currently Available

The Bell 103 and CCITT V.21 modems are very similar. They are
asynchronous modems that can communicate at any rate up to 300 bits per
second.   These modems use frequency shift keyed (FSK) modulation.   They
assign a particular frequency to a logical one value and another frequency
to a logical zero value.   The data stream is transmitted by switching
between the two frequencies as appropriate.   The frequency shift must be
phase continuous which means that the transmitter cannot simply switch
between two free running oscillators.   However, a significant advantage of

513

FSK modulation is that the modem is insensitive to variations in amplitude of the two frequencies.

An additional characteristic of these modems is that they operate in the full duplex mode of operation. This means that they can both transmit and receive information simultaneously on the same pair of wires. Full duplex communication uses frequency domain multiplexing where the originating modem transmits in half of the voice band on the telephone network and the answering modem transmits on the other half. These modems are the simplest modems that are currently widely available.

A step up in performance is provided by the Bell 202 and CCITT V.23 modems. These modems also use FSK modulation but they transmit the data four times as fast. As a consequence the required band width is four times as wide. With the range of frequencies available on the voice band network it turns out that there is not enough room to both transmit and receive simultaneously at this data rate. These modems are therefore designed to transmit or receive but not both at the same time. This is called half duplex operation. Additional complications are introduced in the modem protocol in order to account for turn around delays and to insure that both modems don't try to transmit at the same time. Because of this extra turn around time some modems are provided with a feature called soft carrier turn off. When the modem that is transmitting the data is finished it transmits a tone (typically 900 Hz for a Bell 202) for a short period of time and then it turns off it's transmitter. As soon as the receiving modem detects the soft carrier it immediately shuts off it's receiver even though there is still energy on the line. This can significantly speed up half duplex modems. Since these modems are asynchronous they can actually communicate over the full range from zero to 1200 BPS.

In an attempt to circumvent some of the problems of half duplex operation some modems have a low speed secondary channel. It is also called a reverse channel because it is usually used as a low speed method by which the receiving modem can indicate it's status to the transmitting modem. The CCITT V.23 recommendation specifies a 75 BPS reverse channel while in the United States the data rate can be anything from five BPS to 150 BPS.

Bell 202 compatible modems are not used very much anymore in the United States. Their bit error rate performance is not very good due to the wide bandwidth required and the cost benefit ratio is not very good compared to the other modems available. In Europe, however, CCITT V.23 modems have found a significant market in Viewdata applications where 1200 BPS is the minimum acceptable data rate and low cost is very important. The secondary channel data rate of 75 BPS is typically fast enough to keep up with most people typing on a keyboard.

The Bell 212 and CCITT V.22 compatible modems provide full duplex communication at 1200 BPS. They can do this because they use a different kind of modulation called phase shift keyed (PSK) modulation. The band width required for PSK modulation is only half that required for FSK. As a result two full speed channels can be frequency multiplexed onto one pair of wires. PSK modulation impresses the transmit data onto the phase of a single carrier frequency. Since these modems use four phase PSK·one phase shift actually represents two bits of information and the total bandwidth required is only 600 Hz. Another characteristic of this kind of modulation is that it is differential in nature. The actual information does not depend on the absolute phase of the signal but rather on the phase change between two successive samples. Differential PSK is much easier to implement with only a small penalty in performance.

Both Bell 212 and CCITT V.22 compatible modems have scramblers in their transmitters. These circuits insure that on the average there will be enough phase shifts so that the circuits in the receiver will be able to remain locked onto the transmitted data. They also help insure that certain network tariffs won't be violated.

PSK modulation operates in the synchronous mode which means that data is transmitted at only one specific data rate. A clock signal must be present in both the transmitter and receiver to inform the data terminal equipment when a valid bit is ready to be transmitted or received. These modems also have the ability to transmit data asynchronously. In this case the term asynchronous means that the bit rate remains synchronous but the character rate can vary from zero to 120 characters per second typically. The only requirement is that the character be preceded by a start bit (logic zero) and followed by one or more stop bits (logic one). The modem will also accept a slightly wider range of data rates on the bit rate in asynchronous mode. For example, the Bell 212 can accept a data rate of from 1170 to 1212 BPS. This is accomplished by a converter circuit that will insert or delete stop bits automatically to synchronize the data to the clock in the transmitter. If a stop bit is deleted then the receiver at the remote end must be able to detect the missing stop bit and reinsert it into the demodulated data stream. This is all done transparently to the user. Additional circuits insure that break signals are accounted for and that stop bits are not deleted too often.

The CCITT V.22 recommendation is a super set of Bell 212 modems. In addition to the normal 212 features it also includes half speed, two different kinds of overspeed modes that transmit data at a slightly higher rate and four different character lengths when operating asynchronously. Alternative C of the recommendation also provides a completely different type of bit synchronization as an option.

The Bell 201 and CCITT V.26 modems also use four phase PSK modulation but at a 2400 BPS data rate. Due to bandwidth limitations these modems are half duplex and must make use of the half duplex protocols. In addition, these modems are strictly synchronous in operation. Due to the higher data rate a significant amount of effort is put into the line turn around protocol. The modem must be able to quickly detect a valid carrier and equally important, quickly detect loss of carrier. These modems have been referred to as nothing more that carrier detect circuits with modulators and demodulators attached as an after thought. They are used mostly in financial data transfer applications.

The CCITT V.22 bis recommendation is the first serious attempt to come up with a common voice band modem standard for both the United States and Europe. This standard supports 2400 BPS full duplex operation using a new kind of modulation called quadrature amplitude modulation (QAM). QAM is very similar to PSK modulation but for this modem transmits four bits per baud rather than two. The advantage here is that the bandwidth is still the same as the CCITT V.22 recommendation so full duplex operation is still possible. The trade off is that the modem is more susceptible to noise and phone line distortion and requires a higher level of signal processing to maintain adequate bit error rates.

The CCITT V.22 bis recommendation is similar to the V.22 recommendation with the data rate doubled. In fact, it has a half speed fall back mode that is compatible with the V.22 modem. It also supports both synchronous and asynchronous operation with four character lengths as options in the asynchronous mode and has a scrambler circuit to make the transmitted data appear random. The major difference is the adaptive equalizer that is required for this modem. An equalizer automatically adjusts for phone line distortion and will be described in more detail later.

The Bell 208 and CCITT V.27 compatible modems use eight phase PSK modulation to communicate at 4800 bits per second in the half duplex mode of operation. These modems were originally designed only for use on specially conditioned private phone lines but today are available from several manufacturers for use on dial up networks. They are strictly synchronous and have the normal advantages and limitations of half duplex operation. Eight phase PSK was chosen as the modulation scheme because at the time it provided the best compromise between circuit complexity and performance. Even though these are relatively old standards these modems are still very popular.

The Bell 209 and CCITT V.29 modems were originally strictly for use on private conditioned lines. Versions of these modems are now available for use on dial up phone lines. They typically use some variation of QAM modulation although this depends on the manufacturer and the application. These are 9600 BPS modems that are synchronous and half duplex. The cost of these modems is still quite high but can be expected to drop with further integration.

There are several other kinds of voice band modems that are not included in this brief list but are used in unusual situations. This is particularly true at the higher data rates. However, this list should give the reader an appreciation of the data communication options that are available. More complete information is available in the bibliography [1-14].

2.2. Modem Modulation Techniques

516

There are three common modulation techniques used for voice band modems. These are frequency shift keying, phase shift keying and quadrature amplitude modulation. There are many variations on these techniques but this description will only review the basic schemes. The reader is referred to the bibliography for a more detailed analysis [21,22].

Of the three techniques FSK modulation is the simplest. In this system digital information is transmitted by selecting one of two possible frequencies. For example in the originating Bell 103 modem a logic one is represented by a 1270 Hz tone and a logic zero is represented by a 1070 Hz tone. However FSK modulation is more than just the random selection of two frequencies since the frequency shift has to be phase continuous or else spectral energy outside the normal bandwidth would be transmitted.

The bandwidth required is determined by the peak frequency deviation. For low speed FSK systems this is equal to half the baud rate. A Bell 103 modem has a baud rate of 300 and a peak frequency deviation of 150. The band edges are then the center frequency plus and minus the peak frequency deviation or, in this case, 1020 Hz and 1320 Hz respectively. Figure 1 shows an FSK modulated wave form as well as the the frequency spectrum of the wave form. It should be noted that the wave form can be amplitude modulated as well with no effect on the FSK demodulator. This could be caused by channel distortion and is a significant advantage of FSK over any type of amplitude modulation.

MARK   SPACE

Fig. 1- An Example of FSK Modulation

FSK signals can be demodulated by either coherent or noncoherent means. Noncoherent detection usually uses an envelope detector to convert a frequency difference to an energy level difference which is then converted to the received data stream. During the conversion process some information is lost for the sake of simplicity with a resultant reduction in bit error rate performance. The next section will describe several noncoherent demodulation schemes.

Coherent demodulation is an optimum demodulation process in that no information is thrown away. Ideally, it provides the best possible bit error rate performance at the cost of increased circuit complexity. For practical FSK systems, however, the performance difference between noncoherent and coherent demodulation is typically one dB of signal to noise ratio at a given bit error rate. The next section will also describe coherent demodulation methods using switched capacitors.

Phase shift keyed modulation is a method of transmitting information where the data stream is impressed on the phase of a single carrier frequency. The simplest version allows only one of two alternatives, either a zero degree phase shift (no change) or a 180 degree phase shift. Note that PSK modulation is synchronous in that the phase information of the carrier is only valid at certain discrete times. Even though binary PSK is the simplest phase modulation system practical modems use at least four phase or quaternary phase shift keying (QPSK). This allows one of four possible phase shifts per sample. Note that for this scheme the sample, or baud rate is no longer equal to the bit rate. The CCITT V.22 recommendation, for example, transmits at 1200 bits per second but only 600 baud. The two most common sets of phase assignments are either multiples of 90 degrees beginning at zero degrees or multiples of 90 degrees beginning at 45 degrees. The first set is the simplest to implement but allows a zero degree phase shift as data where the second set always has a phase shift of at least 45 degrees.

Another characteristic of PSK modulation is that practical systems implement it differentially. In pure PSK modulation the phase of the signal is compared with an absolute reference. This turns out to be fairly difficult to do in practice. It has also been demonstrated that any reference tone transmitted with the information bearing signal will always reduce the efficiency of the channel utilization. Differential PSK compares the phase of the current sample of the carrier with the previous sample stored in the demodulator. The difference between the two represents the phase shift which in turn represents the transmitted data stream. The cost of this simpler demodulator is approximately one dB of bit error rate performance.

High performance PSK systems encode more possible phase shifts per baud. Eight phase PSK allows any multiple of 45 degrees and encodes three bits per baud. Some early designs for 9600 bps modems even encoded four bits per baud for differential phase shifts of multiples of 22.5 degrees. Beyond this level other modulation techniques clearly become more suitable.

Like FSK modulation, PSK signals can be demodulated by both noncoherent and coherent methods. In noncoherent systems the phase of the signal is converted to an energy level and then compared to a reference.

Information is again lost in the conversion process. Coherent demodulation preserves all of the information and also results in a better bit error rate performance. Unlike FSK however, here the difference is approximately three dB and is often worth the added circuit complexity.

Figure two shows an idealized result of a QPSK modulated signal. The phase shifts don't actually occur instantaneously but instead occur over the entire sample time. An instantaneous shift would require a wider than necessary bandwidth. The result, however, in terms of information transmitted is the same. Figure two also shows the constellation pattern for a QPSK signal. This pattern shows the phase of the base band signal at the sample time. Note that since this is differential PSK a quadrant one to quadrant two phase shift encodes the same information as, for example, a quadrant four to quadrant one phase shift. A lot of information can be obtained from the shape of the received constellation pattern of a modem in terms of channel noise and distortion.

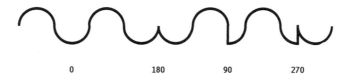

| 0 | 180 | 90 | 270 |

Fig. 2- An Example of **PSK** Modulation

The third major type of modulation used in modern modems is quadrature amplitude modulation (QAM). This technique is very similar to PSK modulation and for four phase PSK it is identical. QAM is implemented with two amplitude modulated signals that are 90 degrees out of phase with respect to each other. If only one bit of information is represented on each signal then the constellation pattern in figure two is the result. PSK modulation represents the constellation pattern in terms of polar coordinates and QAM represents it in terms of rectangular coordinates. The difference between the two modulation systems shows up more clearly in modulation schemes with more bits per baud. Figure 3 shows a 16 point system with four bits per baud for both PSK modulation and QAM modulation. QAM encodes the information into both the amplitude and phase of the carrier.

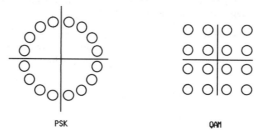

PSK   QAM

Fig. 3- A comparison Between **PSK** and **QAM** Constellation Patterns

Since QAM circuits are fairly complex and difficult to implement they
are generally reserved for high speed voice band modems. At these data
rates bit error rate performance is important and this technique is almost
always implemented coherently.   A later section of the chapter will
describe some general demodulation circuits.

2.3. Telephone Network Characteristics

This section will describe a few of the network impairments that
affect modem performance. The telephone network was originally design for
voice communication and has a number of characteristics that do not impair
speech but can have a significant effect on data.   This is only a brief
overview and more detailed information can be obtained from the references
[24,25].

Telephone line characteristics can be divided into four general
categories which are the passive response characteristics, the noise
characteristics, the steady state impairments and the transient response
characteristics. Each affects a transmitted signal in a different manner

and each must be accounted for in modem design.

The passive response characteristics include primarily amplitude attenuation and envelope delay distortion. Amplitude attenuation is caused by loss in the local loop to the central office and central office to central office attenuation. Typical values at 1000 Hz are 16 dB or less although it can get up to 30 dB or, at 1700 Hz, 35 dB. Daily stability for a particular channel should be within plus or minus four dB although channel to channel values can change significantly. Envelope delay distortion is the frequency derivative of the phase of the signal. Another name for it is group delay distortion and it is caused by the different transmission speed of the frequencies spread across the voice band. This causes the successive symbols to interfere with each other and run together. A typical value is one millisecond or less between any two frequencies in the 800-2300 Hz range.

The hissing background noise sometimes heard on phone lines is referred to as noise. This noise sets the lower limit to the signal level that the telephone network will accept. It is independent of frequency and can extend well beyond the edge of the voice band. When this noise is to be measured it is often passed through a filter that gives it a shape similar to what the human ear can hear. This filter is called a C-message filter. Due to certain characteristics of the network proper noise measurement sometimes requires a tone to be present. This helps set the gain level of certain amplifiers in the signal path. To be measured properly, the tone which is 1004 Hz in the United States is introduced into the channel. Then a C-message filter with a notch at exactly 1004 Hz is connected to stop the tone but pass the resulting noise. This is called a C-notch filter. These are very common noise measurement methods for telecommunication channels.

One of the most common steady state impairments is phase jitter. This is just unwanted random phase modulation of the signal. It has little effect on voice transmission but can have a significant impact on data transmission. Phase jitter can be caused by cross modulation with the network power source and with various ring signals coupling into the channel. Phase jitter that occurs below 20 Hz is also sometimes known as phase wobble. Another steady state impairment is known as frequency offset. Long haul communication systems often use frequency division multiplexing on a common high frequency carrier. When the data signal is separated back out it can have a frequency shift associated with it. In the United States the frequency shift is typically less than five Hz and in Europe it is typically less than six Hz. The design of a modem receiver must account for both this offset and the offset caused by transmitter inaccuracies.

Single frequency interference is an impairment that can occur when the network introduces a tone on the communications channel. During mixing operations it can combine with the carrier frequency and cause both amplitude and phase modulation of the signal. The seriousness of the effect depends on the type of demodulation circuit the modem uses. Another kind of distortion sometimes found on the channel is harmonic distortion.

This is caused by nonlinearities in the amplitude of the signal and can result in integer harmonics of the carrier frequency being generated. This is a significant problem on full duplex modems that are transmitting in the low band and receiving in the high band. The second harmonic of the transmitter could easily be greater in amplitude than the received signal. Amplitude jitter is similar to phase jitter and can be caused by similar impairments. Generally amplitude jitter is not a serious problem for modems because frequency and phase modulated systems can be designed to minimize this effect. Echoes are a steady state impairment that occur at an impedance mismatch in the telephone channel. There are two general types of echoes. Talker echoes are echoes that are heard at the transmitter. These don't affect half duplex modems but can affect full duplex modems. Listener echoes are the type that reflect a multiple number of times and appear at the receiving end of the channel. These echoes can degrade receiver performance. A circuit that can minimize this effect is called an adaptive echo canceler. This is a fairly complex circuit and will be described in a later section.

Impulse noise is the most common type of noise impairment in data transmission today. It appears as short duration bursts with high peaks at random intervals. These impairments will usually cause errors on the data stream for most modems although certain high quality and expensive modems use a technique called data smearing to minimize the damage caused by the impulse. Typical values are 50 impulses per 15 minute interval of 12 dB or more amplitude. Phase or gain hits are also transient response characteristics of the telephone network. These are sudden changes in the phase or amplitude of the signal that last at least four milliseconds. These are caused by the network suddenly switching to another communication channel in the middle of the call.

These are a few of the most common impairments that a modem designer has to understand. If special communication channels like satellites are to be included then their limitations also have to be understood. More information on these additional problems is available in the references [25].

2.4. Review of Switched Capacitor Circuits

Switched capacitor circuit design techniques are a key method of integrating the modems that have been described. These techniques allow analog circuits using MOS processing technologies. Since the major part of this book references switched capacitor circuits this review will be brief and limited to the circuits that are applicable to modems. More information is also available in the references [23].

Three different implementations of an active RC integrator are shown in figure 4. The first design is a typical discrete implementation. The time constant is equal to the resistance times the capacitance. This version is not usually suitable for integration both because the resistor is very large and because the resistor and capacitor values' will not track very well. It does have one advantage over a switched capacitor realization, however, in that it does not have any aliasing problems. There are no sample clocks in this circuit and the high frequency response is monotonically decreasing. It's big application is as an anti-aliasing filter for input and output signals.

The second realization is an idealized switched capacitor integrator. C1 and the two switches replace the resister in the sampled data domain. The switches are clocked on opposite phases at the same frequency so switch one is on or switch two is on but never both simultaneously. The circuit operation is as follows. Switch one is closed first and the input capacitor C1 is charged to the input voltage. Switch one is then opened and switch two is closed. The charge on C1 and C2 is then shared so that the input node to the op amp is brought back to the voltage reference. Switch two is then opened and the cycle repeats. Note that this action causes a specific amount of charge per time to flow though the circuit and that it is proportional to the input voltage. The result is just the equivalent of resistance in the sampled data domain. This circuit is not practical because any capacitance in parallel to C1 is added to the circuit causing erroneous results.

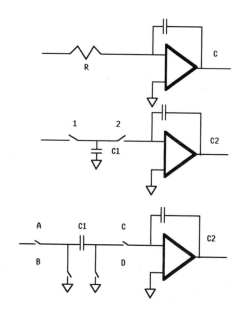

Fig. 4- Three Integrator Circuits

The third circuit in figure 4 is a practical realization of a switched capacitor integrator. Four switches are required instead of just two but the circuit is no longer sensitive to parasitic capacitance. Note that the switches are labeled A, B, C and D. Assume initially that A and C are clocked on phase one and B and D are clocked on phase two. During phase one the input voltage is again stored on C1. During phase two the input capacitor C1 is then discharged completely. Since the capacitor is discharged and the op amp input node is at virtual ground parasitic capacitance on any of these signals will have no effect on the overall performance of the integrator. This circuit is called a negative integrator because the output signal is inverted by the op amp.

A non-inverting integrator can be built with the same circuit by just changing the switch phases. Switches A and D are now clocked on phase one and switches B and C are clocked on phase two. The input capacitor C1 is charged in one direction and discharged in the other. In combination with the op amp the signal is inverted twice resulting in a non-inverting output.

The integrator is a fundamental building block to switched capacitor filters. Since most modems take advantage of the classic filter design techniques using these circuits they won't be reiterated here. Only the unique characteristics of these circuits that are significant to modems will be described. There are several basic signal processing elements that can be implemented with just the integrator by itself. One of the simplest is a frequency mixer.

A mixer is a nonlinear circuit element that combines two signals. Specifically, if one input signal is at a frequency F1 and the other input signal is at a frequency F2 then the output signal is the sum of two frequencies which are F1+F2 and F1-F2. A typical application is mixing a base band data signal with a carrier frequency. The basic switched capacitor integrator in figure 4 provides all the elements necessary for a full wave mixer. The input frequency F1 can be just the input to the integrator and the other input frequency F2 can be implemented by appropriately switching the input capacitor C1. Recall that a negative integrator has switches A and C clocking on phase one and switches B and D clocking on phase two. In the same manner a positive integrator clocks A and D on phase one and B and C on phase two. This is equivalent to multiplying the input signal by plus or minus one. If the circuit is switched between positive and negative integration at a particular frequency, say F2, then the result is full wave mixing and the sum and difference frequency products.

A variation on this technique also provides full wave rectification. If the output of the integrator is monitored for positive going transitions with the result controlling the switching logic then the signal level at the output can always be made positive just by multiplying the input by plus or minus one at the appropriate times. A fairly good precision rectifier can be made this way.

The two significant limitations to this method are caused by DC offsets and by unwanted mixing products. If there is any DC component to the input signal F1 then this component will be multiplied by F2 and show up at the output along with the sum and difference products. Also, any asymmetries in the switching circuits, caused by layout variations, for example, will result in an F1 component showing up at the output. Due to the nonlinear nature of the switching transistors used in these circuits DC offset effects are almost always present in the signal path and have to be accounted for. This is typically accomplished by referencing two signals to each other so that they have the same offsets or by simply keeping the signal large enough so that the offsets are too small to worry about. Large capacitors relative to the switch size also help to minimize this effect. In the worst case an external reference signal can be made available to tune out offsets with a potentiometer.

The other significant limitation is caused by mixing products. Ideally, there are only two input frequencies, F1 and F2. Actually, there are a number of frequency components that are being mixed together. The most obvious one in switched capacitor circuits is the sample clock. This frequency is typically not related in any way to F1 or F2. Also, since it is the rate that switches turn on and off it looks like a square wave with all the odd harmonics that result. F2 is also a square wave and mixes all of it's odd harmonics. F1 in modem applications is the information carrying wave form and is not a frequency but rather a band of frequencies. Finally, add in the frequencies that result from the offsets described earlier. When all of these are mixed together randomly the result is often nothing more than wide band noise. If there is any noise on the power supplies or capacitively coupling into the circuit then these simply add to the problem.

Another common circuit function is summation. A typical realization is shown in figure 5. Note that in this case the integrator also has a resistive equivalent in the feedback path through C4 as well. The DC gain is set by the capacitive ratios of C1 to C4 and C3 to C4 respectively. This circuit will work as a summer as long as all three capacitors clock the analog signals on the same phase.

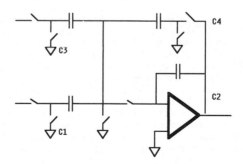

Fig. 5- Am Example of a Multi-Input Integrator

If the two inputs of the summation circuit are tied to a common node and the logic controlling the switching of the input capacitors is independently selected then this circuit can also act as a gain control element for the analog signal. The gain is digitally controlled and can be implemented in two different ways. If the input capacitors C1 and C3 are similar in size then the user would simply switch in one or the other signal path but not both. If the capacitor sizes differed significantly then C3 would be added to the value of C1 rather than replacing it. This would allow closer tracking of gain levels as well as require a smaller over all amount of capacitance. The circuit can be extended to an arbitrary number of parallel input capacitors. Practical implementations usually require both kinds of digital capacitor selection methods.

One final point should be noted. Since the gain level is set by the ratio of input to feedback capacitance it might be assumed that capacitor C4 can be adjusted just as easily as the input capacitors. This is not true because the ratio of C4 to C2 sets the frequency response of the circuit and any modification to the size of C4 would also require modification of C2, doubling the complexity of the circuit.

Another common circuit element that is very useful is the comparator. It is often nothing more than an op amp without a compensation capacitor. In modems comparators are useful for detecting both absolute and relative signal levels. They can also provide hysteresis and, in general, are useful where a one bit D/A converter is needed. Comparators have one significant advantage over op amps in system level designs. Since the output has only one bit of information the comparator can be multiplexed to perform several different system functions. The advantage is a lower overall chip size although some care must be taken with parasitic capacitance. Many people over several years of effort have tried to multiplex op amps almost uniformly without success. Usually problems with DC offsets and mixing products become too difficult to solve.

These are all the basic building blocks that are necessary to design and build voice band modems. Two key ideas should be remembered at this point. The first is the advantage of simplicity. The more specialized the circuit element is the more certain will be an unexpected response. With one exception arbitrarily complex signal processing functions can be implemented with the elements described. The exception is the multiplication of two time varying wave forms. The need for this and some possible realizations will be described in a later section. The second key point is that all these circuits operate with a two phase clock. Some fairly clever signal processing functions have been implemented in the past with higher level multiple phase clocking but on a system level design the additional problems associated with extra clock logic, DC offsets and especially mixing products usually preclude this type of design.

3 FSK MODEM DESIGN

3.1. FSK Modulation Techniques

Frequency shift keyed modulation is a very simple function to implement. The transmit frequency can be derived from a digital divider counting down from the master clock frequency of the chip. The only variations are in how much harmonic distortion is allowed in the transmitted wave form. There are two common techniques currently in use.

The simplest FSK modulator is just a digital circuit that generates a square wave at the transmitted frequency. The digital data stream to be transmitted is simply an input to the divide circuit to change the divide ratio as it is counting down. An advantage to this approach is that a phase continuous frequency shift falls out automatically. A disadvantage is that the master clock frequency has to be an integer multiple of both the mark and space frequencies or else the transmitted frequencies won't be at quite the correct value. Most modem specifications allow the transmitted frequency to vary by plus or minus two Hz.

A more significant problem with this approach is that the square wave is made up of the fundamental and all the odd harmonics. Unless filtered out these harmonics can violate telephone network frequency limits. Therefore this transmitter is usually followed by a high order noncritical filter to reduce the harmonics to an acceptable level. Typically this filter will require a lot of chip real estate, especially if several different carrier frequencies are being generated. Another problem with this approach is that all of the harmonics will mix with the sample clocks causing mixing products that could interfere with the remote modem's receiver.

A common technique to get around these problems is to generate a square wave at an harmonic multiple of the transmitted frequency rather than the frequency itself. The most common harmonic is sixteen times the fundamental. This signal is then used to clock a D/A converter that is made up of circuit elements like those shown in figure 5. The advantage is that the first harmonics that have to be filtered are the fifteenth and seventeenth harmonics. A much smaller, simpler filter is required. There are other variations that take advantage of special circumstances but these two techniques will usually satisfy the needs of most designers.

3.2. FSK Demodulation Techniques

FSK demodulation provides some more interesting applications of

527

switched capacitor circuits. There are three popular techniques in use that are integrated versions of classic discrete designs. These are delay discrimination, phase locked loops and mark/space differential demodulation. Each of these techniques will be described along with their advantages and limitations.

Delay discrimination is the first demodulation technique to be described. A block diagram is shown in figure 6. The carrier frequency received from the telephone line is first filtered to eliminate any out of band noise introduced by the network. The signal is then limited resulting in a square wave at the carrier frequency with a variable period. The advantage of limiting is that now the key part of the demodulator can be implemented digitally. The signal then goes to two separate circuits. One circuit is a delay element that delays the signal by one quarter cycle of the center frequency between the mark and space frequencies. This can be implemented in any number of ways including a CCD delay line or a shift register. The output of the delay element is then mixed digitally with the original signal resulting in the base band data stream and a double frequency component. After the double frequency component is filtered out by a low pass filter the base band signal is limited resulting in the original transmitted data stream.

An interesting variation is that the delay element need not actually delay the original signal at all. The delay element can be replaced by a retriggerable monostable multivibrator with a period equal to the equivalent delay time. If the period is reasonably well controlled then the result will be the same level of performance.

Fig. 6- FSK Delay Discrimination

There are several advantages to this approach. It is predominantly digital and therefore simple to design, the base band filter is not critical and can be fairly small in chip area. For these reasons it is by far the most popular demodulation technique and is used in a number of integrated designs [15,17,18]. It does have some limitations, however. It is a noncoherent technique and will not perform as well as some coherent designs. More importantly, since it throws so much information away it is very sensitive to network impairments including phase jitter and harmonic distortion. In addition the monostable multivibrator version usually requires an external potentiometer to adjust the period of the equivalent delay. This uses up a chip interface pin and introduces an additional aging and temperature sensitivity factor.

A second type of FSK demodulation technique that used to be popular uses a phase locked loop. This is a coherent demodulation technique and can in theory provide close to optimum performance. An example of this demodulation method is shown in figure 7. The received signal is again filtered and then passes through an automatic gain control (AGC) circuit. In a discrete realization the signal could just be limited but in a switched capacitor environment the resulting harmonics would mix with the sample clocks and generate too many mixing products. The AGC can be designed with the circuit elements described earlier in the normal manner. Note that the complete phase locked loop is implemented with two of the basic integrator elements. The first integrator provides both the phase comparator or mixing function and the loop filter function to remove the double frequency component. The output of the loop filter is limited with a comparator and provides the base band data. The output is also the error signal for the voltage controlled oscillator (VCO).

The VCO is another integrator with the output compared to a reference voltage. When the comparator switches and the integrator is reset and begins integrating again. The rate that this occurs is proportional to the input voltage resulting in VCO action. The VCO output also provides the reference frequency at the input to the phase locked loop.

In this application a phase locked loop looks better on paper than it does in the real world. As mentioned earlier, a coherent demodulation technique only provides a one dB improvement in performance. Also, within the constraints of switched capacitors the circuit is difficult to design. It's biggest limitation, however, deals with the system constraints of phase locked loops. With the frequencies normally selected for FSK modems the loop bandwidth has to be very wide. A narrow loop would not be able to maintain lock over the mark/space frequency range. A wide band width means that the loop is susceptible to noise and in the end has no advantage over a noncoherent technique. One method of overcoming this limitation is to center the loop on just the mark carrier frequency. The loop would then treat the space frequency as noise and the demodulator would be nothing more than a loop lock indicator. One of the early modem designs used this technique. Because of these limitations the phase locked loop demodulator should only be used in situations where it's special characteristics are really necessary. Additional information is in the references [26].

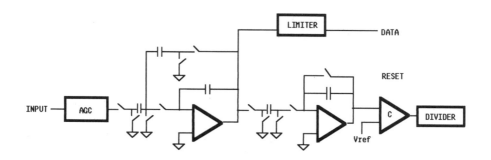

Fig. 7- A Phase Locked Loop Demodulator

The third type of demodulation technique is the mark/space differential demodulator. A block diagram is shown in figure 8. This demodulation technique is also incoherent but it has several advantages over the delay discriminator. The received signal is again filtered as in the other schemes. Since most of the circuits are analog rather than digital an automatic gain control circuit is required here. Once the signal level has been set it follows two separate paths into mark and space band pass filters. These are low order band pass filters with several special requirements. The group delay of the signal through both filters must be the same and it must be less than the baud time of the modem. Also, the gain of the signal through both filters must be the same. The key to these filters is that one is set on the first Bessel side band of the mark frequency and the other is set on the first Bessel side band of the space frequency. For a 300 baud modem these are the center frequency plus and minus 150 Hz. The output of the filters is full wave rectified and then summed differentially. A frequency difference at the input is converted to an amplitude difference at the output. The result is filtered with a base band low pass filter and limited, becoming the demodulated data stream.

530

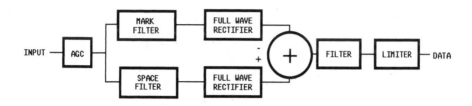

Fig. 8- Mark/Space Differential Demodulation

This circuit technique has been around for several years but has not been used much in discrete designs. The reason is that the components usually could not be matched well enough through both parallel signal paths. This is where the advantages of switched capacitor circuits stand out. Since everything is on the same integrated circuit the capacitor ratios can track each other very well making mark/space differential demodulation a useful design alternative. Even though noncoherent, the bit error rate of this method is generally quite good and it is used in popular FSK modem designs [16,19].

3.3. FSK Filter Requirements

Filters that are designed for use in modems have several special requirements over filters that are used in voice applications. The most obvious constraint is the phase response of the filter. Since the human ear is not sensitive to the phase of a signal voice applications need only be concerned with amplitude response. An example is the elliptic filter in a codec. In general, an elliptic filter is not suitable for use in a modem because of the phase distortion at the band edge. The derivative of phase with respect to frequency is called group delay and is a measure of how fast a signal actually passes through a filter. If the filter does not

531

have a linear phase response then the group delay through the pass band of the filter is a function of the frequency. This can result in intersymbol interference which is another way of saying that the successive baud samples start smearing together.

Another characteristic of receive filters in modems is the trade off between internal and external noise. External noise is the normal out of band interference that the filter is supposed to reject. Internal noise is the noise floor of the filter caused by the circuit itself. If the filter is designed to amplify the incoming signal then proper selection of where the gain should be in the filter is determined in part by the type of noise encountered. If the received signal level is low then the signal to noise ratio (SNR) at the input of the filter will not be very large. As the signal is filtered each stage of the filter raises the noise floor level and the resulting signal to noise ratio is reduced. If the gain is put at the end of the filter then both the noise and the signal are amplified maintaining a low SNR at the output. If, on the other hand, the signal is amplified at the input to the filter then a large SNR is maintained through the filter.

In spite of it's advantages the amplification of a signal should not always be at the input of the filter. Full duplex modems frequency multiplex a large transmitted signal with a small received signal. If the received signal is amplified before the transmitted signal is filtered out then the transmitted signal can saturate the receive filter with harmonics and mixing products overshadowing the received signal. For this reason the designer of a modem filter will usually start by rejecting the adjacent channel before trying to do anything else. Next will come the amplification and only then the actual shaping of the filter. Note that this problem does not apply to half duplex modems because there is no adjacent channel on the line.

The general shape of the receive filter depends on the type of modem, the data rate and whether it is half or full duplex. In spite of these differences there are some general rules of thumb that apply in modem design at low and moderate data rates. First, of course, the high frequency spectral energy limitations of the network apply. In the U.S. energy above 40 kHz must be no higher than -50 dBm. In Europe, the requirement is -80 dBm above 50 kHz. The receive filter is concerned with these limits because of the possibility of mixing products at these frequencies with the sample clocks.

Generally, noise in the voice band below 3.4 kHz should be rejected by at least 40 dB. However, because of the allowed transmitter signal level, the adjacent channel in full duplex modems should be rejected by at least 60 dB. The actual value for a particular application depends on the desired bit error rate for a given demodulator scheme.

Another characteristic in the design of the receive filter in voice band modems is called statistical equalization. As the signal is transmitted over the network the the impairments described earlier distort both the amplitude and phase characteristics of the signal. The receive

532

filter can have some distortion built into it to reconvert the signal back to it's original form. This type of equalization is statistical because the receive filter is designed around the average phone line distortion. Equalization is not necessary at low speeds but is useful at moderate and high data rates.

3.4. FSK Signal Detection

Proper detection of FSK signals buried in noise is one of the most difficult problems in the design of integrated switched capacitor modems. Part of the problem depends on the kind of telephone network interface used. There are two generally accepted means of connecting to the telephone network. They are through an acoustic coupler or directly through a network interface circuit. Each type of connection presents it's own problems.

An acoustic coupler uses the actual telephone itself. The modem transmitter drives a speaker in contact with the microphone of the telephone handset and the modem receiver is connected to a microphone in contact with the handset speaker. The major advantage to this approach is that the PTT approval requirements are simpler. The major disadvantage to this approach is that the microphones will pick up ambient noise and cause a higher bit error rate. Errors are also caused by harmonics in the side tone path of the telephone. Generally, acoustic couplers are useful up through 1200 bits per second. Above that speed a direct connection must be used.

A direct electrical connection can be made between the modem and the network through an interface circuit. This is a much better connection and results in significantly improved performance. The purpose of the interface circuit is to insure that the modem and the network are protected from each other. To insure that the circuit used will do this the PTTs require certification of any equipment to be connected to the network before it can be used. This can be a fairly expensive and time consuming process.

There are three basic techniques in use for detecting an FSK modulated signal. The simplest is to use just an energy detector. The circuit will turn on with any in band noise or signal above a preset level. The energy detector can be nothing more than a full wave rectifier and a low pass filter. The output can be compared against a voltage reference determining the presence of a signal. The weakness of this approach is that the modem will cheerfully try to demodulate impulse noise, somebody's voice or anything else coming out of the network. Also, there is usually enough ambient noise so that the carrier detect circuit of an acoustically coupled modem would never turn off.

The next step up in complexity is a mark carrier detector. This circuit uses an energy detect circuit in combination with the received

533

data. A requirement during modem call initiation protocols is that the modem must receive a valid mark carrier for a set length of time, typically 100 to 200 milliseconds for full duplex modems and ten to 100 milliseconds for half duplex modems. In FSK systems noise on the line will generally not meet this requirement. Once the carrier detect circuit turns on then any in band energy will keep it on. While still limited this is the minimum acceptable carrier detect circuit for acoustic couplers.

A more sophisticated carrier detect circuit can be used with the mark/space discriminator described earlier. If the mark filter has more energy that the space filter then a valid mark signal is present. If the opposite is true then a valid space signal is present. If, however the energy in both filters is approximately the same then the modem must be receiving noise rather than a valid carrier. A circuit to determine this condition can be easily designed and the resulting modem will never try to demodulate noise at any time during the call.

Carrier detect timing is a factor that is significant in half duplex modems. The total data throughput is controlled to a significant extent by the turn around time of a half duplex modem. For this reason a great deal of effort is put into the detection of a valid carrier in a short amount of time. This is especially true at the higher data rates. As mentioned earlier, a Bell 201 compatible modem operating at 2400 bits per second in the half duplex mode has often been described as a carrier detect circuit with a modulator and demodulator thrown in as an after thought.

4 PSK MODEM DESIGN

4.1.PSK Modulation Techniques

Since PSK modems are more complex than FSK modems there are not as many different types of designs available. As described in an earlier section, PSK modems can often be treated as QAM modems. This is specifically true with the Bell 212 and CCITT V.22 compatible recommendations. An advantage of this approach is that once a QAM modem is defined, it is a straight forward if not simple matter to increase the number of points on the constellation pattern and transfer data at higher rates. The designer can, for example, evolve up to the V.22 bis recommendation.

An example of a QAM modulation scheme is shown in figure 9. This example is suitable for use on the V.22 recommendation and has been used in a switched capacitor design [20]. The data to be transmitted is treated as two biphase channels in quadrature. The data is first encoded into differential phase shifts which are just values of plus and minus one in the base band. Each base band is then shaped for both phase and amplitude response by a low pass filter. This is a critical filter and is a key part of the modulator. The two biphase base band signals are then mixed with the carrier frequency at zero and 90 degrees respectively which adds the quadrature component. The two biphase channels are then summed resulting in one quadrature phase signal. Note that these are just the basic switched capacitor circuit elements described earlier. Both mixers and the summer can be implemented with just one op amp. Finally, the transmit signal passes through a band pass filter that provides final shaping at the carrier frequency. Ideally, the sample clock should be at the same frequency for all of the circuit elements and at an integer multiple of the carrier frequency.

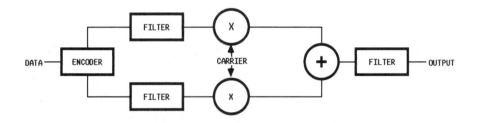

Fig. 9- QAM Modulation Scheme

This basic technique can be extended in a straight forward manner. Some modems add 45 degrees to the differential phase shifts to insure that a phase shift is present at every baud. The additional phase shift can be added in the mixers by some additional logic. Additional points in the QAM constellation pattern can be added by allowing the base band to assume values other than plus or minus one. This circuit is not suitable for eight or sixteen phase PSK systems however and a more conventional phase shift system must be used.

4.2.PSK Demodulation Techniques

535

There are two common techniques used to demodulate phase shifted signals. The simplest one is based on a delay demodulation technique similar to the one used in FSK modems. A block diagram of a delay demodulator is shown in figure 10. The incoming signal is filtered as usual and then limited since the scheme is mostly digital. The signal is again delayed and then exclusively ored with itself in one path and with a 90 degree phase shifted version of itself in the other path. The carrier is again treated as two biphase channels in quadrature. The result is two biphase base band signals which are filtered with a low pass filter, limited and then decoded into the original data stream. Note that in this case the decoder must have a sample clock input since the data is only valid at the baud time. The sample clock is derived from the received data with switched capacitor circuits that will be described in the next section.

The strengths and weaknesses of this circuit are similar to the equivalent FSK circuit. The technique is very simple to design and implement but it doesn't perform very well. If this scheme is used then a lot of attention must be paid to the design of the receive filter to insure adequate performance.

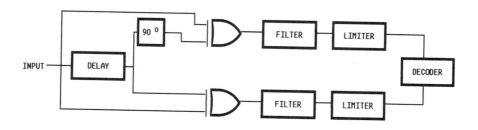

Fig. 10- PSK Delay Demodulator

PSK or QAM modulation can also be demodulated coherently using a Costas phase locked loop. Unlike FSK systems, PSK performance can be

536

significantly better using coherent techniques. A block diagram of this technique is shown in figure 11. The received carrier signal is filtered and amplified by an AGC. It is then mixed with the output of a Voltage Controlled Oscillator (VCO) both at zero and at 90 degrees. The signal is again being divided into two biphase channels. The arm filters are low pass filters that eliminate the double frequency component and are usually fairly critical in this application since they shape the base band data. The signals are then limited and sampled by the received baud clock in the decoder circuit. The result is the original data stream. The signal from the arm filters is mixed and summed and passed into the loop filter. This is a low pass filter with a very low cut off frequency, typically ten to thirty Hz. The result is an error signal that is proportional to the average phase error of the loop and that drives the VCO.

This is a fairly difficult circuit to design. The signal gain of each element in the loop is critical and any significant offsets or mixing products can destroy the performance of the circuit. Mixing products are an especially difficult problem in this circuit because of all the different frequencies that are present in the elements. Those elements that work on the base band data should be clocked at a multiple of the data and those elements that work on the carrier frequency should be clocked at a multiple of the carrier frequency. It may be necessary to separate the various sections of the circuit with an antialias filter. Additional information can be obtained from the references [26]. In any event, every aspect of this circuit should be planned from the beginning of the design.

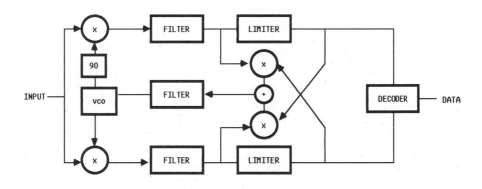

Fig. 11- A Costas Phase Locked Loop Demodulator

There are two common techniques in use in modems for recovering the received data clock. In both cases it is recovered from the base band data. The most common technique is just a simple phase locked loop. All that is needed is one mixer, one loop filter and one VCO, none of which are very critical. The VCO can be the integrator type described earlier. The loop bandwidth has to be fairly wide in this case to allow the wide range of clock rates that are available as options on the various modems. Also, a wide bandwidth loop will lock quickly on the received data, a necessity in half duplex modems.

The other technique was popular several years ago and is not in use very much any more but can still be found in special applications. It can only be used on modems that always have a phase shift on every baud sample, typically a modem with an additional 45 degree phase shift. The circuit splits the incoming signal into two separate paths. One path filters out the carrier and the high band of frequencies and the other path filters out the carrier and the low band of frequencies. These two signals are then mixed in a balanced modulator resulting in sum and difference products. The sum product is filtered out and the difference product is just the received baud rate.

4.3. PSK Filtering and Carrier Detection Methods

The overall constraints that apply in FSK filtering circuits also apply in PSK circuits. They are just a little more critical. The noise rejection and adjacent channel rejection requirements are the same although for a given data rate PSK systems usually have a better bit error rate than FSK systems. This is because of the reduced bandwidth requirement. Also, where equalization is optional on most FSK systems it is usually required on PSK or QAM systems. A small variation on the group delay of a Bell 212 equivalent modem can substantially reduce it's performance.

The carrier detect circuits are a little more limited on PSK circuits. First, very few PSK modulated systems will perform over an acoustic coupler so most designs are limited to a direct electrical connection. Second, many systems are limited to a mark carrier detection scheme because they support remote digital loop back protocols which are sensitive to specific data that must be detected in addition to a mark carrier. High speed half duplex modems must not detect anything other than a specific carrier or else the performance level will degrade.

5 ADVANCED CIRCUIT CONCEPTS

Two additional circuit concepts will be briefly described in this section. These are adaptive equalizers and adaptive echo cancelers. they are normally used only in high performance, high data rate systems because of their complexity although at least one 1200 bit per second modem chip has an adaptive equalizer in it [20]. After both signal processing techniques have been described this section will finish up with a description of some special circuit elements needed for these applications.

5.1.Adaptive Equalization

Adaptive equalization is required for high speed modems because the nonlinear distortion on a telephone network causes intersymbol interference. Successive baud samples begin to overlap and cause data errors. This is caused by both amplitude and group delay distortion. Because this distortion is easiest to deal with in pulse amplitude modulated systems, most modem equalizers are implemented in the base band circuits of the demodulator. Figure 12 shows a block diagram of an adaptive equalizer.

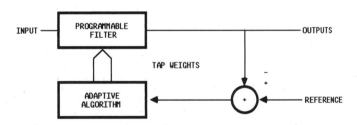

Fig. 12- An Adaptive Equalizer Block Diagram

In a PSK system there are typically two sets of filters, one for each biphase channel. The input is the filtered base band data and the output goes into the decoder to become the demodulated data stream. The reference input is from the other channel that is 90 degrees out of phase. Intersymbol interference causes the signal to couple into the other channel and any correlation between the two channels is subtracted out through the reference input. The programmable filter is usually implemented as a finite impulse response filter because it is unconditionally stable and easy to design. There are several different algorithms in common use. The most popular is the LMS algorithm which correlates on a least mean square error basis. This is an iterative technique that requires the multiplication of two time varying signals. This is sometimes difficult to do in switched capacitor technology but has been implemented in integrated form [27]. Another iterative algorithm is decision feed back equalization. This is a similar scheme except that it has feed back as well as feed forward equalization and the only multiplication requirement is by plus or minus one. This is the technique used on a 1200 BPS modem chip [20]. There are two distinct limitations to these techniques. First, the convergence rate is generally slow. This is a problem for half duplex modems that must converge quickly. Second, the convergence rate is dependent on the signal power level and the signal to noise ratio. There are three newer equalization techniques that avoid these problems.

The first one is sampled matrix inversion. This is a digital technique that calculates the inverse of the input signal covariance matrix. It requires a lot of digital signal processing but since no iterative operations are performed it can adapt very quickly. It is described in the references [29]. Another option is an adaptive lattice filter where each weight is related to a local error rather than a global one. Since local errors are not as interrelated it can converge much faster in some applications. It is referenced in [32]. A third technique is adaptive Kalman filters. These use recursive structures rather than transversal ones and require fewer tap weights to equalize a signal. However, this realization has to be very accurate and can result in unstable signals. Some good references for adaptive equalization are [28,30,31].

5.2. Adaptive Echo Canceling

The adaptive echo canceling problem is very similar to the adaptive equalization problem. A block diagram of the echo canceler is shown in figure 13. This circuit has two different applications related to modems. The first is the elimination of echoes in half duplex modem operation. These echoes occur at any impedance mismatch in the channel. More recently these circuits are allowing full duplex modem operation by adaptively canceling the transmitter. This way both the transmitter and receiver can use the full bandwidth of the channel. An example is the CCITT V.26 ter recommendation.

540

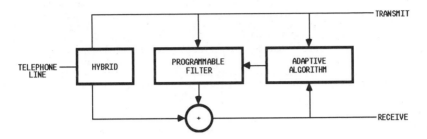

Fig. 13- An Adaptive Echo Canceler Block Diagram

All of the filters and algorithms described for adaptive filters also apply to adaptive echo cancelers. The most popular circuit technique is again the LMS algorithm. A switched capacitor integrated circuit that uses both adaptive filtering and adaptive echo canceling is described in the references [27]. The most successful technique for multiplying two time varying signals in switched capacitor technology has been the multiplying digital to analog converter (MDAC). This allows one signal to remain analog but requires that the other be converted to digital format. With clever design the penalty can be minimized in a system level application.

6 FUTURE TECHNOLOGY DIRECTIONS

6.1.<u>Analog vs Digital</u>

In telecommunications this question has been around as long as there have been integrated circuits. Both design technologies have their advantages. Switched capacitor circuits more closely match the existing technology in telecommunications because they are integrated realizations of existing discrete designs and can draw on a large body of theory. Digital circuits on the other hand are simpler to design and enjoy a substantial momentum in the integrated circuit industry. Analog circuits can generally implement a function in less chip real estate than digital because more than one bit of information can be carried on a chip geometry.

Digital circuits can often multiplex several functions together more easily because it is a more indirect model of the telecommunications function. Digital circuits can be shrunk directly, analog circuits can often implement the same function without having to be shrunk at all. Analog circuits are more susceptible to device and process noise, digital circuits are susceptible to round off error. Shrinking both technologies results in higher frequency applications. Digital circuits can be designed more quickly but analog circuits cost less on a per function basis. And so it goes.

There is one conclusion that proponents of both systems usually agree on. Future telecommunication systems that are implemented on single integrated circuits will almost certainly depend heavily on both technologies.

6.2. The "Ultimate" Voice Band Modem

Recently CCITT study group XVII made available for technical review a draft recommendation for a voice band modem that can communicate at 9600 bits per second in the full duplex mode over a dial up telephone line. This is the V.32 recommendation. The modem will make use of all of the techniques described in this chapter and then some. A forward error correction technique called Trellis coding is used to help improve it's performance at a given signal to noise ratio. The modulation scheme is a version of QAM that uses five bits per baud. The baud rate is 2400 baud so the actual bit rate is 12 kBPS in each direction. One of the five bits is used for error correction so the user data rate is 9600 BPS. The constellation pattern actually has 32 points. Since a lot of bandwidth is required in each direction the modem takes advantage of adaptive echo canceling rather than frequency multiplexing. Adaptive equalization will also be required. More information is available in the references [33].

This is not a trivial design task in either discrete or integrated form. The modem will be fairly expensive initially but as the technologies are made available with advances in switched capacitor and digital signal processing circuits this modem should become the standard telecommunications device for data into the 1990s.

7 CONCLUSION

This chapter began by reviewing the basics of telecommunications. Included was a brief description of the voice band modems that are popular in both the United States and in Europe. This was followed by a review of FSK and PSK modulation techniques. It was noted that QAM modulation is very similar to PSK modulation. A number if different telephone network impairments that affect the performance characteristics were described next. Good modem designs should be able to minimize the effect of the impairments. The final review was of basic switched capacitor circuit techniques. The emphasis was on those functions that are specific to modems rather than a review of general filter theory.

An analysis of various FSK modulation and demodulation techniques was described next. Both coherent and noncoherent circuits were included. This was followed by PSK and QAM modulation methods and how they might be realized with switched capacitors. Next was a brief description of the adaptive signal processing problem applied to equalization and echo canceling. Some relative merits of analog and digital design methodology were followed by a review of a new modem standard that will push the limits of voice band telecommunications. A list of references is provided at the end of the article for those readers who want more information.

8 REFERENCES

[1] K. OLSON, K. O'DONNELL, "Update: Modems Part I", Digital Design, September, 1976.

[2] K. OLSON, K. O'DONNELL, "Update: Modems Part II", Digital Design, October, 1976.

[3] S. DAVIS, "Modems: Their Operating Principles and Applications", Computer Design, September, 1973.

[4] C. BOUSTEAD, "Modems Keep Things Quiet When Ma Bell Gets Noisy", Electronic Products, May, 1973.

[5] M. HIMMELFARB, "Modems", Digital Design, March, 1973.

[6] R. TOOMBS, "Ordering Data Couplers", Data Communications, July, 1975.

[7] T. MAZZARINI, "Specify Low Speed Modems Properly", Electronic Design, April, 1978.

[8] H. ANDREWS, "Let's take the mystery out of modems", The Electronic Engineer, July, 1972.

[9] S. LEVINE, "Focus on modems and multiplexers", Electronic Design, October, 1974.

[10]DATA DECISIONS, "Modem Survey", Datamation, October, 1981.

[11]D. GIBSON, "Designer's Guide To: Data Modems", EDN, March, 1980.

[12]J. EDLIN, "A Micro Age View Of Networking", Data Communications, May, 1981.

[13]J. GAUDREAULT, "Microprocessor Based 212-type Modem Designs", MIDCON/82, November, 1982.

[14]CCITT, "Data Communication Over The Telephone Network, Recommendations Of The V Series", International Telecommunication Union, 1981.

[15]A. MENDELSOHN, "Monolithic Modems Marry Micros To Landlines", Electronic Products, January, 1983.

[16]K. HANSON, "Chip Set Opens Public Networks To Low-cost Data Communications", Electronic Design, November, 1981.

[17]R. MACDONALD, M. GODDARD, ET AL, "Single-chip 1200-b/s Modem Melds U.S. and European Standards", Electronics, January, 1983.

[18]B. JORDAN, T. DUGAN, "Single-chip Modem Links To PC In Many Ways",

Electronic Design, September, 1983.

[19]K. HANSON, "Single-chip Modem Slashes Space Requirements", Computer Design, March, 1983.

[20]K. HANSON ET AL, "A 1200b/s QPSK Full Duplex Modem", ISSCC Conf. , February, 1984.

[21]P. PANTER, Modulation, Noise, And Spectral Analysis, McGraw-Hill, 1965.

[22]P. PEEBLES, Communication System Principles, Addison-Wesley, 1976.

[23]P. ALLEN, E. SANCHEZ-SINENCIO, Switched Capacitor Circuits, Van Nostrand Reinhold, 1984.

[24]BELL SYSTEM TECHNICAL REFERENCE, "Data Communications Using The Switched Telecommunications Network", May, 1971.

[25]D. TUGAL, O. TUGAL, Data Transmission Analysis, Design, Applications, McGraw-Hill, 1982.

[26]F. GARDNER, Phaselock Techniques, John Wiley & Sons, 1966.

[27]E. SWANSON ET AL, "A Fully Adaptive Transversal Canceler and Equalizer Chip", ISSCC Conf., February, 1983.

[28]C. COWAN, P. GRANT, "Adaptive Filters Await Breakthroughs", MSN, August, 1981.

[29]F. KRETSCHMER ET AL, "A Digital Open-loop Adaptive Processor", IEEE Transactions on Aerospace and Electronic Systems, January, 1978.

[30]C. COWAN, P. GRANT,"Adaptive Processing Improves Filters and Antennas", MSN, July, 1981.

[31]S. QURESHI, "Adaptive Equalization", IEEE Communications Magazine, March, 1982.

[32]E. SATORIUS, S. ALEXANDER, "Channel Equalization Using Adaptive Lattice Algorithms", IEEE Transactions on Communications, June, 1979.

[33]E. MIER, "A Standard For Faster Data Communications", Data Communications, February, 1984.

16

ECHO CANCELLERS: THEIR ROLE AND CONSTRUCTION

E. J. Swanson
AT&T Bell Laboratories
2525 N. 12th Street
Reading, Pennsylvania 19604

1   INTRODUCTION

Echo cancellers are adaptive transversal filters. As such, their impulse responses (tap weights) are not set at the silicon foundry. Chances are remote indeed that any two echo cancellers will ever have exactly the same transfer function.

Adaptive filters have dramatically increased their market share in the monolithic filter business. They avoid many of the compromises inherent in fixed transfer function filters. Higher performance systems can result. Adaptive filters can frequently eliminate the need for expensive manual adjustments, both during initial system setup and throughout the system life cycle.

In the next section, we examine a specific application for an adaptive filter. This will serve as a brief introduction to the least mean square (LMS) adaptive filter algorithm [1]. Other applications for adaptive filters will be discussed in Section 3, with emphasis on analog/digital tradeoffs. Wide dynamic range analog CMOS echo canceller designs are the subject of Section 4.

The ubiquitous 2-to-4 wire interface is one of the first application areas for echo cancellers. An idealized view of the interface is shown in Figure 1.

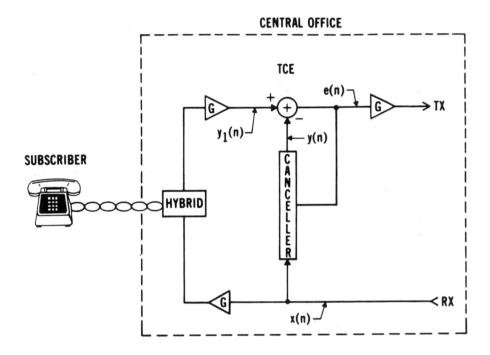

Fig. 1 - 2-to-4 wire interface

In the subscriber loop plant two-wires must accommodate both talking and listening. Once inside the central office, however, the two directions are separated into transmit (TX) and receive (RX) paths. An ideal hybrid provides complete isolation between the two directions. The subscriber listens to all of the received signal. The transmitted signal contains only the subscriber's message. No echo canceller is needed with an ideal hybrid.

Of course, the ideal hybrid has never been invented — and with just cause. The two-wire loop plant is incredibly diverse. Different cable lengths, wire gauges, bridge taps, and plenty of other factors vary the effective impedance seen by the hybrid. Echo return loss (ERL) is a measure of the isolation between TX and RX paths provided by the hybrid. Worst case ERLs are only 10 dB or so. RX energy that returns via the TX path to the far-end customer is called echo.

The gain and filtering blocks labelled "G" in Figure 1 cause poor ERL to be dangerous. These blocks provide gain to compensate for the attenuation in the two-wire loop. If the isolation provided by the hybrid is poor, the four-wire loop can break into oscillation, or "sing."

The job of the echo canceller is to remove energy in the transmit path that is correlated with the receive path. Uncorrelated energy, the subscriber's speech, is not attacked. Referring again to Figure 1, if the response of the echo canceller to an impulse at $x(n)$ can converge to match the impulse response from $x(n)$ to $y_1(n)$, no echo energy will remain at $e(n)$.

The output of a sampled-data echo canceller is given by

$$y(n) = \sum_{i=0}^{N-1} a_i^{(n)} x(n-i)$$

The tap weights $a_i^{(n)}$ define the impulse response of the filter. The superscript reflects the changing of the filter's transfer function with time. N is the order of the transversal filter. Obviously, if the trans-hybrid impulse response duration exceeds the total delay available in the canceller, the residual energy cannot be eliminated. The adaptive filter may even diverge.

The tap weight update algorithm is simple and intuitive:

$$a_i^{(n+1)} = a_i^{(n)} + \alpha x(n-i)e(n)$$

Non-linear variants of this algorithm abound [2], and while they may converge more slowly, they are equally robust. For example:

$$a_i^{(n+1)} = a_i^{(n)} + \text{sgn } \{x(n-i)e(n)\}$$

To develop your intuition, consider the following 4th order circuit.

$$y(n) = a_0^{(n)} \, x(n) + a_1^{(n)} \, x(n-1) + a_2^{(n)} \, x(n-2) + a_3^{(n)} \, x(n-3)$$

For n = 0, suppose e(n) is positive while the samples in the filter have the signs given below:

Sample	Sign	Update and Direction
x(0)	+	a_0, +1
x(1)	−	a_1, −1
x(2)	+	a_2, +1
x(3)	−	a_3, −1

With e(0) positive, we know that y(0) was not positive enough. Coefficients are updated in the direction that would have produced a more positive echo estimate, GIVEN THE SAME FILTER DATA. Each coefficient is adjusted as though it alone is responsible for all of the error. Wrong update decisions are made, but the overall transfer function trend is in the right direction.

If the subscriber is talking as the filter adapts, the sign e(n) will depend not just on the trans-hybrid path, but on this uncorrelated energy as well. Indeed, if the canceller has converged perfectly, e(n) is determined exclusively by the subscriber's transmitted signal. The erroneous updates average to zero, however, and result in the filter transfer function meandering about its target value. Convergence may be slowed down markedly by the presence of this so-called "double-talk."

Adaptive filters are vulnerable to another problem: they can only cancel the energy they see. If narrowband data constitutes the input x(n), the canceller will work very nicely over the data frequency range. Unfortunately, the canceller is liable to do just about anything at other frequencies. It may even provide gain, promoting rather than hindering 4-wire loop oscillation. Broadband signals like white noise or speech are ideal training signals for echo cancellers.

Voiceband transversal filters can be implemented via either analog sampled-data or digital signal processing techniques. A host of engineering tradeoffs are part of such decisions. Low-volume, long impulse response applications favor digital techniques [3]. The perfect analog memory will never be invented. Wide dynamic range cost-and-power-sensitive applications still favor analog CMOS. This advantage is shrinking, however, with digital design rules and advancing high-resolution CODEC technology.

549

3   OTHER APPLICATIONS FOR ECHO CANCELLERS

3.1 High Speed Data Communications

With ever-increasing demand for higher-speed data communication over
two-wire loops, the use of adaptive filtering has become unavoidable. Full
duplex data rates of up to 500 kHz will be possible through use of echo
cancellers.

A conventional frequency division multiplexed, full duplex data system
is shown in Figure 2. Since cable losses increase with $f^{.5}$ or so, problems
with the hybrid are severe. With 40 dB of cable loss combined with only
10 dB of trans-hybrid loss, echo energy dominates the desired signal at the
hybrid transmit output. In FDM systems such as the one shown, a high pass
filter is used to clean up the hybrid. Roughly 20 dB of signal-to-echo (or
noise) ratio is required at the filter output.

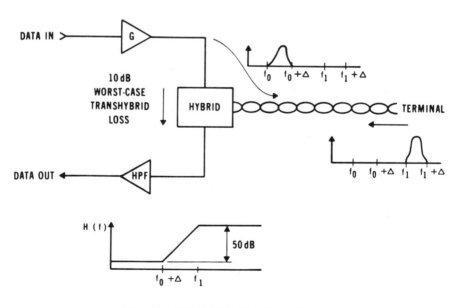

Fig. 2 — FDM Full Duplex Data System

For a full duplex data rate of Δ, FDM systems for two-wire transmission require 2Δ of cable bandwidth. To keep the high pass filter order reasonable, even more cable bandwidth is needed for practical systems. Time division multiplexed (or ping-pong) systems can emulate full duplex operation, but they require over 2Δ of loop bandwidth for analogous reasons.

Through use of an echo canceller in the same topology as Figure 1, two directions of data transmission can be isolated, even if they occupy the same frequency bands. Each direction of transmission can use the full cable bandwidth, offering better than a factor of two improvement in performance.

When comparing echo cancellers for this application with their voiceband counterparts, several generalizations can be made. First, sample rates must be much higher. Fortunately, the canceller input is a high-level digital signal; analog memory is replaced by a shift register. Because the canceller must supply 50 dB of loss to the echo signal, high resolution tap weights are required. The Agazzi architecture [4] for this type of echo canceller is discussed in Section 4.

3.2 Decision Feedback Equalizers

In the high data rate market, not only do we have poor hybrids to contend with, we also have poor cables. Bridge taps can corrupt a terminal's data signal, giving rise to intersymbol interference. Decision feedback equalizers (DFEs) are adaptive filters ideally suited for cleaning up such interference. These equalizers can be applied to TDM, FDM, or echo canceller hybrid systems.

Think of a decision feedback equalizer as an echo canceller that literally operates on its own input rather than on some echo path. Instead of an echo estimate, an estimate of the intersymbol interference error caused by the N-1 preceding data bits is produced by the DFE. This error is subtracted off before a new data bit decision is made. Adaptation is driven by the fact that intersymbol interference errors are correlated with previously detected data bits, while new data is uncorrelated.

A block diagram of a decision feedback equalizer appears in Figure 3. Operation of the equalizer requires the following steps:

1/   Clock shift register $(S_0 \rightarrow S_1, \; S_1 \rightarrow S_2,$ etc.)

2/   Compute Σ

3/ $V'_{OUT} = V_{OUT} - \Sigma$

4/ $S_0 = \text{sgn} \; (V'_{OUT})$

5/ $e = V'_{OUT} - A_0 S_0$

6/   Update tap weights: $a_i^{(n+1)} = a_i^{(n)} + \text{sgn} \; (S_i e)$

7/   Go to #1

Note that S_i here is the contents of the ith shift register cell. Figure 4 shows example waveforms [5] in a converged DFE.

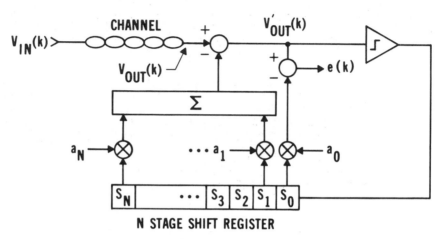

Fig. 3 - DFE Block Diagram

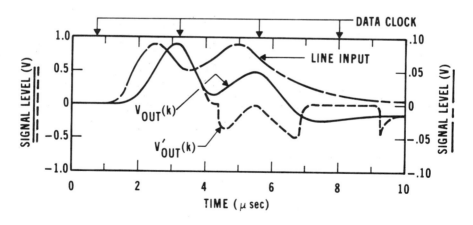

Fig. 4 - DFE Waveforms

Circuit designs for high data rate decision feedback equalizers are quite similar to those for Agazzi cancellers. Zero delay tap and update circuits are modified, but much of the analog core remains intact.

3.3 Acoustic Noise Filters

Ever-improving cost-performance of digital circuits is leading to the use of echo cancellers to enhance room acoustics or to cancel unwanted noise.

Many audiophiles with digital audio systems would be ecstatic over the prospect of transforming their living room acoustics to match those of Carnegie Hall. Acoustic mapping filters, as much echo generators as echo cancellers, use a microphone and/or training signals to determine the speaker-to-audiophile impulse response [6]. This response can then be mapped to an arbitrary response; one location in any room can have sublime acoustics.

Reverberation time constants are very long, perhaps on order 30K samples for high-quality audio. Digital transversal filters and high-resolution analog-to-digital converters are required. Consumer markets for state-of-the-art data conversion products will not be limited to DACs.

Another intriguing application for acoustic filtering is in the active suppression of unwanted noise. Adaptive filters have successfully achieved 15 dB of noise reduction in the interiors of turboprop passenger planes [7]. A pictorial view of this system appears in Figure 5.

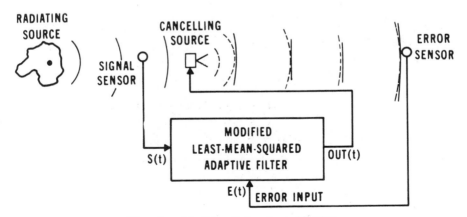

Fig. 5 - Adaptive Noise Cancellation

3.4 Television Ghost Cancellers

"Ghosts" represent one of the most significant remaining defects in television picture quality. Video sampling rates are on the order of 10 MHz. Ghost delays require adaptive filters with impulse responses extending out hundreds of sample periods. Delay-line complexities rule out switched-capacitor techniques; CCDs or high-throughput digital signal processing is required.

The ghost cancelling job is by no means easy, but the problem is rendered less intractable by the nature of the ghosts themselves. Ghosts represent pure delays, and very few canceller tap weights are non-zero. Taps not corresponding to ghost delays need not be accumulated, reducing the number of multiply-accumulate operations to only several per sample period.

Another factor simplifying the design of these cancellers is the limited dynamic range of video signals (<45 dB). 8-bit digital arithmetic should be adequate for consumer applications. Nevertheless, throughput requirements are so formidable that monolithic cancellers are just beginning to appear.

4 ANALOG CMOS ADAPTIVE FILTER DESIGNS

4.1 Voiceband Echo Cancellers

Voiceband filter applications are characterized by low sample rates (f_s = 8 kHz) and wide dynamic range (>80 dB). Adaptive hybrid balancing circuits for most 2-to-4 wire interfaces require modest impulse response duration (3 msec). Low leakage currents for floating nodes in analog MOS technology [8] allow for cost-effective sampled-data echo cancellers [9].

4.1.1. Overall Circuit Topology – A classic transversal filter architecture appears in Figure 6. Analog samples propagate down a delay-line, with multipliers at each delay tap.

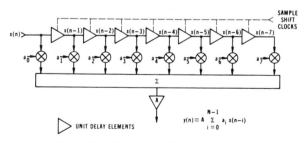

Fig. 6 – Classic FIR Architecture

554

The classic approach fails to provide small, low-noise monolithic filters for two reasons. First, each time an analog sample is moved, noise is added to it. Total noise power increases linearly with delay. Second, the high cost in silicon area of multipliers demands a multiplexed configuration. A more reasonable circuit topology appears in Figure 7.

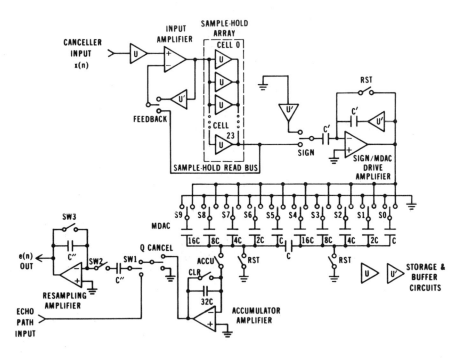

Fig. 7 - Voiceband Transversal Filter Topology

A sample-hold array replaces the delay-line. Analog samples are not moved; rather, tap weights are addressed as required to implement the convolution summation. In each sample interval the oldest sample in the sample-hold array is overwritten.

For an echo canceller operating at f_s = 8 kHz, multiplexing one multiplier to serve every sample-hold cell is attractive. For a 24th-order canceller, the effective multiplication rate is only 192 kHz, well within the speed range of 10V analog CMOS technology. Tap weights are stored digitally, and a capacitor array MDAC realizes multiplication as a digitally-controlled analog attenuation.

4.1.2. <u>Sample-hold Design</u> – One design for a sample-hold unit cell appears in Figure 8. A source-follower buffer, switch, and storage capacitor form a nearly unity-gain feedback path around the opamp in the write mode. Opening the switch "freezes" the analog sample on the storage capacitor for the duration of the impulse response.

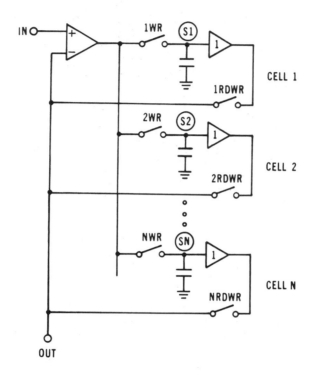

Fig. 8 – Conventional Sample-hold

While the sample-hold delay-line synthesis minimizes thermal noise, it imposes severe matching constraints on the individual storage cells. Small differences in hold-steps from cell-to-cell give rise to in-band tones of fixed pattern noise. These tones appear at multiples of f_s/N. Matching of hold-steps depends on gate-to-drain capacitance and channel charge splitting effects that produce cell-to-cell 1σ variation of order $300\mu V$. Unless corrected, fixed pattern noise will dominate the idle channel noise of sample-hold based transversal filters.

One approach to minimizing cell-to-cell hold-step mismatches is to employ a second feedback loop to cancel them [10]. The modified unit cell is shown in Figure 9 with a timing diagram. The WR1 interval is a conventional write operation. Opening of the WR1 and AWR1 switches results in a feedthrough mismatch which the opamp will correct through the WR2 path. C_B is chosen to be much smaller than C_S or C_C. Thus, overall loop gain for the WR1 and WR2 loops will be similar to that shown in Figure 10.

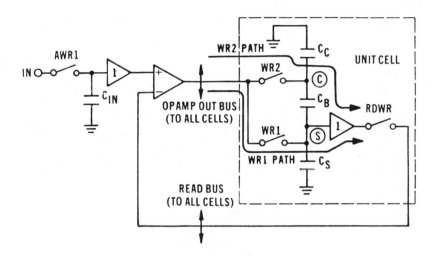

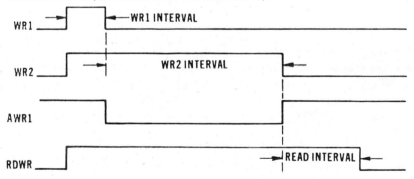

Fig. 9 - Dual Feedback Sample-Hold

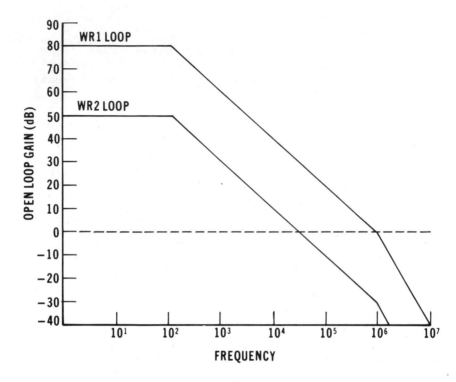

Fig. 10 - Feedback Loop Gain

At the end of the WR2 interval, the unit cell output voltage will match the "reference feedthrough" on C_{IN}. Opening of the WR2 switch generates another feedthrough on $C_C \sim C_S$, but mismatches here are divided down by the C_B to C_S ratio. Another advantage of the WR2 loop lies in its bandlimiting of the opamp thermal noise. Overall sample-hold thermal noise power approximates kT/C_{IN}. Measured total dynamic range of these dual-feedback sample-hold arrays exceeds 100 dB.

During read mode, coupling into the floating storage node can produce severe PSRR problems. The coupling path is illustrated in Figure 11 for a simple buffer. As shown, the enhancement/depletion cascode arrangement eliminates this problem and elegantly implements C_B as C_{GD1}.

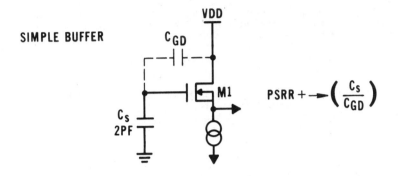

SIMPLE BUFFER

$$PSRR + \longrightarrow \left(\frac{C_S}{C_{GD}} \right)$$

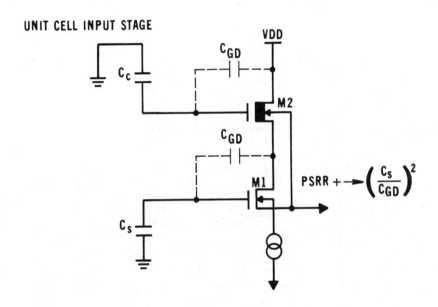

UNIT CELL INPUT STAGE

$$PSRR + \longrightarrow \left(\frac{C_S}{C_{GD}} \right)^2$$

Fig. 11 – Hold Mode PSRR

4.1.3. <u>MDAC/Accumulator Design</u> - The multiplexed MDAC shown in Figure 7 operates N times each sample interval. "Front end" sample-hold and MDAC buffer amplifier noise sources are attenuated by the tap-weight spectrum. Accumulator amplifier noise and offset are accumulated independent of tap weights, however. Even with careful design of this amplifier, its noise dominates the filter.

Minimization of accumulator amplifier noise demands a low noise density, low noise bandwidth amplifier. Large compensation capacitance is combined with high input stage transconductance. 5nV/rt.Hz thermal noise spectral density can be achieved, but the resulting dc gain is too low (worst case ‘60 dB) if a scaled minimal opamp design is used. An improved $(g_m/g_o)^3$ amplifier design appears in Figure 12. A high-swing, geometric ratio cascode input stage allows performance of this design to rival that of bipolar linear circuits. Open-loop gain and phase characteristics for the amplifier driving an MDAC load appear in Figure 13.

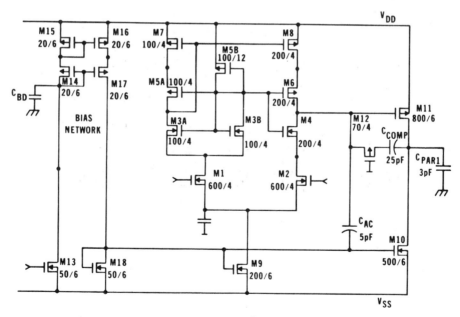

Fig. 12 - $(g_m/g_o)^3$ Amplifier

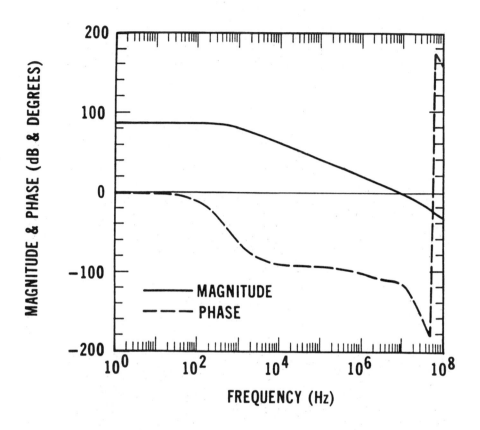

Fig. 13 - Amplifier Performance

4.1.4. <u>Performance Results for Voiceband Filters</u> – Conventional comparator designs and digital circuitry convert the core transversal filter into an adaptive echo canceller. Echo and error signals are shown in Figure 14 for a typical two-wire loop application. The residual error is nearly independent of two-wire loop characteristics. Uniform transmission quality is yet another advantage of adaptive filtering.

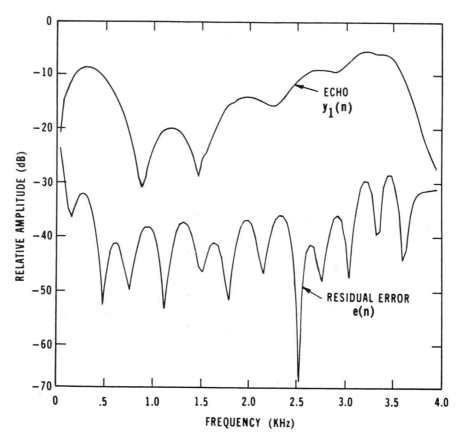

Fig. 14 – Echo Canceller Performance

The same core analog technology can be used to produce digitally-programmed analog filters. Measured and ideal results for an eighth-order transversal equalizer appear in Figure 15. The canceller and equalizer in series achieve a dynamic range of 84 dB. Die area for the complete canceller/equalizer, including a high-level digital control interface, is 34 mm^2.

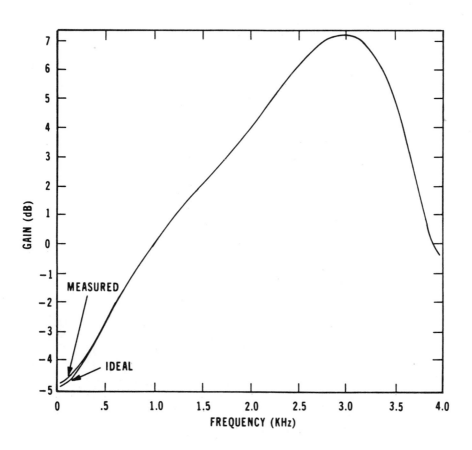

Fig. 15 - Programmable Equalizer Performance

4.2 Digital Subscriber Loop Cancellers

As mentioned in Section 3.1, echo cancellers can improve digital
transmission over two-wire loops. These cancellers have digital inputs,
and the delay line is a simple digital shift register. The echo path
begins with an alternate bipolar waveform sent to the hybrid and contains
the unknown trans-hybrid impulse response. The canceller must provide over
50 dB of echo suppression across the band, attenuating the echo to a level
well below the received data.

The high sample rates of these circuits demand echo estimates at up to
a 500 kHz rate. At only 2μsec per sample, the transversal filter
accumulation must be implemented as a flash parallel sum. Fortunately, the
digital delay line and alternate bipolar coding combine to simplify the
multiplication immensely.

$$\sum_{i=0}^{N-1} a_i^{(n)} \, x(n-i) \rightarrow \sum_{i=0}^{N-1} a_i^{(n)} \, (\pm 1)$$

A two-phase non-overlapping clock and SPDT switches implement the plus
or minus accumulation. The result is the Agazzi topology shown in
Figure 16.

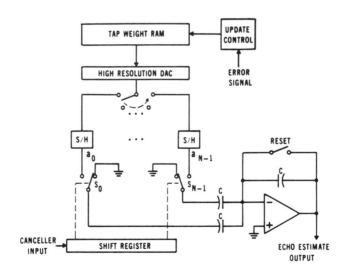

Fig. 16 - Agazzi Canceller Topology

564

In this circuit, the tap weights are converted to analog and stored in sample-holds. Switches S_0 through S_{N-1} move left-to-right or vice versa depending on their corresponding shift register data bits.

Key circuits in this transversal filter are the DAC and the accumulator amplifier. Very accurate, 50 dB return loss echo estimates require high resolution tap weights. Absolute tap weight accuracy is not required since the coefficients are imbedded in the adaptive filter feedback loop. A variety of high-resolution inherently monotonic DAC architectures will suffice [11, 12].

The accumulator amplifier must settle to .1% accuracy in about 1μsec. Note that the capacitive feedback topology shown in Figure 16 is gain of N. Lightly compensated $(g_m/g_o)^3$ amplifiers have proven adequate for this application.

The echo canceller itself produces estimates that are linear combinations of its previous data inputs. Harmonic distortion in the echo path cannot be attacked by a simple canceller. Additional special taps can correct for non-linearities in the echo path; such techniques are described elsewhere [4].

5   CONCLUSION

The role of echo cancellers in a variety of applications has been examined. The usefulness of adaptive filters in extending the performance of a wide variety of systems has resulted in their capture of an increasing share of the monolithic filter business.

Two examples of analog sampled-data echo cancellers are described. For wide dynamic range voiceband applications, nearly ideal adaptive transversal filters can be realized. Last generation two-wire loop plant data transmission systems will exploit echo cancellers to extend range and/or increase speed.

Whether constructed in the analog or digital domain, if you build an echo canceller, beware that the experience may be habit forming. Adaptive filters are simply much more interesting than their fixed transfer function counterparts. The systems they make possible are likewise more flexible, more exciting. The high-levels of integration required to implement adaptive filters promise ever increasing applications for VLSI in telecommunications.

6   REFERENCES

[1] M. M. Sondhi, "An Adaptive Echo Canceller," BSTJ, 56, March 1967.

[2] M. M. Sondhi, D. Mitra, "New Results on the Performance of a Well-Known Class of Adaptive Filters," Proceedings of the IEEE, 64, November 1976.

[3] D. L. Duttweiler, Y. S. Chen, "Single-Chip VLSI Echo Canceller," BSTJ, 59, February 1980.

[4] O. Agazzi, D. A. Hodges, D. G. Messerschmitt, W. Lattin, "Echo Canceller for a 80 kbs Baseband Modem," ISSCC Digest, 25, 1982.

[5] Y. Hino, et al., "A Burst-Mode LSI Equalizer with Analog-Digital Building Blocks," ISSCC Digest, 27, 1984.

[6] R. K. Randall, private communication.

[7] L. A. Poole, G. E. Warnaka, R. C. Cutter, "Adaptive Cancellation of Acoustic Noise," ICASSP Digest, 1984.

[8] A. B. Grebene, Bipolar and MOS Analog Integrated Circuit Design, Wiley, New York, 1984, p. 716.

[9] E. J. Swanson, et al., "A Fully-Adaptive Transversal Canceller and Equalizer Chip," ISSCC Digest, 26, 1983.

[10] Hans Lie, private communication.

[11] B. Fotouhi, D. A. Hodges, "High Resolution A/D Conversion in MOS/LSI," IEEE JSSC, SC-14, December 1979.

[12] H. S. Lee, D. A. Hodges, "Self-Calibration Technique for A/D Converters," IEEE CAS, CAS-30, March 1983.

17

DIGITAL TELEVISION

Rainer Schweer[*], Peotr Baker[**],
Thomas Fischer[*], and Heinrich Pfeifer[*]

[*]Concept Engineering
ITT-Intermetall
Hans-Bunte Str. 19
7800 Freiburg
West Germany

[**]Filter Design
Consultants Limited
Park Road, Allington
Lincolnshire NG32.2EB
U.K.

1   INTRODUCTION

Digital television is a vast field, actually too large to be treated completely in a single chapter. Thus we will focus on digital baseband processing within a TV receiver and its realization with integrated circuits. For the realization we will concentrate on a system which is already available. References of the relevant literature and hints to competetive system solutions will be included.

The normal reader is not expected to be familiar with TV concepts. So at the beginning of each chapter we will point out the different problems without going too far into tedious details which are only of interest to TV specialists. The different solutions and their realizations on silicon will be considered afterwards.

The TV receiver or its monitor is thought of as a center piece in a home communication system. Therefore the digital TV fits very well in this environment and furthermore enables us to get improved picture quality using digital enhancement methods.
From Table 1.1 one can deduce that a video system differs from normal telecommunication applications. One of the main aspects is that telecommunication systems must react to a large number of interactive users, whereas a home system is under the control of one person only.
Major constraints are set by the commercial environment. You have to make it inexpensive and they must work without any maintanance in the field. But the greatest challenge is the huge amount of data.

Let us look at the recommended A/D conversion rate of the video signal of 8 Bit. It is 13.5 MHz. Considering a simple digital TV system in terms of MIPS (million of instructions per second) to compare it to normal computer efficiency, leads to a demand which is above that available even from the fastest mainframe.

	PROCESSING SPEED	SYSTEM STRUCTURE	PRODUCTION VOLUME	USER NUMBER IN PARALLEL
TELECOMMUNICATION	SLOW-FAST	MULTI PATH	MEDIUM	VERY HIGH
DIGITAL - TV	VERY FAST	PARALLEL	HIGH	VERY FEW
MAINFRAME COMPUTER	VARIABLE AS FAST AS POSSIBLE	HIERARCHICAL	MEDIUM	FEW
HOME COMPUTER	MEDIUM	HIERARCHICAL	MEDIUM	VERY FEW

Tab. 1.1 System Comparison

E. g. the video processor unit performs, besides other functions, 500 million/sec additions and 20 million/sec multiplications. For the storage of one picture (elimination of flickereffects e. g.) 4 MBit must be written into RAM in 20 msec. It can be easily calculated how much film material can be digitally stored on todays storage media.
These demands lead at once to extremely dedicated signal processing units. Therefore the system design aspects have to be considered carefully to get a flexible system.

After system definition we will look at the different interfaces and system connections. The description of the control unit will be followed by a discussion of the deflection and video processing units and the audio processing section.
The teletextchip, which provides the entry into new communication services will be discussed. Finally future trends and challenges will be considered.

2   SYSTEM AND INTERFACES

2.1 Problems

The system design must be performed with regard to future systems, where the monitor or TV set is the center of a home communication system.
The different stages and functions of a TV receiver are shown in Fig. 2.1. At the front end is the tuning system which for now and the near future will only be a digitally controlled system, whereas the IF stages are completely in the reach of digital systems in the next years. The system of today covers the blocks surrounded by dashed line.

568

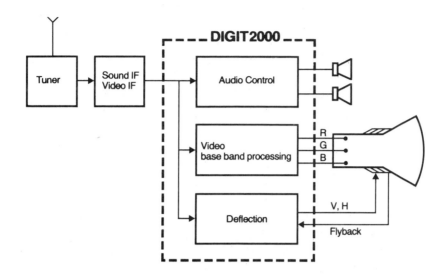

Fig. 2.1 Block diagram of a TV receiver

Future tasks and interconnections have been gathered in Fig. 2.2.

Fig. 2.3 shows the main points of the different TV norms. For detailed specifications see [1-2]. One has to keep in mind that due to the historical development and market forces the TV system is well suited to analog techniques. Furthermore, the consumer is accustomed to a mature system with some nonlinear corrections (flesh tone correction, gamma correction, etc.). Additional features inherent to a normal TV screen have to be provided (picture geometry corrections, fly-back control, etc.). Thus the developed digital TV has to compete with a well introduced product with many years of manpower in its development and state of the art components.

2.2   System Aspects

One of the first things which has to be resolved is the choice of clock frequency. The choice is not only dependent on limits due to the Nyquist theorem, but also due to dependences on clock frequency versus horizontal line frequency versus chroma subcarrier frequency. In some cases the chroma subcarrier frequency varies against the line frequency. Taking a quartz stabilized fixed clock would be a solution. But then you collect all the disadvantages of the system comparison made in Table 2.1.

	VIDEO PROCESSING	SAMPLE FREQUENCY	DEFLECTION	INTERFACE TO FRAME STORE	FEATURE CAPABILITIES CRT-CORRECTION
FREQUENCY LOCKED TO HORIZONTAL LINE	COMPLEX	REL. LOW	EASY	EASY	RESTRICTED
FREQUENCY LOCKED TO COLOR SUBCARRIER	EASY	HIGH	COMPLEX	INTERPOLATOR	POSSIBLE

Tab. 2.1 Comparison of different system clocks

The choice was 4 times colorsubcarrier for A/D and D/A conversion and system clock frequency of the digital signal-processing systems. This resulted in a sampling frequency of 17.7 MHz (PAL), 14.4 MHz (NTSC), 17.14 MHz (SECAM). Due to the technology (HMOS), only parallel and pipelined processing of operations related to each sample (filters, multipliers, code converters etc.) is used. The clock frequency is higher than the Nyquist theorem requires, but phase locking this sampling clock to the color burst (the reference transmitted by TV station) at 45 degrees results in an inherent separation of the R-Y and B-Y components of the composite video signal. Another important situation where high sampling frequency is advantageous is in a Video-tape-recorder where the degradations of the video signal are considerably large. This system profits from the fact that the subcolor carrier regeneration is treated with much more emphasis, and furthermore, oversampling (signal frequency 0-3 MHz only) can be used for noise reduction.

Due to the high speed and need of parallel processing the multiplexing of busses is not possible. Using the flyback times, however, will give some time division multiplexing opportunity which has been used for pin count reduction.

2.3 System Partitioning and Interfaces

The partitioning of the system is straightforward according to the different tasks: deflection, Video signal processing, conversion etc. The chip set consists of:

Video Codec Unit	VCU
Video Processor Unit	VPU
Deflection Processor Unit	DPU
Audio Processor Unit	APU
Audio A/D Converter	ADC
Central Processor Unit	CPU

TV-environment

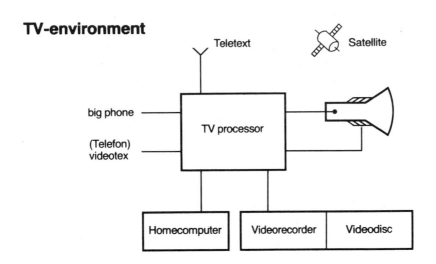

Fig. 2.2 TV Environment

TV Norms:
Differencies and
Similarities

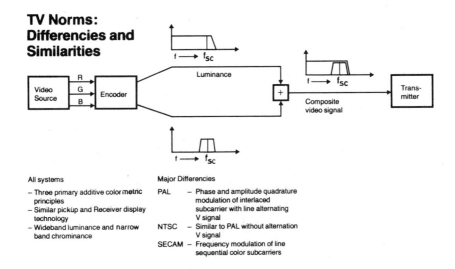

All systems

– Three primary additive color metric
  principles
– Similar pickup and Receiver display
  technology
– Wideband luminance and narrow
  band chrominance

Major Differencies

PAL – Phase and amplitude quadrature
      modulation of interlaced
      subcarrier with line alternating
      V signal

NTSC – Similar to PAL without alternation
       V signal

SECAM – Frequency modulation of line
        sequential color subcarriers

Fig. 2.3 TV norms differencies and similarities

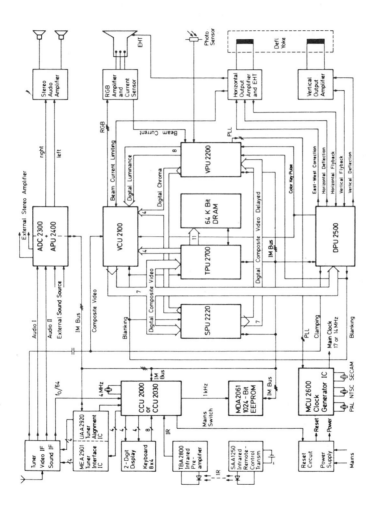

Fig. 2.4 Block diagram of a color TV receiver

572

and optional

Teletext Processor Unit	TPU
NTSC Comb Filter Video Processor	CVPU
SECAM Chroma Processor	SPU
Automatic Picture Control Unit	APCU

All ICs but the VCU are fabricated in a 3 μm N-channel MOS technology. The effective channel lengths are 2.5 μm. The VCU and ADC are bipolar as are a couple of smaller circuits that assist the digital signal processors, that is color PLL and clock driver and infrared preamplifier. Fig. 2.4 shows the block diagram of a receiver built with the described chip set.

The CCU consists of the well-known 8049 8-bit microcomputer and some additional circuitry on chip for tuner control and infrared command receptions, (user interface). Here all incoming commands and information from the signal processing units will be interpreted and harmonized with the CRT-type or Video norm and the according settings which are given to the different units.

As the course is intended to deal with MOS-circuits the VCU is discussed only in short. Its block diagram is given in Fig. 2.5. It is intended for the conversion of an analog composite video signal into Gray coded data and further for the reconversion of the processed video signals followed by an RGB matrix and output amplifiers with programmable gain and offset and some additional circuitry which is of importance for the CRT.

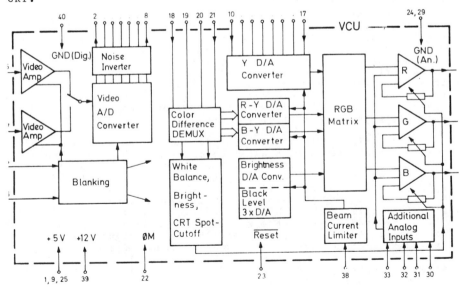

Fig. 2.5 Block diagram of the video conversion unit

573

The A/D converter is of the flash type. Thus it is important to have as few bits as possible. For a slowly varying video signal, 8 bits are required. In order to achieve 8-bit picture resolution using a 7-bit converter a trick is used: during every alternating line the reference voltage of the A/D converter is changed by an amount corresponding to one half of a least significant bit. In this procedure, a grey value located between two 7-bit steps is converted to the next lower value during one line and to the next higher value during the next line. The two grey values on the screen are averaged by the viewers eye, thus producing the impression of grey values with 8-bit resolution.

3   VIDEO SIGNAL PROCESSING

3.1   Overview and Coding Principles

The technical basis of the different color TV norms lie in the fundamentals of the science of colorimetry. With the concept of hue, saturation and brightness a norm compatible to the monochrome TV could be introduced without increase of the transmission channel bandwidth.

In the following discussion we choose the standard PAL as a vehicle for digital TV realization and point out afterwards or in parallel necessary modifications for other norms. The standard RGB-signals of a video signal source representing the red, green, and blue information of a scene on a point by point basis as determined by the scanning rates have a large bandwidth, (0-5MHz), and have to be transmitted from the source to the receiver. Different methods could be adopted. Employing three different channels or one common channel time multiplexed.

The discussion of the latter method was revived for the introduction of direct broadcasting transmission systems, as it is well suited to modern digital systems. This method avoids the drawbacks of the formerly adopted frequency interlacing systems for Color TV which had been implemented with regard to the analog techniques of those days.

The effect that the eye is more sensitive against brightness changes than against color changes has been known for long time. The color acuity of the eye decreases as the size of the viewed object decreases. Thus the three signals have been split into two different signals, the luminance signal with high bandwidth, which contains the brightness information including signal details, and the chrominance signal with low bandwidth, which is added in such a way that it is ignored by monochrome receivers. In Fig. 3.1 a typical FBAS line signal is depicted. The different abbreviations are included for clarity. In standard PAL the chrominance information is carried simultaneously as phase and amplitude modulation of the color subcarrier, which was chosen to lie around 4.3 MHz. From Fig. 3.2 the modulation scheme can be derived. Further, the phase of the V=R-Y signal is altered from line to line by 180 degree. Thus the PAL chroma expression is given by

$$C = U/2.03 \sin \omega_{sc} \, t + V/1.14 \cos \omega_{sc} t$$

where $U=B-Y$, $V=R-Y$, and ω sc is the color subcarrier frequency.

Typical FBAS line

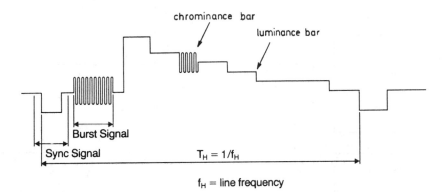

Fig. 3.1 Typical FBAS-Signal during one horizontal line

The phase reversal was applied to cancel errors due to differential phase distortion of the color modulation sidebands. This cancellation can be performed by the viewers eyes (simple PAL), but is commonly performed via a so called PAL-Delay, which is a 1H-Delay.
With the line to line averaging a cancellation of the phase error is achieved. The saturation is somewhat reduced, but this error is less visible. The subcarrier frequency is chosen to be $f_{sc} = 1135/4 \, f_H + 1/2 \, f_V$, thus getting a mostly interlaced signal as depicted in Fig. 3.3. The detailed structure can be found there also.

After this excursion into TV principles we present the realization for the video signal processing. We start with a functional description and discuss afterwards the most important tool the digital filters. In particular, the PAL VPU luminance and chrominance filters will be discussed.

3.2 Functional Description

Fig. 3.4 shows the block diagram of the video processing circuitry. All the decoding and filtering circuitry is situated on the VPU. That is:

- a code converter

575

Chrominance coding principle

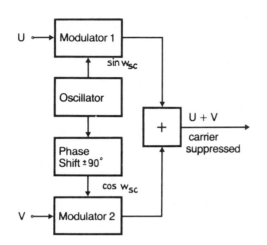

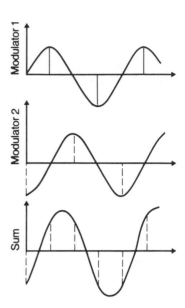

Fig. 3.2 Chrominance coding principle

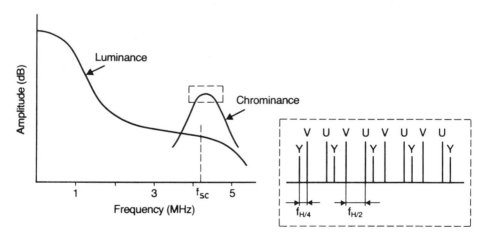

Fig. 3.3 Frequency relationship of modulated chrominance and luminance

- the chroma bandpass filter
- the variable peaking filter with chroma trap
- a contrast multiplier with limiter for the luminance signal
- all color signal processing circuits such as automatic color control (ACC), color killer, Pal identification, decoder with Pal compensation or NTSC comb filter, hue correction working on PAL and NTSC
- a color saturation multiplier with multiplexer for the color difference signals

There are two main signal paths, the luminance channel and the chrominance channel.

In Fig. 3.5 the amplitude of the video signal in the different stages is shown. This is an important aspect to be looked at especially at the points of nonlinear processing, where overshoot or undershoot may occur.

3.2.1 The Luminance Channel - This channel starts with the chroma trap, followed by the peaking which will be discussed at length in the following. Behind the luminance filter 8 bits are used to carry the luminance signal, thus the overshoots caused by the peaking filter would not lead out of range as can be seen in Fig. 3.5.

The contrast multiplier which follows the peaking circuit is combined with a limiter, limiting the luminance signal if its amplitude becomes too high. The contrast is adapted to ambient lighting by means of a photo sensor connected to the chip. In this process, the signal generated by the photo sensor is first digitized and then, during vertical flyback, transferred to the central processor. After calculation of the needed contrast settings with the central processor the values are presented back to the VCU.

3.2.2 The Chrominance Channel - The chroma channel starts with a filter too, the chroma bandpass. The filter is followed by the automatic color control (ACC), which measures the burst amplitude, as the burst signal can be looked at as a good reference. If the burst signal is too weak the chrominance channel will be switched off (color killer). In the other cases the chroma channel will be influenced accordingly. The PAL flip-flop which is synchronized by the burst is also situated here.

The chrominance signal is decoded by an AM Synchronous Demodulator, giving the color difference signals R-Y and B-Y. The decoder also contains the PAL compensation circuitry which uses an adressable RAM for signal storage of one line. The delayed and the non-delayed signal are added according to the PAL coding scheme to compensate for phase-angle dependent errors. In NTSC operation, the PAL compensation circuit works as a comb filter, improving cross color suppression.

The digital color difference signals R-Y and B-Y are now routed to one color saturation multiplier at a rate of 2*fsc. This is possible because the bandwidth of the chrominance

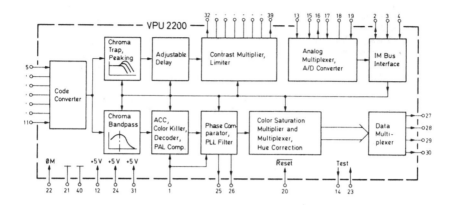

Fig. 3.4 Block diagram of the video processing unit

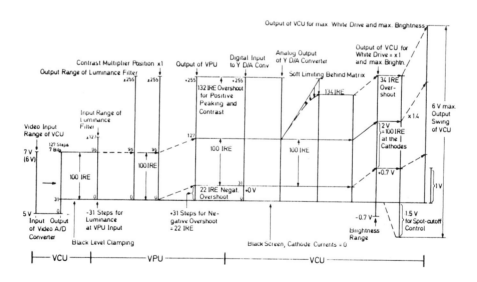

Fig. 3.5 Luminance level diagram

channel is low, (0-1.1 MHz). Therefore color subcarrier
frequency is now used as the sampling frequency. The decrease
in sampling frequency also permits the multiplexed multiplier to
reduce chip size and pin count.
    Hue correction is applied in the saturation multiplier
according to the formula

$$(R-Y) = (R-Y) * f_{SC} * \cos \alpha - (B-Y) * f_{SC} * \sin \alpha$$

$$(B-Y) = (R-Y) * f_{SC} * \sin \alpha + (B-Y) * f_{SC} * \cos \alpha$$

The possible resolution of the angle rotation is three degrees
and the range is from 0 to 360 degrees.

3.2.3 Phase Comparator - An additional circuit in the chroma
channel after the PAL compensator is the phase comparator, which
is part of a phase locked loop. It consists of a simple digital
comparator triggered by the color key signal which is supplied
from the deflection processor. The output signal due to input
phase deviation is shown in Fig. 3.6. This signal drives the
PLL compensation filter and in turn a D/A converter and the
system VCO. Thus completing the loop.

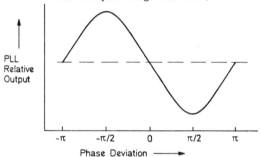

Fig. 3.6 Phase Comparator Output

3.3 Video Filters

    The principal requirement of these filters, given that the
video spectral energy distribution takes the form shown in
Fig. 3.7, is that the luminance filter extracts luminance
information and conditions it while rejecting chrominance, and
that the chrominance filter extracts chroma information while
rejecting luminance. As we shall see linear phase and picture
edge enhancement capabilities are important conditioning
properties of these filters.

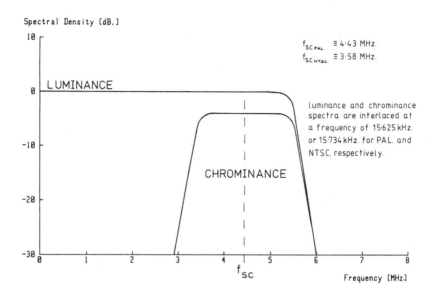

Fig. 3.7 Video Signal Spectral Distribution

For digital synchronous demodulation of the chrominance signal in either PAL or NTSC systems the practical ideal sampling frequency is four times color subcarrier frequency (Fig. 3.2). Bearing in mind the required spectral bandwidth, (5.5 MHz), and taking the Nyquist theorem into account, a filter system sampling frequency of 4 * fsc is also quite useful. Furthermore, it can be shown that many simplifications result with regard to filter synthesis whith such a choice.

Examples of this include the ability to generate arithmetically symmetric bandpass and bandstop filter approximations for chrominance signal processing. These are economic in terms of multipliers, adders and, consequently silicon area.

These filters which have a sampling frequency of 17.7 MHz are implemented in NMOS technology and thus give problems in achieving the required processing speed. To this end filters have to be realized using special pipelining and elementary systolic processing techniques, as will be discussed. Such techniques permit the elemental processing networks to process data at clock frequency. A final point to be made is:

constraints of linear phase and small silicon area indicate that finite impulse response, FIR, filter structures are particularly, although not exclusively, useful for our application.

3.3.1 Video Filter Structures and Techniques - Implementing video frequency digital filters in NMOS VLSI creates its own particular set of problems and solutions. Primarily, one is limited by technology in producing filter response approximations, and the associated networks, at high frequencies, and in a given small area of silicon. These limitations have led to the following set of techniques and constraints being applied.

(i) Pipelining:

This is applied over two domains. Firstly, in the transfer function domain, the filter circuit layout is very similar to the network schematic, having a one-to-one correspondance, to aid realization except for included redundancy. This permits pipelining in the conventional sense. Secondly, in the arithmetic domain, a bit wise pipeline exists from the lsb to msb. In this case the particular calculation on the m'th bit of a word, at a particular time instant, is delayed by m time samples. This method can be considered to be an elementary form of systolic processing, [3]. Both forms of pipelining will be considered in a little more detail, further on in this chapter.

(ii) Simple multipliers:

These take the form of a single hard-wired arithmetic shift, or shift and add operation of the form:

$$\sum_{i=1}^{L} \pm 2^{p_i}; \; p_i \text{ integer}$$

where L is as small as possible and the range of the p_i is as small as possible. In both cases this is to minimize silicon area.

(iii) Finite impulse response approximations:

FIR filter approximations are particularly well suited to our requirements because they are ameanable to the methods of pipelining employed. They are also suitable due to linear phase and constant group delay constraints, which are easily satisfied. Indeed, given we are to use FIR filters, these constraints are advantageous because the number of multipliers is generally reduced, [4], for a given approximation. Finally, FIR filter approximations, when implemented using non-recursive techniques, have the property that they do not generate so called limit cycle noise.

(iv) Cascade networks:

Cascading of relatively small subnetworks to provide a complete filter permits localized and silicon area efficient design, by reducing the amount of metallization and associated path lengths. From a point of view of synthesis, cascade filter design is also simpler (bearing in mind the silicon size and range of multipliers available). This is because the approximation can be broken down into a series of smaller problems and tackled as such. This in turn reduces design time and appears to lead to the selection of low complexity subnetworks.

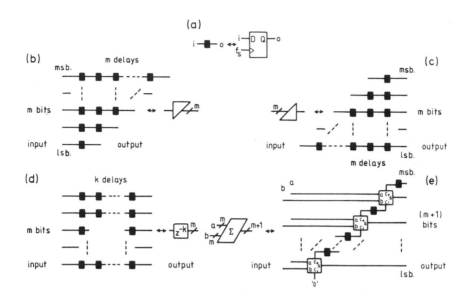

Fig. 3.8 Elemental building blocks for function-bit pipelined structure. (a) one-one comparison of a single bit unit-delay, (b) bit pipeline converter, (c) bit pipeline inverter, (d) K unit-delay, (e) bit pipelined adder.

Consideration will now be given to methods of pipelining employed. The input data word to a filter is bit wise skewed by a pipeline converter of the form shown in Fig. 3.8b. This operator is applied solely to permit pipelined carry propagation during addition.

To illustrate the principle, a partial network for pipelined addition is given in Fig. 3.9. This network has two pipeline converters, a bit pipelined adder and inputs a(n) and b(n), where n is the discrete time sample number. As the lsbs $a_1(n)$ and $b_1(n)$ are added by adder#1 the sum and a carry output are generated. The carry output propagates through the unit delay to the carry input of adder#2. At the same time unit-delayed $a_2(n)$ and $b_2(n)$ are presented to adder#2. Thus these quantities are added in the correct time frame and a further sum and carry output are generated. This operation continues toward the msb and further partial summations follow.

As shown, partial sums $S_k(n-k)$ are generated at the time $a_k(n)$ and $b_k(n)$ are presented at the pipeline converter input. Reference to the bit pipeline inverter of Fig. 3.8c shows this network has the ability to reorder $S_k(n-k)$ to give a truely parallel sum S(n), with only a pure processing time delay. To illustrate more fully how the elemental networks of Fig. 3.8 are combined within a typical filter, the derivation of a subnetwork, (which is incorporated within the PAL luminance

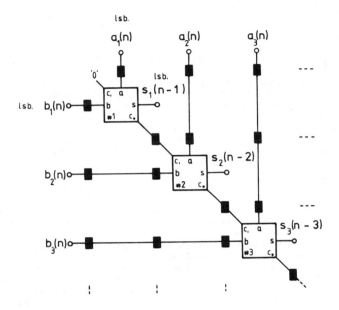

Fig. 3.9 Bit pipelined converter and adder connection example

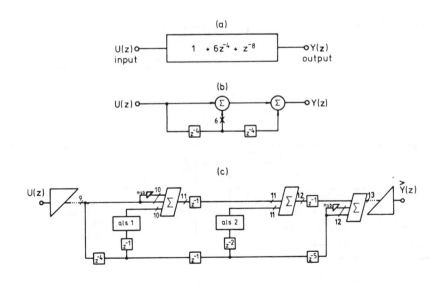

Fig. 3.10 Subnetwork derivation. (a) Transfer function,
(b) Structure Derivation, taking silicon layout into account,
(c) pipelined structure.

filter), is given in Fig.   3.10.   Two   factors   of   importance
within Fig. 3.10c are:

firstly, the input is a pipelined version of U(z). This is
derived either directly from the bit pipeline converter or
alternatively from the output of a preceding bit pipelined
filter or algebraic subnetwork.

And secondly, integer multiplication by a factor of six is
achieved   by arithmetic left shift operations, (als.) and
addition. The inherent delay in the als. stages are to allow
the signal to be time frame synchronized with the other signals
they are to be added to.

To isolate the partial results of the multiplication and to
make the network realizable in NMOS, unit-delays are inserted
between the adder stages. A further point of interest is that
the bit pipelined adders have to be sufficiently wide to accept
the maximum input. To cope with this, the smaller word has to
be expanded to the maximum input word length by padding out its
msb, to extra inputs, via a pipeline converter as before, to
place the signals in the correct time frame.

Finally, from a point of view of practical implementation
the actual wordlengths shown will totally avoid overflow. This
constraint although necessary is very expensive to apply in
terms of silicon area, and extremely pessimistic bearing in mind
a filtering operation is being performed. Word length can be
usually reduced in theory and even further in practice, [5],
without overflow occuring.

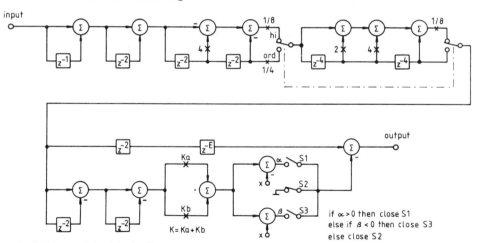

note: E is the equalising delay for the
nonlinear network in the lower path.

Fig. 3.11 PAL VPU luminance filter structure

584

3.3.2 PAL VPU Luminance Filter - The luminance filter has a large number of requirements and constraints placed upon it. Some of the main requirements and our solution to these will now be summarized.

(i) The filter must have a bandstop characteristic about the color carrier frequency, of 4.43 MHz and a lowpass characteristic after this. This response is supplied by the first two subnetworks of the filter shown in Fig. 3.11. They suppress the chroma signal and high frequency noise. Additionally they provide interpolation for dynamic input signals.

(ii) The filter should have a symmetric peaking response. That is, when a step is applied to the filter, the step response should have an identical preundershoot and overshoot characteristic. Such responses are given in Fig. 3.12 and Fig. 3.13. This form of response enhances transition sharpness of the visual information presented on the television screen, giving an implied increased spatial bandwidth. Such a requirement places the constraint of linear phase and constant group delay on the complete filter. Peaking is obtained by the lower subnetwork cascade of Fig. 3.11. This network in effect performs a double differentiation function and hence enhances rapid edge transitions.

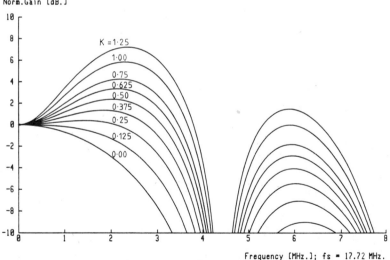

Fig. 3.12 PAL VPU luminance filter magnitude response, with switch in position "ord".

585

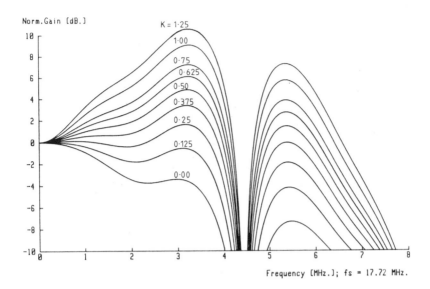

Fig. 3.13 PAL VPU luminance filter magnitude response, with switch in position "hi".

The degree to which enhancement takes place is controlled primarily by the varaible gain factor K. The effect of variation in K is shown in the frequency domain and time domain responses of Fig. 3.12 to Fig. 3.15. The frequency domain responses show the differentiating effect quite well by the increasing positive slope with increasing K.

(iii) The effect of peaking introduces undesired visual noise enhancement to almost constant grey levels on television screen, which is disturbing to the eye and has to be removed. This removal is accomplished by the nonlinear subnetwork incorporated within the filter. Its transfer characteristic is piecewise linear with a deadband of +-x about zero input.

The deadband can be adjusted, to suppress low level noise by varying the input paramter x. However, larger amplitude edge transition information can easily break the deadband and pass through.

(iv) Zero frequency gain must be invariant of mode or peaking selection, to maintain constant grey level. In variable simple coefficient filters this is problematic because gain compensation can rarely be performed using simple coefficients. In this filter structure the problem is overcome by controlling the section with no zero frequency gain contribution.

(v) Zero frequency gain must be of the form 2^m, m integer. The reason for this being that when a gradual noise free ramp is applied at the filter input, that is, the input changes in single quantization steps of Q giving a difference at each point in time of Q, Q, Q, etc. If the filter is not constrained in this way then, the output has uneven changes in Q steps.

586

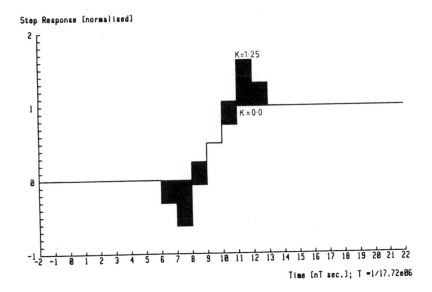

Fig. 3.14 PAL VPU luminance filter step response, with switch
in position "ord", for K = 0.0 to 1.25

This is a correlated noise mechanism. A good example of the
noise mechanism occurs if the zero frequency gain is 3/2. Then
for a Q, Q, Q etc. difference in input, the output changes by
Q, 2Q, Q, 2Q, Q, etc. This shows the effect of not obeying the
gain constraint appears subjectively to increase the visual
quantization noise, by one bit, for a given filter output
wordlength.
(vi) Ordering of the network could in a sense be arbitrary,
because no lsb. quantization is applied within the network.
However, ordering is applied with view to minimizing network
complexity and silicon area.
     Our solution filter is switchable to increase the bandwidth
of peaking, as can be seen if the magnitude responses of Fig.
3.14 and Fig. 3.15 are compared. This is desirable to increase
the horizontal spatial resolution on the television screen.
However, having achieved this, it is apparent that for excessive
peaking the cost is that the chrominance signal can leak into
the Luminance channel causing undesirable effects.

     SECAM requires that a larger stopband bandwidth exists,
primarily because the FM chrominance information in a SECAM
signal extends lower in frequency than in a PAL system. The
NTSC system tends to rely more on higher peaking (increased K)
to subjectively improve the television picture.

3.3.3 PAL VPU Chrominance Filter – Chrominance filter requirements and constraints and our filter solution to them are summarized as follows:

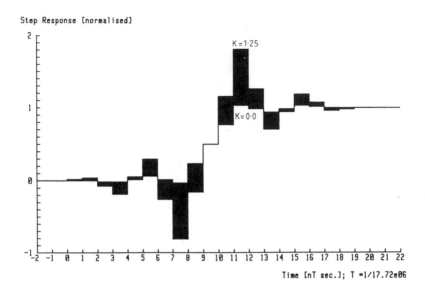

Step Response [normalised]

Time [nT sec.]; T =1/17.72e06

Fig. 3.15 PAL VPU luminance filter step response, with switch in position "hi", for k=0.0 to 1.25

(i) When coded rectangular chrominance signals are demodulated, it is important, for visually subjective reasons that the rising and falling edges have a mirror image symmetric response. To achieve this the chrominance filter must be linear phase.
(ii) For the case of gradually changing color it is important to have a midband gain of 2^m, m integer. The reasoning for this is the same as for the luminance filter.
(iii) It is necessary to have the filter passband switchable, see Fig. 3.16, to compensate the preceding IF section filter. Where, the IF section could have either a surface acoustic wave filter, (flat passband selected), or a lumped element passive filter, (equalizing passband selected).
(iv) The stopband response attenuation of the chrominance filter needs to be of the order of 40 dB in the lower band to suppress, to within an lsb, feedthrough of the baseband luminance signal into the demodulator and subsequently to the TV screen. Also, a reasonably large attenuation is required in the upper band to suppress higher frequency noise being imaged down to the baseband after demodulation and again being presented to the TV screen.

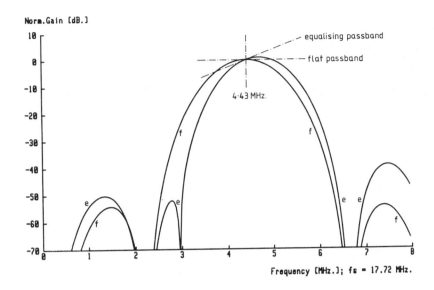

Fig. 3.16 PAL VPU chrominance filter magnitude response

All the above requirements have been achieved with the network of Fig. 3.17. This filter is novel due to its simplicity, in that only simple multiplication factors of +-1 are used.

The filter, with a change in sampling frequency, can be used in an NTSC system without modification. For SECAM however, requirements are quite different from those of PAL. Consequently a completely different solution exists. The SECAM chrominance filter must compensate both phase and amplitude of the transmitted signal, [6]. Also in an analog TV the chrominance filter compensates the IF filter response. In a digital TV concept the first two problems can be overcome by the use of an infinite impulse response, IIR, filter matched to the phase and amplitude of a SECAM analog prototype filter. The third problem can be overcome by the inclusion of a variable passband equalizer. Our present filter solution to this problem is given in Fig. 3.18.

This consists of a cascade of FIR equalizing subnetworks, providing a gain gradient of 6.37 (b1+b2) dB/MHz, (at 4.43 MHz), see Fig. 3.19, and a cascade of FIR and IIR subnetworks forming the bandpass network, see Fig. 3.20. The whole filter including IIR subnetworks is bit pipelined. Limitations exist with regard to the implementation of bit pipelined IIR subnetworks. This is due to what can be the inherent long processing delay in the bit pipeline causing a timing conflict with the requirement of short delay for recursive processing and particularly because of the inclusion of shift and add

589

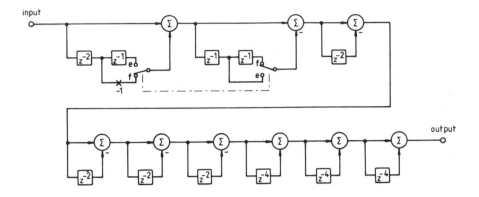

note: e ⇒ equalising passband response.
      f ⇒ flat passband response.

Fig. 3.17 PAL VPU chrominance filter structure

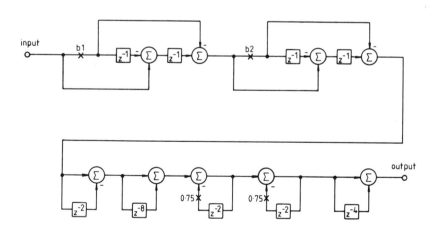

Fig. 3.18 SECAM VPU chrominance filter structure

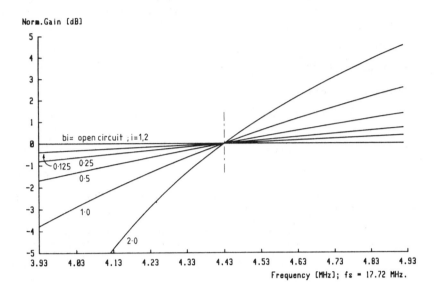

Fig. 3.19 SECAM VPU chrominance filter

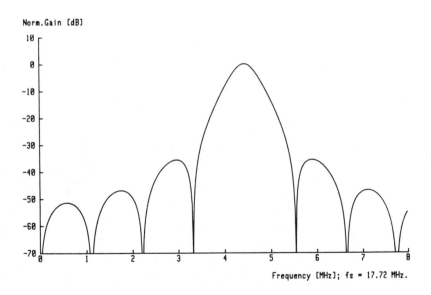

Fig. 3.20 SECAM VPU chrominance filter

591

multiplication delays. In our case the delay of two samples and a coefficient arithmetic right shift of two places to achieve a multiplier of 0.75 are compatable.

4   SCAN CONTROL PROCESSING

4.1 Tasks and Problems

The deflection circuit has the following tasks:

- horizontal synchronization.
- vertical synchronization.
- video signal clamping.
- generation of the horizontal and vertical sync and monitor correction signals.
- incorporation of monitor transfer function for picture contents depedent corrections.

    The latter two tasks represent the first step to the combination of a digital TV plus a digital controlled monitor.

    The deflection circuit delivers the scan control signals for the monitor. These must be synchronized to the incoming video signal, and incorporate monitor or type dependent corrections.
    Therefore they can be looked at as frame points. The problem which arises in some systems is that the system clock is not phase locked to a multiple of the horizontal clock frequency. In standard color TV transmission systems the color subcarrier is locked to the horizontal frequency.
    From this point of view, if we generate the sync-signal (D/A-conversion) with a one Bit converter - which can be a clean digital output - we have a systematic uncertainty of +- 1/2/fs, which is +-28 ns in PAL-systems. The same argument holds for the detection of the fly back pulse when a single one-bit comparator is used. The incoming video signal and therefore the inserted sync signal too has a better resolution (7 Bit) and is bandlimited. Thus the extraction of the sync shoulder can be performed with better resolution.
    A 56 ns clock period corresponds at a 66 cm diameter tube, to approximately 0.5 mm on the tube horizontal direction. This is a width which can be neglected, if you look at VTR generated pictures or normal TV pictures. The disturbing effect is that this phase or displacement shift can and does occur due to noise mechanisms from frame to frame. Thus generating a moving impression, especially at edges. This oscillating or fringing can be recognized at good (66 cm) monitors and good TV signals down to a tenth of the actual pixel width. The reason is that the eye is very sensitive to movements in a quiet picture.
    The flyback signal is not bandlimited. Thus the detection of the rising edge resolution has to be performed either with a higher system clock or with an antialising filter and more than a one bit A/D converter resolution.
    The preceding comparison (Tab. 2.1) of horizontal frequency - versus subcarrier frequency coupled sampling clock,

appears in a different aspect with regard to the monitor control and correction.

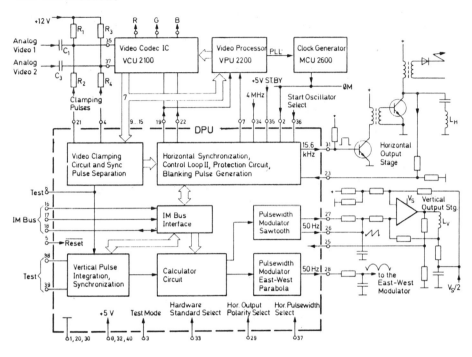

Fig. 4.1 Block diagram of the deflection processor

4.2 Deflection Processor Unit (DPU)

The deflection processor unit shown in Fig. 4.1 contains the circuit solutions to these problems. The most interesting part besides the clamping circuit is the horizontal synchronization. The block diagram is shown in Fig. 4.2. It comprises two different phase comparators and two filters to fulfill the above described requirements.

Two different operation modes exist, the so called non locked mode, and the locked mode. In the latter case there is a fixed ratio between the color subcarrier frequency and the horizontal frequency. In this mode the Influence of the PLL filter on the programmable divider is switched off. Therefore, interfering noise and pulses cannot influence horizontal deflection, if they occur during the sync signal.

But nevertheless phase comparator two is not disconnected and the influence of the monitor inherent distortions and signal distortions are measured by means of a balanced gate delay line, to get a time resolution of the leading edge up to +-3.5 ns.

With the same resolution the leading edge of the outgoing sync
signal can be adjusted.

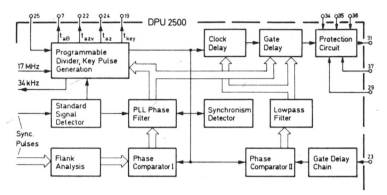

Fig. 4.2 Principle of the horizontal synchronization

In the unlocked mode one has to think of an internal
running horizontal frequency which is calculated up to fractions
of a clock cycle. The difference against the incoming signal
can be calculated by analyzing the Sync signal up to the same
resolution (bandlimited). The difference will be fed to a
lowpass filter which has variable coefficients (VTR, Standard
signals). Its result (msb) is routed to the programmable
divider to close the PLL loop. The remainder (less than one lsb
of the divider) is presented to the clock and gate delay
circuitry. Here due to the influence of flyback this value will
be corrected and via the programmable gate delay presented to
the monitor.

The influence of the flyback is also filtered and the
filter coefficients can be changed under program control, too.

4.3   Picture Geometry Correction

The monitor consists of a relatively flat screen. This
results in picture geometry distortion, which looks if
completely uncorrected like a pin cushion.

The distortion in the horizontal direction is called
east-west distortion and the distortion in the vertical
direction is call North south distortion. Corrections can be
applied by modulating the horizontal and vertical deflection
signals. The corresponding signals which are known as east west
parabola and vertical sawtooth (see Fig. 4.3) are generated as
pulse width modulated signals. A lowpass filter must be
provided at the output pin for analog reconstruction filter.

Fig. 4.3 East-west parabola and vertical sawtooth

5   AUDIO PROCESSING

A TV without sound would lead us back into the days of old movies. This is undesirable, so an audio section has been included. And as more and more functions have to be performed here (pilot detection, pseudo stereo etc.) a special signal processor has been developed.

The audio signal processing is distributed between two chips, the Audio conversion unit, Fig. 5.1 and the Audio Processor Unit (APU) including the D/A conversion, Fig. 5.5.

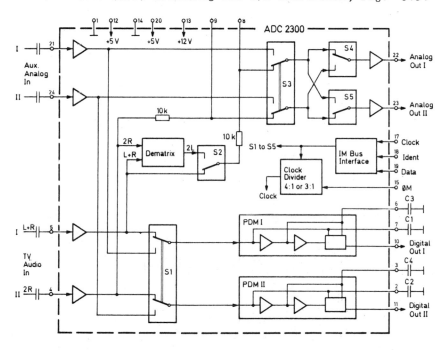

Fig. 5.1 Block Diagram of the Audio ADC

595

5.1 A/D Conversion

The main part of the Audio ADC consists of a Sigma-Delta Modulator. This can be described as a one bit converter with error feed back. The clock frequency must be much higher than required by the Nyquist limit if a high bit resolution has to be achieved. The realization is shown in Fig. 5.2a. It consists of an RC-network connected as an integrator. Its output voltage is fed to a comparator, which has a clocked output. The Q-output is treated as an analog signal fed back to the RC-network and also used as a digital signal which is routed to the APU. The left part of Fig. 5.2b. shows a typical timing diagram with zero input and the right part shows the situation with a positive input. Whereas in the left part the number of plus and minus pulses are equal in the right side the number of positive (density) pulses is considerably larger.

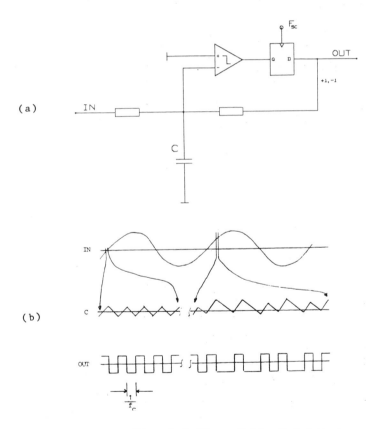

(a)

(b)

Fig. 5.2 Sigma-Delta Modulator

The main property of the pulse stream is that its average is a direct representation of the input signal. Thus the pulse stream could be reconverted using an analog lowpass filter. Obviously the pulse stream is not free of noise. The signal to noise performance over a given baseband is as follows:

$$S/N \ (ratio) \ = \ \frac{3}{4 \ pi} \cdot \left(\frac{f_c}{f_0} \right)^{3/2} \tag{7}$$

The performance can be enhanced with the implementation of the sigma-delta modulator in a feed back loop of a high gain opamp as shown in Fig. 5.3. The baseband component, (after being low pass filtered), is subtracted from the input signal. This results in an error signal which is amplified by the said amplifier and then connected to the input of the former circuit. S/N is increased by the amount of loop gain.

The overall noise cannot be reduced by this method, however the baseband noise is reduced at the expense of increase in out of band noise. The resulting equation for the signal to noise performance of the complete pulse-density modulator is then:

$$S/N \ (dB) \ = \ 20 \ log \left[\frac{3}{4\pi} \left(\frac{f_c}{f_0} \right)^{\frac{3}{2}} \right] \ + \ a$$

with an outer loop gain in dB.

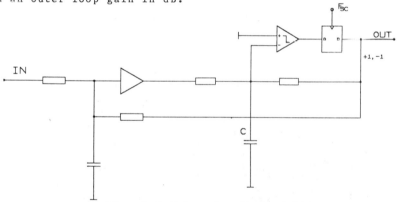

Fig. 5.3 Pulse-density modulator

5.2 Decimation, Processing, and Reconversion of Audiosignals

The PDM-pulse stream is routed to a lowpass conversion filter, which decreases the sampling frequency and increases the number of bits in parallel. Furthermore the conversion filter must suppress the out of band noise so that no downfolding of disturbing noise can occur. Such a filter consists of a cascade of moving time averages. Fig. 5.4. After conversion the audio

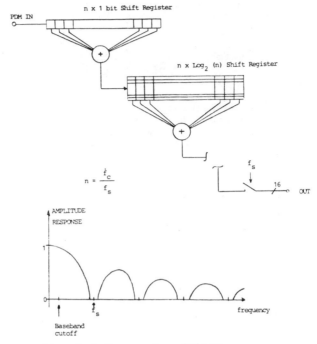

Fig. 5.4 Conversion filter

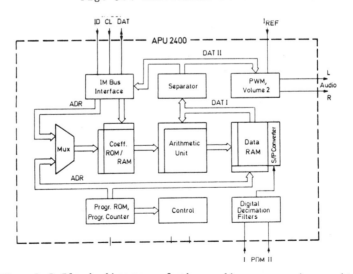

Fig. 5.5 Block diagram of the audio processing unit

598

information can be further processed by the APU, a programmable
real time processor Fig. 5.5.

Here the different additional filters and parameter
settings can be controlled by software. In a sampling period of
28.9μs one has to do all the nescessary multiplications,
additions, and data transfers, which can easily amount to over
100 operations. This leads to a cycle time (multiplication
included) of less than 280 ns.

The solution was to implement a Signal processor with an
appropriate instruction set. E. g.

MUAD	multiply and add
MUSU	multiply and subtract
CLSA	multiply and compare whether less or equal ACCU
etc.	

With a processor like this e. g. the stereo dematrixing has
been programmed, Fig. 5.6.

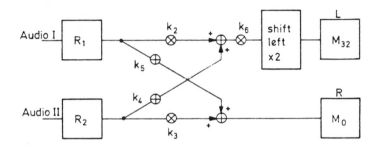

Fig. 5.6 Programming Structure for the Stereo Dematrix

The output of the APU consists of a PWM (pulse-width modulator)
running with a 16 fs clock rate. With a frequency of fs/16 a 4
Bit Data stream can be converted directly. To get an increase
in the signal to noise ratio a recursive error feedback filter
is implemented. In this way a signal to noise ratio of 75 dB
can be achieved.

For volume control the pulse height of the PWM output can
be programmed. Finally the analog signal is reconstructed by a
simple analog lowpass filter.

6   TELETEXT PROCESSING

With the start of new services delivered by TV signals or
Telephone connection the incorporation of decoders into the
Digital TV was a must. The first chip available for this
application is the teletext processor which extracts the digital
coded data out of the data input stream, processes it and stores
it on command in memory or displays it.

Working systems show some sensitive points of the existing

decoder, (distortion due to echoes or weak signals) . Due to this the consumer has a good picture quality but incorrect detection of the parallel transmitted teletext.

This case was or is circumvented in some countries by transmitting in parallel the teletext on all channels, especially when additional transmitters are in the signal path. In future however, this cannot be accepted because teletext is intended for subtitles in connection with particular and different programmes.

Another point of interest is the decoder capability of recognizing higher levels of teletext which will be transmitted in future. We start with a level 1 decoder which represents the standard teletext at present. Higher level teletext comparable to the videotext are under development.

6.1   Data Aquisition

After the code converter (Gray to binary) the FBAS signal is routed through a filter and slicer to detect the medium level between zeroes and ones, this is to reject low frequency noise influence. It is followed by the ghost compensation circuit, which comprises a transversal filter with variable tap values determined by a correlator output derived from the correlation of the ideal reference signal and the processed waveform. This is shown in Fig. 6.1.

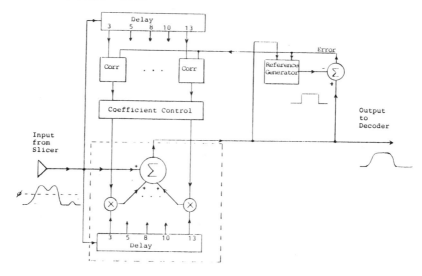

Fig. 6.1 Equalizer of the teletext processor

The amplitude value for the high value of the reference waveform is derived from the average of the high level in the teletext signal, the same holds for the low level. The

600

information high or low is given by the slicer output.

For a detailed discussion of the algorithm involved see [8]. The method leads to good results when the ones and zeroes are distributed randomly. In today standards the eighth bit is reserved as parity bit, therefore a zero to one or one to zero change in a line is guaranteed after 14 bits.

Criteria for the convergence of the coefficients can be found in [9].

the multipliers (tapweights) cycle time has to be equal to system cycle time which equals 56 ns for PAL, thus the coefficients were only realized as single shift stages, due to chip area constraints. This led to the following coefficients: 1, 0.5, 0.25, 0.125, 0.05625, 0.028125, (plus minus each). At the first sight this seems inadequate, but, nevertheless, the logarithmic nature of the possible coefficients is well suited for this purpose since it gives a large dynamic range. Only a digital decision is needed (high or low), therefore the analog value can have a relatively large variation. Furthermore, an additional strategy for the finding of the best coefficients has been implemented, described in the following. Important for this strategy are the following facts and ideas:

The incoming teletext samples are correlated continuously with the ideal signal, whereas the updating of the tapweights is less frequent. They are latched in at the beginning of each new vertical blanking interval (VBI). Finding a clock run in pattern (always preceding a teletext line) activates the correlator activity.

Starting with an ideal tapweight which cannot be realized with our coefficients, we observe the change of the tapweights and make a statistical observation.

The value which has the majority under these nominations representing the best choice, will be taken if a selected page is detected. This equalizer allows reliable decoding of the binary data of teletext signals with 20% eye height, whereas 60% are required but not always delivered.

6.2 Display and Storage

If 6 VBIs have passed without detection of clock run in, all values are set back to overcome possible divergence or false convergence.

Fig. 6.2 shows the system interconnections of the Teletext IC and additional multi page memory. In this application some pages can be reserved for prepared information like menues, etc. Many applications can be implemented in this way.

The Standard for Teletext, defining different quality level (L1-L5) includes in L1 fixed character positions (24-25 lines, 40 rows) 7 bit plus parity for every character with additional sequential attributes (boxing, flashing, etc.) Thus a low-cost one-bit organized dynamic RAM can be used as page store. For higher-level or full-channel teletext the transmission bandwidth between TTP and memory must be increased, due to parallel attributes (L2-L5, ANTIOPE) or pixel oriented information (L5).

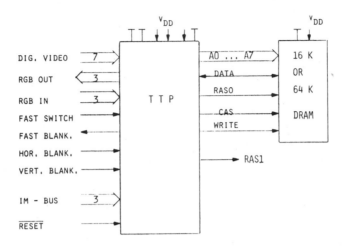

Fig. 6.2 Digital teletext system

7   FUTURE TRENDS

The future of the various television systems is not as determined as most people see it today. Flat display technology could change the whole market. Direct broadcasting systems, maybe with other video norms e. g. MAC-C or the so called high definition TV (HDTV), are at the verge of introduction for the consumer. Not to speak of Pay TV and scrambling techniques.

All these systems have one thing in common. They need at least one frame store or will greatly profit from the use of one.

Besides these high flying future quality improvements today's pictures can be improved by digital means. There are two areas of consideration

(i) Video processing enhancement

Due to the frequency interlacing of chrominance and luminance there are interferences visible: the so called cross luminance (leakage of chrominance into luminance) and cross chrominance (leakage of luminance into chrominance).

reduction of areaflicker and lineflicker via scan conversion and according processing.

(ii) CRT corrections digital geometry and convergence corrections.

7.1 Video Processing Enhancement

In the chapter about video processing, we presented today's solution for video processing and enhancements. Further improvements for the extraction of luminance and chrominance information can be achieved with comb filtering techniques but are fairly restricted to certain textures or still pictures.

One of the most disturbing effects (area flicker) can only be overcome in today's CRT-technology by scan conversion (50 Hz to 75 Hz or 100 Hz). The so called critical flicker frequency for bright areas is somewhere between 70 and 80 Hz. This leads to the demand for a frame store (6.8 MBit digital norm for professional equipment; components stored, Y, U, V) with cycle time of write (16 bit 74 ns) and read (32 bit 74 ns) for 100 Hz Display. This would be capable of excellent picture quality for most future consumer applications, besides HDTV.

Due to the today's standards a scan conversion must be in front of the frame store. This incorporates movement detectors, a comb filter, two dimensional filters (to increase spatial resolution) or even three dimensional filters (including frame to frame filtering). This represents a wide field and great challenge for future developments. Many studies in this field have been reported. But there remains much work to be done before these methods are in the reach of consumer application. Digital CRT correction solutions are certain to be introduced in the next generation of Digital Television.

7.2 CRT Corrections

If the CRT has a spherical form and the RGB guns start from one point a perfect picture would be seen. But as the screen is relatively flat, the picture geometry is distorted and as the three electron beams with the RGB informations respectively have different paths through the electrical and magnetic fields, they have a diverging deflection speed, leading to convergence errors (color bars at steep black white jumps).

Today's color TV (CTV) receivers apply correction by the modulation of the deflection current (geometry corrections) or by alteration of magnetic fields (convergence). The horizontal corrections can be easily performed via digital means. This leads to better picture sharpness (less correcting fields or currents) and better correction possibilities which are of great importance for TVs used as a monitor. Whereas in a natural picture, geometry distortions and small convergence errors are subjectively hardly visible, this situation is changed when columns of numbers or still rectangular structures are depicted.

In Fig. 7.1 the principle problems which can be overcome by digital means are depicted. In Fig. 7.2. the corrected signal is depicted. The underlying correction scheme is the same for geometry and convergence correction.

In the right part of the figure you can see the altered signal in an anlog sense. The correction is not achieved by changing the deflection, (changing the horizontal speed of the beam), but by means of a time compression of the videosignal to take into account the distortions. As can be seen this time

603

compression must be performed depending on the horizontal position and vertical position. Thus it has to be performed in a digital system with every sample different.

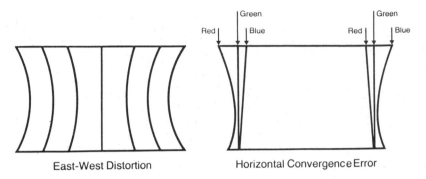

East-West Distortion   Horizontal Convergence Error

Fig. 7.1 Typical convergence and east west distortion

In Fig. 7.3 the proposed correction method is depicted. The signal will be resampled; that means interpolated at the appropriate times. An interpolation of zero order, that means taking the next possible value, can be achieved via a memory with different read and write addresses. In Fig. 7.4 are typical memory adressing schemes to achieve the corrections depicted. The addressing of the read pointer is to be modulated according to the correction. If the write pointer will be modulated the 'read' and 'write' are interchanged and mirrored at 45 degrees. The interpolation of the values between clockcycles for better approximation will be done in the first case in the read path and in the latter case in the write path.

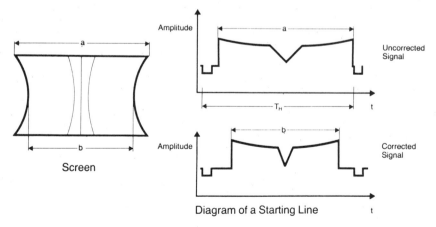

Screen   Diagram of a Starting Line

Fig. 7.2 Correction scheme

Care must be take because in general there is no

604

reconstruction filter after the D/A converter (sinx/x correction). The D/A converter produces in general a rectangular signal. That means higher frequencies corresponding to edges will be attenuated.

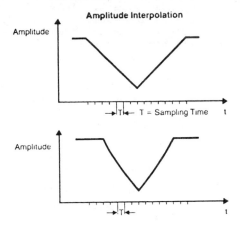

Fig. 7.3 Digital time base correction

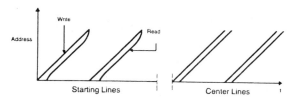

Fig. 7.4 Typical memory addressing

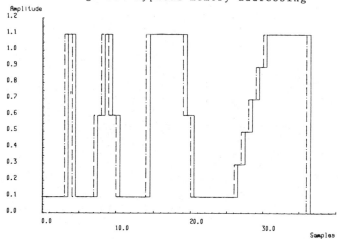

In this application the phase with respect to the sampling frequency is of importance giving different amplitude values. For a quiet scene and line coupled systems this would be of no interest. But small changes (a little noise) lead to a phase jump from frame to frame. The case that a shift of a half sample has to be performed is shown in Fig. 7.5. It is the difference between dashed line and solid line. The starting signal jump does not occur due to the Nyquist limit, but is a practial case for synthetic digital signal.

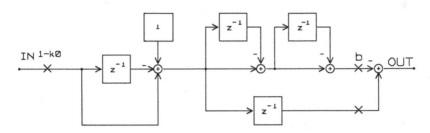

Fig 7.6 Interpolation and peaking filter structure

The shift will be provided by interpolation and peaking with the structure depicted in Fig. 7.6; The result for $K_0=0.5$ and $b=0.2$ in the time domain are shown in Fig. 7.7 and Fig.7.8.

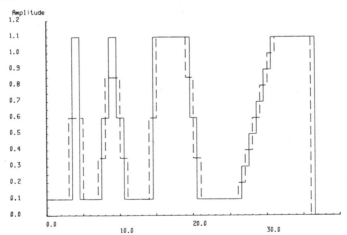

Fig. 7.7 Half sample shift, $k_0=0.5$, $b=0.0$

The phase delay and the amplitude response for different k_0 can be found in Fig.7.9 and Fig.7.10. For the relevant frequencies (up to 6 MHz for PAL) the phase delay error is considerably less than 1/10 sample unit. Thus only the resulting amplitude error in Fig. 7.10 can be of importance.

606

This can be accomplished by the shown peaking filter. The peaking value is depending of the interpolation value k_0 via the formula:

$$b = b_0 * k_0$$

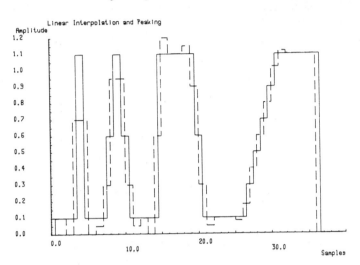

Fig. 7.8 Half sample shift with peaking value $b_0 = 0.2$

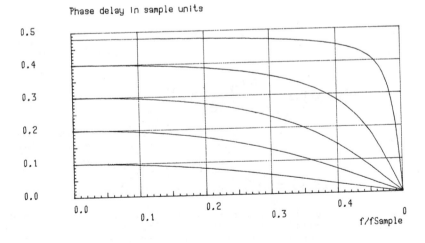

Fig. 7.9 Phase delay for $k_0 = 0.48, 0.4, 0.3, 0.2, 0.1$.

However, these signals refer to regions where the luminance

607

values or color values change rapidly. In areas like this the resolution of the eye is very poor and therefore this error is undetectable. Furthermore there is another effect. The situation on the screen is influenced by the video amplifiers and the beam diameter.

This can be simulated in order to calculate the screen nescessary D/A conversion clockrate and interpolation order or the nescessity of a reconstruction filter.

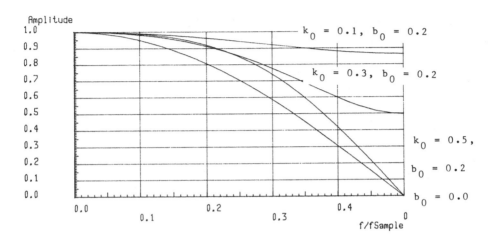

Fig. 7.10 Magnitude response of the interpolation and peaking filter

The beam is modelled Gaussian shape and the beam current is integrated at every spot. With $i(t,x)$ as current per pixel per time the light intensity $h(x,t)$ per pixel can be written as

$$h(x) = \int_{t}^{t=T_b} i_e(x,t)$$

where $T_b = 1/f_v$ denotes the picture display time, the inverse of the vertical frequency. For one pixel in one line it follows

$$i_e(x,t) = i(t)\ e^{-((x(t) - x(t_0))^2}$$

where $i(t)$ is a rectangular function see Fig. 7.5, x corresponds to the horizontal difference. Taking 10% as beam diameter shown in Fig. 7.10. The spot diameter can be correlated to the sample number. That is the distance of the spot diameter versus the distance the spot center moved during a certain number of samples. The simulation results are shown in Fig. 7.12-7.14.

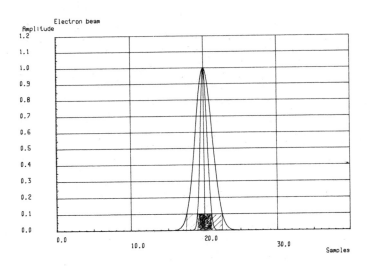

Fig. 7.11 Electron beam shape

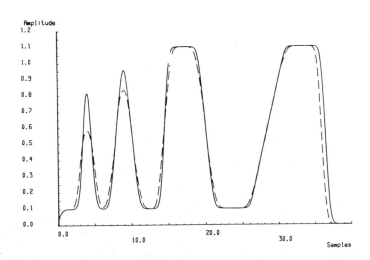

Fig. 7.12 Linear interpolation no peaking, spot diameter = 2.0
* sample

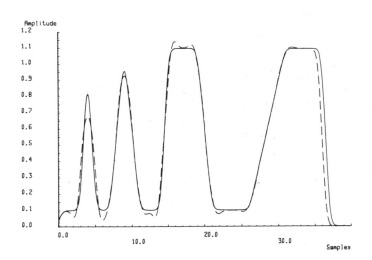

Fig. 7.13 Linear interpolation and peaking, spot diameter = 2.0
* sample

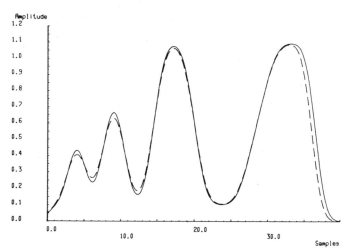

Fig. 7.14 Linear interpolation no peaking, spot diameter = 5.0
* sample

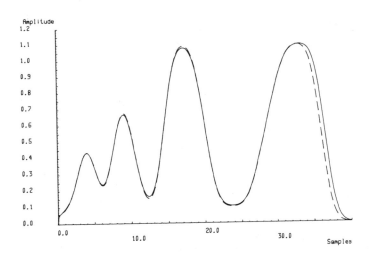

Fig. 7.15 Linear interpolation and peaking, spot diameter = 5.0 * sample

For a 66 cm diameter screen with standard scan this predicts superior quality for a conversion rate of 35 MHz and linear interpolation. That means following the VPU by an interpolator with a sampling rate of 35 MHz is therefor desirable. Taking into account the resolution ability of the human eye the standard conversion rate of 17.7 MHz is sufficient for standard high quality TV screens. A feasability study showed the convenience for integration and the circuit is under development.

8   CONCLUSION

The digital TV system is a good example of the problems the design engineer is involved with. It has shown that more and more system knowhow is incorporated on chip. Thus cooperation of silicon foundries and the chip buyers will clearly be altered. One way will be that the customers will use silicon foundries making their own design. That means they must gain knowledge of the IC design. Many levels can be thought of: custom design, user libary, etc.

The other way arround is that the designer will get system knowledge. The last situation will be preferred for consumer applications, as a great number of companies will use such chips in great numbers.

9.0 REFERENCES

[1] D. H. PRITCHARD, and J. J. GIBSON, "Worldwide Color Television Standards - Similarities and Differences" J. Soc. Motion Pict. Telev. Eng., vol. 89, pp. 111-120, Feb. 1980.

[2] Normen fuer Schwarzweiss- und Farbfernsehen, Funktechnische Arbeitsblaetter FS 01, Funkschau, Vol. 7 1975.

[3] U. WEISER and A. DAVIS, "A Wavefront Notation Tool for VLSI Array design" in VLSI Systems and Computations, Springer Verlag. 1981.

[4] L. RABINER and B. GOLD, "Theory and Application of Digital Signal Processing", Prentice Hall, 1975.

[5] L. JACKSON, "On the Interaction of Rounoff Noise and Dynamic Range in Digital Filters", Bell Systems Technical J., vol. 49 pp. 159-184, Feb. 1970.

[6] SECAM Colour TV System, in Compange Francaise De television Laboratoire et Usine: Rue Ernest Cognacq 92-Levallois-Perret

[7] H. INOSE et al., "A telemetering system by code modulations -Delta-Sigma modulation", IRE trans. on Space electronics and Telemetry. Sept. 1962.

[8] J. O. VOORMAN, P. J. SNIJDER, P. J. BARTH, and J. S. VROMANS, "A One-Chip Automatic Equalizer for Echo Reduction in Teletext", IEEE Transactions on Consumer Electronics in Vol CE-27, No. 3, pp 512-529.

[9] W. CICIORA, G. SPRIGNOLI, and W. THOMAS, "A Tutorial on Ghost Cancelling in Television Systems", IEEE Trans. on Consumer Electronics, Vol CE-25, No. 1, pp 9-44.

10.0 ACKNOWLEDGEMENT

The authors wish to thank their colleagues in the concept engineering group and the MOS-design group for valuable discussions and are particularly grateful for the help of Mr. Elmis, Mr. Flamm, Mr. Mehrgardt, and Mr. Freyberger.